AS 13

14 —

From Calculus to Cohomology

Top

From Calculus to Cohomology

de Rham cohomology and characteristic classes

Ib Madsen and Jørgen Tornehave

University of Aarhus

CAMBRIDGE
UNIVERSITY PRESS

PUBLISHED BY THE PRESS SYNDICATE OF THE UNIVERSITY OF CAMBRIDGE
The Pitt Building, Trumpington Street, Cambridge CB2 1RP, United Kingdom

CAMBRIDGE UNIVERSITY PRESS
The Edinburgh Building, Cambridge, CB2 2RU, United Kingdom
40 West 20th Street, New York, NY 10011-4211, USA
10 Stamford Road, Oakleigh, Melbourne 3166, Australia

First published 1997

Printed in the United Kingdom at the University Press, Cambridge

A catalogue record for this book is available from the British Library

Library of Congress Cataloguing in Publication data

Madsen, I. H. (Ib Henning), 1942-
 From calculus to cohomology: de Rham cohomology and
 chracteristic classes / Ib Madsen and Jørgen Tornehave.
  p.   cm.
 Includes bibliographical references and index.
 ISBN 0-521-58059-5 (hc). -- ISBN 0-521-58956-8 (pbk).
 1. Homology theory. 2. Differential forms. 3. Characteristic classes.
 I. Tornehave, Jørgen. II. Title.
QA612.3.M33 1996
514'.2--dc20 96-28589 CIP

ISBN 0 521 58059 5 headback
ISBN 0 521 58956 8 paperback

Contents

PREFACE

This text offers a self-contained exposition of the cohomology of differential forms, de Rham cohomology, and of its application to characteristic classes defined in terms of the curvature tensor. The only formal prerequisites are knowledge of standard calculus and linear algebra, but for the later part of the book some prior knowledge of the geometry of surfaces, Gaussian curvature, will not hurt the reader.

The first seven chapters present the cohomology of open sets in Euclidean spaces and give the standard applications usually covered in a first course in algebraic topology, such as Brouwer's fixed point theorem, the topological invariance of domains and the Jordan–Brouwer separation theorem. The next four chapters extend the definition of cohomology to smooth manifolds, present Stokes' theorem and give a treatment of degree and index of vector fields, from both the cohomological and geometric point of view.

The last ten chapters give the more advanced part of cohomology: the Poincaré–Hopf theorem, Poincaré duality, Chern classes, the Euler class, and finally the general Gauss–Bonnet formula. As a novel point we prove the so called splitting principles for both complex and real oriented vector bundles.

The text grew out of numerous versions of lecture notes for the beginning course in topology at Aarhus University. The inspiration to use de Rham cohomology as a first introduction to topology comes in part from a course given by G. Segal at Oxford many years ago, and the first few chapters owe a lot to his presentation of the subject. It is our hope that the text can also serve as an introduction to the modern theory of smooth four-manifolds and gauge theory.

The text has been used for third and fourth year students with no prior exposure to the concepts of homology or algebraic topology. We have striven to present all arguments and constructions in detail. Finally we sincerely thank the many students who have been subjected to earlier versions of this book. Their comments have substantially changed the presentation in many places.

Aarhus, January 1996

1. INTRODUCTION

It is well-known that a continuous real function, that is defined on an open set of $\mathbb{R}$ has a primitive function. How about multivariable functions? For the sake of simplicity we restrict ourselves to smooth (or C^∞-) functions, i.e. functions that have continuous partial derivatives of all orders.

We begin with functions of two variables. Let $f:U \to \mathbb{R}^2$ be a smooth function defined on an open set of $\mathbb{R}^2$.

$(a,b) \mapsto (f_1(a), f_2(b))$

Question 1.1 Is there a smooth function $F:U \to \mathbb{R}$, such that

(1) $$\frac{\partial F}{\partial x_1} = f_1 \quad \text{and} \quad \frac{\partial F}{\partial x_2} = f_2, \quad \text{where } f = (f_1, f_2)?$$

Since

$$\frac{\partial^2 F}{\partial x_2 \partial x_1} = \frac{\partial^2 F}{\partial x_1 \partial x_2}$$

we must have

$P \Leftrightarrow q$

$P \Rightarrow q$ nec.

$q \Rightarrow p$ suff.

(2) $$\frac{\partial f_1}{\partial x_2} = \frac{\partial f_2}{\partial x_1}.$$

The correct question is therefore whether F exists, assuming $f = (f_1, f_2)$ satisfies (2). Is condition (2) also sufficient?

Example 1.2 Consider the function $f:\mathbb{R}^2 \to \mathbb{R}^2$ given by

$$f(x_1, x_2) = \left(\frac{-x_2}{x_1^2 + x_2^2}, \frac{x_1}{x_1^2 + x_2^2} \right).$$

It is easy to show that (2) is satisfied. However, there is no function $F:\mathbb{R}^2 - \{0\} \to \mathbb{R}$ that satisfies (1). Assume there were; then

$$\int_0^{2\pi} \frac{d}{d\theta} F(\cos\theta, \sin\theta) d\theta = F(1,0) - F(1,0) = 0.$$

$x_1 = \cos\theta, \quad x_2 = \sin\theta$

On the other hand the chain rule gives

$$\frac{d}{d\theta} F(\cos\theta, \sin\theta) = \frac{dF}{dx} \cdot (-\sin\theta) + \frac{\partial F}{\partial y} \cdot \cos\theta$$
$$= -f_1(\cos\theta, \sin\theta) \cdot \sin\theta + f_2(\cos\theta, \sin\theta) \cdot \cos\theta = 1.$$

This contradiction can only be explained by the non-existence of F.

Definition 1.3 A subset $X \subset \mathbb{R}^n$ is said to be *star-shaped* with respect to the point $x_0 \in X$ if the line segment $\{tx_0 + (1-t)x \mid t \in [0,1]\}$ is contained in X for all $x \in X$.

Theorem 1.4 *Let $U \subset \mathbb{R}^2$ be an open star-shaped set. For any smooth function $(f_1, f_2): U \to \mathbb{R}^2$ that satisfies (2), Question 1.1 has a solution.*

Proof. For the sake of simplicity we assume that $x_0 = 0 \in \mathbb{R}^2$. Consider the function $F : U \to \mathbb{R}$,

$$F(x_1, x_2) = \int_0^1 [\, x_1 f_1(tx_1, tx_2) + x_2 f_2(tx_1, tx_2)\,]dt.$$

Then one has

$$\frac{\partial F}{\partial x_1}(x_1, x_2) = \int_0^1 \left[f_1(tx_1, tx_2) + tx_1\frac{\partial f_1}{\partial x_1}(tx_1, tx_2) + tx_2\frac{\partial f_2}{\partial x_1}(tx_1, tx_2) \right] dt$$

and

$$\frac{d}{dt} t f_1(tx_1, tx_2) = f_1(tx_1, tx_2) + tx_1\frac{\partial f_1}{\partial x_1}(tx_1, tx_2) + tx_2\frac{\partial f_1}{\partial x_2}(tx_1, tx_2).$$

Substituting this result into the formula, we get

$$\frac{\partial F}{\partial x_1}(x_1, x_2) = \int_0^1 \left[\frac{d}{dt} t f_1(tx_1, tx_2) + tx_2\left(\frac{\partial f_2}{\partial x_1}(tx_1, tx_2) - \frac{\partial f_1}{\partial x_2}(tx_1, tx_2)\right) \right] dt$$
$$= [\, t f_1(tx_1, tx_2)\,]_{t=0}^1 = f_1(x_1, x_2).$$

Analogously, $\frac{\partial F}{\partial x_2} = f_2(x_1, x_2)$.

Example 1.2 and Theorem 1.4 suggest that the answer to Question 1.1 depends on the "shape" or "topology" of U. Instead of searching for further examples or counterexamples of sets U and functions f, we define an *invariant* of U, which tells us whether or not the question has an affirmative answer (for all f), assuming the necessary condition (2).
Given the open set $U \subseteq \mathbb{R}^2$, let $C^\infty(U, \mathbb{R}^k)$ denote the set of smooth functions $\phi: U \to \mathbb{R}^k$. This is a vector space. If $k = 2$ one may consider $\phi: U \to \mathbb{R}^k$ as a vector field on U by plotting $\phi(u)$ from the point u. We define the *gradient* and the *rotation*

$$\text{grad}: C^\infty(U, \mathbb{R}) \to C^\infty(U, \mathbb{R}^2), \qquad \text{rot}: C^\infty(U, \mathbb{R}^2) \to C^\infty(U, \mathbb{R})$$

by

$$\text{grad}(\phi) = \left(\frac{\partial \phi}{\partial x_1}, \frac{\partial \phi}{\partial x_2} \right), \qquad \text{rot}(\phi_1, \phi_2) = \frac{\partial \phi_1}{\partial x_2} - \frac{\partial \phi_2}{\partial x_1}$$

Note that rot ∘ grad = 0. Hence the kernel of rot contains the image of grad,

$$\text{Ker(rot)} = \text{Kernel of rot}$$
$$\text{Im(grad)} = \text{Image of grad}$$

Since both rot and grad are linear operators, Im(grad) is a subspace of Ker(rot). Therefore we can consider the quotient vector space, i.e. the vector space of cosets $\alpha + \text{Im(grad)}$ where $\alpha \in \text{Ker(rot)}$:

(3) $$H^1(U) = \text{Ker(rot)}/\text{Im(grad)}.$$

Both Ker(rot) and Im(grad) are infinite-dimensional vector spaces. It is remarkable that the quotient space $H^1(U)$ is usually finite-dimensional.

We can now reformulate Theorem 1.4 as

(4) $$H^1(U) = 0 \quad \text{whenever } U \subseteq \mathbb{R}^2 \text{ is star-shaped.}$$

On the other hand, Example 1.2 tells us that $H^1(\mathbb{R}^2 - \{0\}) \neq 0$. Later on we shall see that $H^1(\mathbb{R}^2 - \{0\})$ is 1-dimensional, and that $H^1(\mathbb{R}^2 - \bigcup_{i=1}^{k}\{x_i\}) \cong \mathbb{R}^k$. The dimension of $H^1(U)$ is the number of "holes" in U.

In analogy with (3) we introduce

(5) $$H^0(U) = \text{Ker(grad)}.$$

This definition works for open sets U of $\mathbb{R}^k$ with $k \geq 1$, when we define

$$\text{grad}(f) = \left(\frac{\partial f}{\partial x_1}, \ldots, \frac{\partial f}{\partial x_n}\right).$$

Theorem 1.5 *An open set $U \subseteq \mathbb{R}^k$ is connected if and only if $H^0(U) = \mathbb{R}$.*

Proof. Assume that $\text{grad}(f) = 0$. Then f is locally constant: each $x_0 \in U$ has a neighborhood $V(x_0)$ with $f(x) = f(x_0)$ when $x \in V(x_0)$. If U is connected, then every locally constant function is constant. Indeed, for $x_0 \in U$ the set

$$\Gamma = \{x \in U | f(x) = f(x_0)\} = f^{-1}(f(x_0)).$$

is closed because f is continuous, and open since f is locally constant. Hence it is equal to U, and $H^0(U) = \mathbb{R}$. Conversely, if U is not connected, then there exists a smooth, surjective function $f: U \to \{0,1\}$. Such a function is locally constant, so $\text{grad}(f) = 0$. It follows that $\dim H^0(U) > 1$. □

The reader may easily extend the proof of Theorem 1.5 to show that $\dim H^0(U)$ is precisely the number of connected components of U.

We next consider functions of three variables. Let $U \subseteq \mathbb{R}^3$ be an open set. A real function on U has three partial derivatives and (2) is replaced by three equations. We introduce the notation

$$\text{grad}\colon C^\infty(U, \mathbb{R}) \to C^\infty(U, \mathbb{R}^3)$$
$$\text{rot}\colon C^\infty(U, \mathbb{R}^3) \to C^\infty(U, \mathbb{R}^3)$$
$$\text{div}\colon C^\infty(U, \mathbb{R}^3) \to C^\infty(U, \mathbb{R})$$

for the linear operators defined by

$$\text{grad}(f) = \left(\frac{\partial f}{\partial x_1}, \frac{\partial f}{\partial x_2}, \frac{\partial f}{\partial x_3} \right)$$

$$\text{rot}(f_1, f_2, f_3) = \left(\frac{\partial f_3}{\partial x_2} - \frac{\partial f_2}{\partial x_3}, \frac{\partial f_1}{\partial x_3} - \frac{\partial f_3}{\partial x_1}, \frac{\partial f_2}{\partial x_2} - \frac{\partial f_1}{\partial x_2} \right)$$

$$\text{div}(f_1, f_2, f_3) = \frac{\partial f_1}{\partial x_1} + \frac{\partial f_2}{\partial x_2} + \frac{\partial f_3}{\partial x_3}.$$

Note that $\text{rot} \circ \text{grad} = 0$ and $\text{div} \circ \text{rot} = 0$. We define $H^0(U)$ and set $H^1(U)$ as in Equations (3) and (5) and

(6) $$H^2(U) = \text{Ker}(\text{div})/\text{Im}(\text{rot}).$$

Theorem 1.6 *For an open star-shaped set in $\mathbb{R}^3$ we have that $H^0(U) = \mathbb{R}$, $H^1(U) = 0$ and $H^2(U) = 0$.*

Proof. The values of $H^0(U)$ and $H^1(U)$ are obtained as above, so we shall restrict ourselves to showing that $H^2(U) = 0$. It is convenient to assume that U is star-shaped with respect to 0. Consider a function $F : U \to \mathbb{R}^3$ with $\text{div}\, F = 0$, and define $G : U \to \mathbb{R}^3$ by

$$G(\mathbf{x}) = \int_0^1 (F(t\mathbf{x}) \times t\mathbf{x})\, dt$$

where $\times$ denotes the cross product,

$$(f_1, f_2, f_3) \times (x_1, x_2, x_3) = \begin{vmatrix} e_1 & f_1 & x_1 \\ e_2 & f_2 & x_2 \\ e_3 & f_3 & x_3 \end{vmatrix}$$
$$= (f_2 x_3 - f_3 x_2, f_3 x_1 - f_1 x_3, f_1 x_2 - f_2 x_1).$$

Straightforward calculations give

$$\text{rot}(F(t\mathbf{x}) \times t\mathbf{x}) = \frac{d}{dt}\left(t^2 F(t\mathbf{x}) \right).$$

Hence

$$\text{rot}\, G(\mathbf{x}) = \int_0^1 \frac{d}{dt}\left(t^2 F(t\mathbf{x}) \right) dt = F(\mathbf{x}). \qquad \square$$

If $U \subseteq \mathbb{R}^3$ is not star-shaped both $H^1(U)$ and $H^2(U)$ may be non-zero.

e.g. 1.7

Example 1.7 Let $S = \left\{ (x_1, x_2, x_3) \in \mathbb{R}^3 \big| x_1^2 + x_2^2 = 1, x_3 = 0 \right\}$ be the unit circle in the (x_1, x_2)-plane. Consider the function

$$f(x_1, x_2, x_3)$$
$$= \left(\frac{-2x_1 x_3}{x_3^2 + \left(x_1^2 + x_2^2 - 1 \right)^2}, \frac{-2x_2 x_3}{x_3^2 + \left(x_1^2 + x_2^2 - 1 \right)^2}, \frac{x_1^2 + x_2^2 - 1}{x_3^2 + \left(x_1^2 + x_2^2 - 1 \right)^2} \right)$$

on the open set $U = \mathbb{R}^3 - S$.

One finds that $\mathrm{rot}(f) = 0$. Hence f defines an element $[f] \in H^1(U)$. By integration along a curve γ in U, which is linked to S (as two links in a chain), we shall show that $[f] \neq 0$. The curve in question is

$$\gamma(t) = \left(\sqrt{1 + \cos t}, 0, \sin t \right), \quad -\pi \leq t \leq \pi.$$

Assume $\mathrm{grad}(F) = f$ as a function on U. We can determine the integral of $\frac{d}{dt} F(\gamma(t))$ in two ways. On the one hand we have

$$\int_{-\pi+\epsilon}^{\pi-\epsilon} \frac{d}{dt} F(\gamma(t)) dt = F(\gamma(\pi - \epsilon)) - F(\gamma(-\pi + \epsilon)) \to 0 \quad \text{for } \epsilon \to 0$$

and on the other hand the chain rule gives

$$\frac{d}{dt} F(\gamma(t)) = f_1(\gamma(t)) \cdot \gamma_1'(t) + f_2(\gamma(t)) \cdot \gamma_2'(t) + f_3(\gamma(t)) \cdot \gamma_3'(t)$$
$$= \sin^2 t + 0 + \cos^2 t = 1.$$

Therefore the integral also converges to 2π, which is a contradiction.

e.g. 1.8

Example 1.8 Let U be an open set in $\mathbb{R}^k$ and $X: U \to \mathbb{R}^k$ a smooth function (a smooth vector field). Recall that the *energy* $A_\gamma(X)$, of X along a smooth curve $\gamma: [a, b] \to U$ is defined by the integral

$$A_\gamma(X) = \int_a^b \langle X \circ \gamma(t), \gamma'(t) \rangle dt$$

where $\langle , \rangle$ denotes the standard inner product. If $X = \mathrm{grad}(\Phi)$ and $\Phi\gamma(a) = \Phi\gamma(b)$, then the energy is zero, since

$$\langle X \circ \gamma(t), \gamma'(t) \rangle = \frac{d}{dt} \Phi(\gamma(t))$$

by the chain rule; compare Example 1.2.

2. THE ALTERNATING ALGEBRA

Let V be a vector space over $\mathbb{R}$. A map

$$f : \underbrace{V \times V \times \cdots \times V}_{k \text{ times}} \to \mathbb{R}$$

is called k-linear (or multilinear), if f is linear in each factor.

Definition 2.1 A k-linear map $\omega \colon V^k \to \mathbb{R}$ is said to be alternating if $\omega(\xi_1, \ldots, \xi_k) = 0$ whenever $\xi_i = \xi_j$ for some pair $i \neq j$. The vector space of alternating, k-linear maps is denoted by $\mathrm{Alt}^k(V)$.

We immediately note that $\mathrm{Alt}^k(V) = 0$ if $k > \dim V$. Indeed, let $e_1, \ldots, e_n$ be a basis of V, and let $\omega \in \mathrm{Alt}^k(V)$. Using multilinearity,

$$\omega(\xi_1, \ldots, \xi_k) = \omega\left(\sum \lambda_{i,1} e_i, \ldots, \sum \lambda_{i,k} e_i \right) = \sum \lambda_J \omega(e_{j_1}, \ldots, e_{j_k})$$

with $\lambda_J = \lambda_{j_1,1} \ldots \lambda_{j_k,k}$. Since $k > n$, there must be at least one repetition among the elements $e_{j_1}, \ldots, e_{j_k}$. Hence $\omega(e_{j_1}, \ldots, e_{j_k}) = 0$.

The symmetric group of permutations of the set $\{1, \ldots, k\}$ is denoted by $S(k)$. We remind the reader that any permutation can be written as a composition of transpositions. The transposition that interchanges i and j will be denoted by (i, j). Furthermore, and this fact will be used below, any permutation can be written as a composition of transpositions of the type $(i, i+1)$, $(i, i+1) \circ (i+1, i+2) \circ (i, i+1) = (i, i+2)$ and so forth. The sign of a permutation:

(1) $$\mathrm{sign} \colon S(k) \to \{\pm 1\},$$

is a homomorphism, $\mathrm{sign}(\sigma \circ \tau) = \mathrm{sign}(\sigma) \circ \mathrm{sign}(\tau)$, which maps every transposition to -1. Thus the sign of $\sigma \in S(k)$ is -1 precisely if σ decomposes into a product consisting of an odd number of transpositions.

Lemma 2.2 If $\omega \in \mathrm{Alt}^k(V)$ and $\sigma \in S(k)$, then

$$\omega\big(\xi_{\sigma(1)}, \ldots, \xi_{\sigma(k)}\big) = \mathrm{sign}(\sigma)\omega(\xi_1, \ldots, \xi_k).$$

Proof. It is sufficient to prove the formula when $\sigma = (i, j)$. Let

$$\omega_{i,j}\big(\xi, \xi'\big) = \omega\big(\xi_1, \ldots, \xi, \ldots, \xi', \ldots, \xi_k\big),$$

with ξ and ξ' occurring at positions i and j respectively. The remaining $\xi_\nu \in V$ are arbitrary but fixed vectors. From the definition it follows that $\omega_{i,j} \in \mathrm{Alt}^2(V)$. Hence $\omega_{i,j}(\xi_i + \xi_j, \xi_i + \xi_j) = 0$. Bilinearity yields that $\omega_{i,j}(\xi_i, \xi_j) + \omega_{i,j}(\xi_j, \xi_i) = 0$. $\square$

Example 2.3 Let $V = \mathbb{R}^k$ and $\xi_i = (\xi_{i1}, \ldots, \xi_{ik})$. The function $\omega(\xi_1, \ldots, \xi_k) = \det((\xi_{ij}))$ is alternating, by the calculational rules for determinants.

We want to define the exterior product

$$\wedge \colon \mathrm{Alt}^p(V) \times \mathrm{Alt}^q(V) \to \mathrm{Alt}^{p+q}(V).$$

When $p = q = 1$ it is given by $(\omega_1 \wedge \omega_2) = \omega_1(\xi_1)\omega_2(\xi_2) - \omega_2(\xi_1)\omega_1(\xi_2)$.

Definition 2.4 A (p, q)-shuffle σ is a permutation of $\{1, \ldots, p + q\}$ satisfying

$$\sigma(1) < \ldots < \sigma(p) \quad \text{and} \quad \sigma(p + 1) < \ldots < \sigma(p + q).$$

The set of all such permutations is denoted by $S(p, q)$. Since a (p, q)-shuffle is uniquely determined by the set $\{\sigma(1), \ldots, \sigma(p)\}$, the cardinality of $S(p, q)$ is $\binom{p+q}{p}$.

Definition 2.5 (Exterior product) For $\omega_1 \in \mathrm{Alt}^p(V)$ and $\omega_2 \in \mathrm{Alt}^q(V)$, we define

$$(\omega_1 \wedge \omega_2)(\xi_1, \ldots, \xi_{p+q})$$
$$= \sum_{\sigma \in S(p,q)} \mathrm{sign}(\sigma)\omega_1\big(\xi_{\sigma(1)}, \ldots, \xi_{\sigma(p)}\big) \cdot \omega_2\big(\xi_{\sigma(p+1)}, \ldots, \xi_{\sigma(p+q)}\big).$$

It is obvious that $\omega_1 \wedge \omega_2$ is a $(p + q)$-linear map, but moreover:

Lemma 2.6 *If $\omega_1 \in \mathrm{Alt}^p(V)$ and $\omega_2 \in \mathrm{Alt}^q(V)$ then $\omega_1 \wedge \omega_2 \in \mathrm{Alt}^{p+q}(V)$.*

Proof. We first show that $(\omega_1 \wedge \omega_2)(\xi_1, \xi_2, \ldots, \xi_{p+q}) = 0$ when $\xi_1 = \xi_2$. We let

(i) $S_{12} = \{\sigma \in S(p, q) \mid \sigma(1) = 1, \sigma(p + 1) = 2\}$
(ii) $S_{21} = \{\sigma \in S(p, q) \mid \sigma(1) = 2, \sigma(p + 1) = 1\}$
(iii) $S_0 = S(p, q) - (S_{12} \cup S_{21})$.

If $\sigma \in S_0$ then either $\omega_1(\xi_{\sigma(1)}, \ldots, \xi_{\sigma(p)})$ or $\omega_2(\xi_{\sigma(p+1)}, \ldots, \xi_{\sigma(p+q)})$ is zero, since $\xi_{\sigma(1)} = \xi_{\sigma(2)}$ or $\xi_{\sigma(p+1)} = \xi_{\sigma(p+2)}$. Left composition with the transposition $\tau = (1, 2)$ is a bijection $S_{12} \to S_{21}$. We therefore have

$$(\omega_1 \wedge \omega_2)(\xi_1, \xi_2, \ldots, \xi_{p+q}) =$$
$$\sum_{\sigma \in S_{12}} \mathrm{sign}(\sigma)\omega_1\big(\xi_{\sigma(1)}, \ldots, \xi_{\sigma(p)}\big)\omega_2\big(\xi_{\sigma(p+1)}, \ldots, \xi_{\sigma(p+q)}\big)$$
$$- \sum_{\sigma \in S_{12}} \mathrm{sign}(\sigma)\omega_1\big(\xi_{\tau\sigma(1)}, \ldots, \xi_{\tau\sigma(p)}\big) \cdot \omega_2\big(\xi_{\tau\sigma(p+1)}, \ldots \xi_{\tau\sigma(p+q)}\big).$$

Since $\sigma(1) = 1$ and $\sigma(p+1) = 2$, while $\tau\sigma(1) = 2$ and $\tau\sigma(p+1) = 1$, we see that $\tau\sigma(i) = \sigma(i)$ whenever $i \neq 1, p + 1$. But $\xi_1 = \xi_2$ so the terms in the two sums cancel. The case $\xi_i = \xi_{i+1}$ is similar. Now $\omega_1 \wedge \omega_2$ will be alternating according to Lemma 2.7 below. $\qquad\square$

Lemma 2.7 *A k-linear map ω is alternating if $\omega(\xi_1, \ldots, \xi_k) = 0$ for all k-tuples with $\xi_i = \xi_{i+1}$ for some $1 \leq i \leq k - 1$.*

Proof. $S(k)$ is generated by the transpositions $(i, i+1)$, and by the argument of Lemma 2.2,

$$\omega(\xi_1, \ldots, \xi_i, \xi_{i+1}, \ldots, \xi_k) = -\omega(\xi_1, \ldots, \xi_{i+1}, \xi_i, \ldots, \xi_k).$$

Hence Lemma 2.2 holds for all $\sigma \in S(k)$, and ω is alternating. $\qquad\square$

It is clear from the definition that

$$\left(\omega_1 + \omega_1'\right) \wedge \omega_2 = \omega_1 \wedge \omega_2 + \omega_1' \wedge w_2$$
$$(\lambda\omega_1) \wedge \omega_2 = \lambda(\omega_1 \wedge \omega_2) = \omega_1 \wedge \lambda\omega_2$$
$$\omega_1 \wedge \left(\omega_2 + \omega_2'\right) = \omega_1 \wedge \omega_2 + \omega_1 \wedge \omega_2'$$

for $\omega_1, \omega_1' \in \mathrm{Alt}^p(V)$ and $\omega_2, \omega_2' \in \mathrm{Alt}^q(V)$.

Lemma 2.8 *If $\omega_1 \in \mathrm{Alt}^p(V)$ and $\omega_2 \in \mathrm{Alt}^q(V)$ then $\omega_1 \wedge \omega_2 = (-1)^{pq}\omega_2 \wedge \omega_1$.*

Proof. Let $\tau \in S(p+q)$ be the element with

$$\tau(1) = p+1, \tau(2) = p+2, \ldots, \tau(q) = p+q$$
$$\tau(q+1) = 1, \tau(q+2) = 2, \ldots, \tau(p+q) = p.$$

We have $\mathrm{sign}(\tau) = (-1)^{pq}$. Composition with τ defines a bijection

$$S(p, q) \overset{\cong}{\to} S(q, p); \quad \sigma \mapsto \sigma \circ \tau.$$

Note that

$$\omega_2\left(\xi_{\sigma\tau(1)}, \ldots, \xi_{\sigma\tau(q)}\right) = \omega_2\left(\xi_{\sigma(p+1)}, \ldots, \xi_{\sigma(p+q)}\right)$$
$$\omega_1\left(\xi_{\sigma\tau(q+1)}, \ldots, \xi_{\sigma\tau(p+q)}\right) = \omega_1\left(\xi_{\sigma(1)}, \ldots, \xi_{\sigma(p)}\right).$$

Hence

$$\omega_2 \wedge \omega_1(\xi_1, \ldots, \xi_{p+q})$$
$$= \sum_{\sigma \in S(q,p)} \mathrm{sign}(\sigma)\omega_2\left(\xi_{\sigma(1)}, \ldots, \xi_{\sigma(q)}\right)\omega_1\left(\xi_{\sigma(q+1)}, \ldots, \xi_{\sigma(p+q)}\right)$$
$$= \sum_{\sigma \in S(p,q)} \mathrm{sign}(\sigma\tau)\omega_2\left(\xi_{\sigma\tau(1)}, \ldots, \xi_{\sigma\tau(q)}\right)\omega_1\left(\xi_{\sigma\tau(q+1)}, \ldots, \xi_{\sigma\tau(p+q)}\right)$$
$$= (-1)^{pq} \sum_{\sigma \in S(p,q)} \mathrm{sign}(\sigma)\omega_1\left(\xi_{\sigma(1)}, \ldots, \xi_{\sigma(p)}\right)\omega_2\left(\xi_{\sigma(p+1)}, \ldots, \xi_{\sigma(p+q)}\right)$$
$$= (-1)^{pq}\omega_1 \wedge \omega_2(\xi_1, \ldots, \xi_{p+q}). \qquad\square$$

Lemma 2.9 *If $\omega_1 \in \mathrm{Alt}^p(V)$, $\omega_2 \in \mathrm{Alt}^q(V)$ and $\omega_3 \in \mathrm{Alt}^r(V)$ then*

$$\omega_1 \wedge (\omega_2 \wedge \omega_3) = (\omega_1 \wedge \omega_2) \wedge \omega_3.$$

Proof. Let $S(p,q,r) \subset S(p+q+r)$ consist of the permutations σ with

$$\sigma(1) < \ldots < \sigma(p),$$
$$\sigma(p+1) < \ldots < \sigma(p+q),$$
$$\sigma(p+q+1) < \ldots < \sigma(p+q+r).$$

We will also need the subsets $S(\bar{p},q,r)$ and $S(p,q,\bar{r})$ of $S(p,q,r)$ given by

$$\sigma \in S(\bar{p},q,r) \Longleftrightarrow \sigma \text{ is the identity on } \{1,\ldots,p\} \text{ and } \sigma \in S(p,q,r)$$
$$\sigma \in S(p,q,\bar{r}) \Longleftrightarrow \sigma \text{ is the identity on } \{p+q+1,\ldots,p+q+r\}$$
$$\text{and } \sigma \in S(p,q,r).$$

There are bijections

(2)
$$S(p,q+r) \times S(\bar{p},q,r) \overset{\cong}{\to} S(p,q,r); (\sigma,\tau) \to \sigma \circ \tau$$
$$S(p+q,r) \times S(p,q,\bar{r}) \overset{\cong}{\to} S(p,q,r); (\sigma,\tau) \to \sigma \circ \tau.$$

With these notations we have

$$[\omega_1 \wedge (\omega_2 \wedge \omega_3)](\xi_1,\ldots,\xi_{p+q+r})$$
$$= \sum_{\sigma \in S(p,q+r)} \mathrm{sign}(\sigma) \omega_1\big(\xi_{\sigma(1)},\ldots,\xi_{\sigma(p)}\big)(\omega_2 \wedge \omega_3)\big(\xi_{\sigma(p+1)},\ldots,\xi_{\sigma(p+q+r)}\big)$$
$$= \sum_{\sigma \in S(p,q+r)} \mathrm{sign}(\sigma) \sum_{\tau \in S(\bar{p},q,r)} \mathrm{sign}(\tau) \big[\omega_1(\xi_{\sigma(1)},\ldots,\xi_{\sigma(p)})$$
$$\omega_2(\xi_{\sigma\tau(p+1)},\ldots,\xi_{\sigma\tau(p+q)})\omega_3(\xi_{\sigma\tau(p+q+1)},\ldots,\xi_{\sigma\tau(p+q+r)})\big]$$
$$= \sum_{u \in S(p,q,r)} \big[\mathrm{sign}(u)\omega_1(\xi_{u(1)},\ldots,\xi_{u(p)})\omega_2(\xi_{u(p+1)},\ldots,\xi_{u(p+q)})$$
$$\omega_3(\xi_{u(p+q+1)},\ldots,\xi_{u(p+q+r)})\big]$$

where the last equality follows from the first equation in (2). Quite analogously one can calculate $[(\omega_1 \wedge \omega_2) \wedge \omega_3](\xi_1,\ldots,\xi_{p+q+r})$, employing the second equation in (2). $\qquad\square$

Remark 2.10 In other textbooks on alternating functions one can often see the definition

$$\omega_1 \bar{\wedge} \omega_2(\xi_1,\ldots,\xi_{p+q})$$
$$= \frac{1}{p!q!} \sum_{\sigma \in S(p+q)} \mathrm{sign}(\sigma)\omega_1\big(\xi_{\sigma(1)},\ldots,\xi_{\sigma(p)}\big)\omega_2\big(\xi_{\sigma(p+1)},\ldots,\xi_{\sigma(p+q)}\big).$$

Note that in this formula $\{\sigma(1), \ldots, \sigma(p)\}$ and $\{\sigma(p+1), \ldots, \sigma(p+q)\}$ are not ordered. There are exactly $S(p) \times S(q)$ ways to come from an ordered set to the arbitrary sequence above; this causes the factor $\frac{1}{p!q!}$, so $\omega_1 \bar{\wedge} \omega_2 = \omega_1 \wedge \omega_2$.

An $\mathbb{R}$-algebra A consists of a vector space over $\mathbb{R}$ and a bilinear map $\mu \colon A \times A \to A$ which is associative, $\mu(a, \mu(b, c)) = \mu(\mu(a, b), c)$ for every $a, b, c \in A$. The algebra is called *unitary* if there exists a unit element for μ, $\mu(1, a) = \mu(a, 1) = a$ for all $a \in A$.

Definition 2.11

(i) A graded $\mathbb{R}$-algebra A_* is a sequence of vector spaces $A_k, k = 0, 1 \ldots$, and bilinear maps $\mu \colon A_k \times A_l \to A_{k+l}$ which are associative.

(ii) The algebra A_* is called connected if there exists a unit element $1 \in A_0$ and if $\epsilon \colon \mathbb{R} \to A_0$, given by $\epsilon(r) = r \cdot 1$, is an isomorphism.

(iii) The algebra A_* is called (graded) commutative (or anti-commutative), if $\mu(a, b) = (-1)^{kl} \mu(b, a)$ for $a \in A_k$ and $b \in A_l$.

The elements in A_k are said to have degree k. The set $\mathrm{Alt}^k(V)$ is a vector space over $\mathbb{R}$ in the usual manner:

$$(\omega_1 + \omega_2)(\xi_1, \ldots, \xi_k) = \omega_1(\xi_1, \ldots, \xi_k) + \omega_2(\xi_1, \ldots, \xi_k)$$
$$(\lambda \omega)(\xi_1, \ldots, \xi_k) = \lambda \omega(\xi_1, \ldots, \xi_k), \quad \lambda \in \mathbb{R}.$$

The product from Definition 2.5 is a bilinear map from $\mathrm{Alt}^p(V) \times \mathrm{Alt}^q(V)$ to $\mathrm{Alt}^{p+q}(V)$. We set $\mathrm{Alt}^0(V) = \mathbb{R}$ and expand the product to $\mathrm{Alt}^0(V) \times \mathrm{Alt}^p(V)$ by using the vector space structure. The basic formal properties of the alternating forms can now be summarized in

Theorem 2.12 $\mathrm{Alt}^*(V)$ *is an anti-commutative and connected graded algebra.* $\square$

$\mathrm{Alt}^*(V)$ is called the exterior or alternating algebra associated to V.

Lemma 2.13 *For* 1*-forms* $\omega_1, \ldots, \omega_p \in \mathrm{Alt}^1(V)$,

$$(\omega_1 \wedge \ldots \wedge \omega_p)(\xi_1, \ldots, \xi_p) = \det \begin{pmatrix} \omega_1(\xi_1) & \omega_1(\xi_2) & \cdots & \omega_1(\xi_p) \\ \omega_2(\xi_1) & \omega_2(\xi_2) & \cdots & \omega_2(\xi_p) \\ \vdots & \vdots & & \vdots \\ \omega_p(\xi_1) & \omega_p(\xi_2) & \cdots & \omega_p(\xi_p) \end{pmatrix}$$

Proof. The case $p = 2$ is obvious. We proceed by induction on p. According to Definition 2.5,

$$\omega_1 \wedge (\omega_2 \wedge \ldots \wedge \omega_p)(\xi_1, \ldots, \xi_p)$$

$$= \sum_{j=1}^{p} (-1)^{j+1} \omega_1(\xi_j)(\omega_2 \wedge \ldots \wedge \omega_p)\left(\xi_1, \ldots, \hat{\xi}_j, \ldots, \xi_p\right)$$

where $(\xi_1, \ldots, \hat{\xi}_j, \ldots, \xi_p)$ denotes the $(p-1)$-tuple where ξ_j has been omitted. The lemma follows by expanding the determinant by the first row. $\qquad\square$

Note, from Lemma 2.13, that if the 1-forms $\omega_1, \ldots, \omega_p \in \mathrm{Alt}^1(V)$ are linearly independent then $\omega_1 \wedge \ldots \wedge \omega_p \neq 0$. Indeed, we can choose elements $\xi_i \in V$ with $\omega_i(\xi_j) = 0$ for $i \neq j$ and $\omega_j(\xi_j) = 1$, so that $\det(\omega_i(\xi_j)) = 1$. Conversely, if $\omega_1, \ldots, \omega_p$ are linearly dependent, we can express one of them, say ω_p, as a linear combination of the others. If $\omega_p = \sum_{i=1}^{p-1} r_i \omega_i$, then

$$\omega_1 \wedge \cdots \wedge \omega_{p-1} \wedge \omega_p = \sum_{i=1}^{p-1} r_i \omega_1 \wedge \cdots \wedge \omega_{p-1} \wedge \omega_i = 0,$$

as the determinant in Lemma 2.13 has two equal rows. We have proved

Lemma 2.14 *For 1-forms $\omega_1, \ldots, \omega_p$ on V, $\omega_1 \wedge \ldots \wedge \omega_p \neq 0$ if and only if they are linearly independent.* $\qquad\square$

Theorem 2.15 *Let $e_1, \ldots, e_n$ be a basis of V and $\epsilon_1, \ldots, \epsilon_n$ the dual basis of $\mathrm{Alt}^1(V)$. Then*

$$\left\{\epsilon_{\sigma(1)} \wedge \epsilon_{\sigma(2)} \wedge \ldots \wedge \epsilon_{\sigma(p)}\right\}_{\sigma \in S(p, n-p)}$$

is a basis of $\mathrm{Alt}^p(V)$. In particular

$$\dim \mathrm{Alt}^p(V) = \binom{\dim V}{p}.$$

Proof. Since $\epsilon_i(e_j) = 0$ when $i \neq j$, and $\epsilon_i(e_i) = 1$, Lemma 2.13 gives

$$(3) \quad \epsilon_{i_1} \wedge \ldots \wedge \epsilon_{i_p}\left(e_{j_1}, \ldots, e_{j_p}\right) = \begin{cases} 0 & \text{if } \{i_1, \ldots, i_p\} \neq \{j_1, \ldots, j_p\} \\ \mathrm{sign}(\sigma) & \text{if } \{i_1, \ldots, i_p\} = \{j_1, \ldots, j_p\} \end{cases}$$

Here σ is the permutation $\sigma(i_k) = j_k$. From Lemma 2.2 and (3) we get

$$\omega = \sum_{\sigma \in S(p, n-p)} \omega\left(e_{\sigma(1)}, \ldots, e_{\sigma(p)}\right) \epsilon_{\sigma(1)} \wedge \ldots \wedge \epsilon_{\sigma(p)}$$

for any alternating p-form. Thus $\epsilon_{\sigma(1)} \wedge \ldots \wedge \epsilon_{\sigma(p)}$ generates the vector space $\mathrm{Alt}^p(V)$. Linear independence follows from (3), since a relation

$$\sum_{\sigma \in S(p, n-p)} \lambda_\sigma \epsilon_{\sigma(1)} \wedge \ldots \wedge \epsilon_{\sigma(p)} = 0, \quad \lambda_\sigma \in \mathbb{R}$$

evaluated on $(e_{\sigma(1)}, \ldots, e_{\sigma(p)})$ gives $\lambda_\sigma = 0$. $\qquad\square$

Note from Theorem 2.15 that $\mathrm{Alt}^n(V) \cong \mathbb{R}$ if $n = \dim V$ and, as mentioned earlier, that $\mathrm{Alt}^p(V) = 0$ if $p > n$. A basis of $\mathrm{Alt}^n(V)$ is given by $\epsilon_1 \wedge \ldots \wedge \epsilon_n$. In particular every alternating n-form on $\mathbb{R}^n$ is proportional to the form in Example 2.3.

A linear map $f: V \to W$ induces the linear map

$$(4) \qquad\qquad \mathrm{Alt}^p(f): \mathrm{Alt}^p(W) \to \mathrm{Alt}^p(V)$$

by setting $\mathrm{Alt}^p(f)(\omega)(\xi_1, \ldots, \xi_p) = \omega(f(\xi_1), \ldots, f(\xi_p))$. For the composition of maps we have $\mathrm{Alt}^p(g \circ f) = \mathrm{Alt}^p(f) \circ \mathrm{Alt}^p(g)$, and $\mathrm{Alt}^p(\mathrm{id}) = \mathrm{id}$. These two properties are summarized by saying that $\mathrm{Alt}^p(-)$ is a *contravariant functor*. If $\dim V = n$ and $f: V \to V$ is a linear map then

$$\mathrm{Alt}^n(f): \mathrm{Alt}^n(V) \to \mathrm{Alt}^n(V)$$

is a linear endomorphism of a 1-dimensional vector space and thus multiplication by a number d. From Theorem 2.18 below it follows that $d = \det(f)$. We shall also be using the other maps

$$\mathrm{Alt}^p(f): \mathrm{Alt}^p(V) \to \mathrm{Alt}^p(V).$$

Let $\mathrm{tr}(g)$ denote the trace of a linear endomorphism g.

Theorem 2.16 *The characteristic polynomial of a linear endomorphism $f: V \to V$ is given by*

$$\det(f - t) = \sum_{i=0}^{n} (-1)^i \mathrm{tr}\left(\mathrm{Alt}^{n-i}(f)\right) t^i,$$

where $n = \dim V$.

Proof. Choose a basis $e_1, \ldots, e_n$ of V. Assume first that $e_1, \ldots, e_n$ are eigenvectors of f,

$$f(e_i) = \lambda_i e_i, \quad i = 1, \ldots, n.$$

Let $\epsilon_1, \ldots, \epsilon_n$ be the dual basis of $\mathrm{Alt}^1(V)$. Then

$$\mathrm{Alt}^p(f)\left(\epsilon_{\sigma(1)} \wedge \cdots \wedge \epsilon_{\sigma(p)}\right) = \lambda_{\sigma(1)} \cdots \lambda_{\sigma(p)} \epsilon_{\sigma(1)} \wedge \cdots \wedge \epsilon_{\sigma(p)}$$

and

$$\mathrm{tr}\, \mathrm{Alt}^p(f) = \sum_{\sigma \in S(p, n-p)} \lambda_{\sigma(1)} \cdots \lambda_{\sigma(p)}.$$

On the other hand

$$\det(f - t) = \Pi(\lambda_i - t) = \sum (-1)^{n-p} \left(\sum \lambda_{\sigma(1)} \cdots \lambda_{\sigma(p)}\right) t^{n-p}.$$

This proves the formula when f is diagonal.

If f is replaced by gfg^{-1}, with g an isomorphism on V, then both sides of the equation of Theorem 2.16 remain unchanged. This is obvious for the left-hand side and follows for the right-hand side since

$$\mathrm{Alt}^p\big(gfg^{-1}\big) = \mathrm{Alt}^p(g)^{-1} \circ \mathrm{Alt}^p(f) \circ \mathrm{Alt}^p(g)$$

by the functor property. Hence $\operatorname{tr} \mathrm{Alt}^p\big(g \circ f \circ g^{-1}\big) = \operatorname{tr} \mathrm{Alt}^p(f)$. Consider the set

$$D = \big\{ gfg^{-1} \,\big|\, f \text{ diagonal}, \, g \in GL(V) \big\}.$$

If V is a vector space over $\mathbb{C}$ and all maps are complex linear, then D is dense in the set of linear endomorphisms on V. We shall not give a formal proof of this, but it follows since every matrix with complex entries can be approximated arbitrarily closely by a matrix for which all roots of the characteristic polynomial are distinct. Since eigenvectors belonging to different eigenvalues are linearly independent, V has a basis consisting of eigenvectors for such a matrix, which then belongs to D

For general $f \in \mathrm{End}(V)$ we can choose a sequence $d_n \in D$ with $d_n \to f$ (i.e. the (i, j)-th element in d_n converges to the (i, j)-th element in f). Since both sides in the equation we want to prove are continuous, and since the equation holds for d_n, it follows for f. $\square$

It is not true that the set of diagonalizable matrices over $\mathbb{R}$ is dense in the set of matrices over $\mathbb{R}$ – a matrix with imaginary eigenvalues cannot be approximated by a matrix of the form gfg^{-1}, with f a real diagonal matrix. Therefore in the proof of Theorem 2.16 we must pass to complex linear maps, even if we are mainly interested in real ones.

3. DE RHAM COHOMOLOGY

In this chapter U will denote an open set in $\mathbb{R}^n$, $\{e_1, \ldots, e_n\}$ the standard basis and $\{\epsilon_1, \ldots, \epsilon_n\}$ the dual basis of $\text{Alt}^1(\mathbb{R}^n)$.

Definition 3.1 A *differential p-form* on U is a smooth map $\omega \colon U \to \text{Alt}^p(\mathbb{R}^n)$. The vector space of all such maps is denoted by $\Omega^p(U)$.

If $p = 0$ then $\text{Alt}^0(\mathbb{R}^n) = \mathbb{R}$ and $\Omega^0(U)$ is just the vector space of all smooth real-valued functions on U, $\Omega^0(U) = C^\infty(U, \mathbb{R})$.

The usual derivative of a smooth map $\omega \colon U \to \text{Alt}^p(\mathbb{R}^n)$ is denoted $D\omega$ and its value at x by $D_x\omega$. It is the linear map

$$D_x\omega \colon \mathbb{R}^n \to \text{Alt}^p(\mathbb{R}^n),$$

with

$$(D_x\omega)(e_i) = \frac{d}{dt}\omega(x + te_i)_{t=0} = \frac{\partial \omega}{\partial x_i}(x).$$

In $\text{Alt}^p(\mathbb{R}^n)$ we have the basis $\epsilon_I = \epsilon_{i_1} \wedge \ldots \wedge \epsilon_{i_p}$, where I runs over all sequences with $1 \leq i_1 < i_2 < \ldots < i_p \leq n$. Hence every $\omega \in \Omega^p(U)$ can be written in the form $\omega(x) = \sum \omega_I(x)\epsilon_I$, with $\omega_I(x)$ smooth real-valued functions of $x \in U$.

The differential $D_x\omega$ is the linear map

$$(1) \qquad D_x\omega(e_j) = \sum_I \frac{\partial \omega_I}{\partial x_j}(x)\epsilon_I \ , \ j = 1, \ldots, n.$$

The function $x \mapsto D_x\omega$ is a smooth map from U to the vector space of linear maps from $\mathbb{R}^n$ to $\text{Alt}^p(\mathbb{R}^n)$.

Definition 3.2 The exterior differential $d \colon \Omega^p(U) \to \Omega^{p+1}(U)$ is the linear operator

$$d_x\omega(\xi_1, \ldots, \xi_{p+1}) = \sum_{l=1}^{p+1} (-1)^{l-1} D_x\omega(\xi_l)(\xi_1, \ldots, \hat{\xi}_l, \ldots, \xi_{p+1})$$

with $(\xi_1, \ldots, \hat{\xi}_l, \ldots, \xi_{p+1}) = (\xi_1, \ldots, \xi_{l-1}, \xi_{l+1}, \ldots, \xi_{p+1})$.

It follows from Lemma 2.7 that $d_x\omega \in \text{Alt}^{p+1}(\mathbb{R}^n)$. Indeed, if $\xi_i = \xi_{i+1}$, then

$$\sum_{l=1}^{p+1} (-1)^{l-1} D_x\omega(\xi_l)(\xi_1, \ldots, \hat{\xi}_l, \ldots, \xi_{p+1})$$

$$= (-1)^{i-1} D_x\omega(\xi_i)(\xi_1, \ldots, \hat{\xi}_i, \ldots, \xi_{p+1})$$

$$+ (-1)^i D_x\omega(\xi_{i+1})(\xi_1, \ldots, \hat{\xi}_{i+1}, \ldots, \xi_{p+1})$$

$$= 0$$

because $(\xi_1, \ldots, \hat{\xi}_i, \ldots, \xi_{p+1}) = (\xi_1, \ldots, \hat{\xi}_{i+1}, \ldots, \xi_{p+1})$.

Example 3.3 Let $x_i \colon U \to \mathbb{R}$ be the i-th projection. Then $dx_i \in \Omega^1(U)$ is the constant map $dx_i \colon x \to \epsilon_i$. This follows from (1). In general, for $f \in \Omega^0(U)$, (1) shows that

$$(2) \qquad d_x f(\zeta) = \frac{\partial f}{\partial x_1}(x)\zeta^1 + \cdots + \frac{\partial f}{\partial x_n}(x)\zeta^n$$

with $(\zeta^1, \ldots, \zeta^n) = \zeta$. In other words: $df = \sum \frac{\partial f}{\partial x_i} \epsilon_i = \sum \frac{\partial f}{\partial x_i} dx_i$.

Lemma 3.4 *If* $\omega(x) = f(x)\epsilon_I$ *then* $d_x\omega = d_x f \wedge \epsilon_I$.

Proof. By (1) we have

$$D_x\omega(\zeta) = (D_x f)(\zeta)\epsilon_I = \left(\frac{\partial f}{\partial x_1}\zeta^1 + \cdots + \frac{\partial f}{\partial x_n}\zeta^n\right)\epsilon_I = d_x f(\zeta)\epsilon_I$$

and Definition 3.2 gives

$$d_x\omega(\xi_1, \ldots, \xi_{p+1}) = \sum_{k=1}^{p+1} (-1)^{k-1} d_x f(\xi_k)\epsilon_I\left(\xi_1, \ldots, \hat{\xi}_k, \ldots \xi_{p+1}\right)$$

$$= [d_x f \wedge \epsilon_I](\xi_1, \ldots, \xi_{p+1}). \qquad \square$$

Note for $\epsilon_I \in \mathrm{Alt}^p(\mathbb{R}^n)$ that

$$\epsilon_k \wedge \epsilon_I = \begin{cases} 0 & \text{if } k \in I \\ (-1)^r \epsilon_J & \text{if } k \notin I \end{cases}$$

with r the number determined by $i_r < k < i_{r+1}$ and $J = (i_1, \ldots, i_r, k, \ldots, i_p)$.

Lemma 3.5 *For* $p \geq 0$ *the composition* $\Omega^p(U) \to \Omega^{p+1}(U) \to \Omega^{p+2}(U)$ *is identically zero.*

Proof. Let $\omega = f\epsilon_I$. Then

$$d\omega = df \wedge \epsilon_I = \frac{\partial f}{\partial x_1}\epsilon_1 \wedge \epsilon_I + \cdots + \frac{\partial f}{\partial x_n}\epsilon_n \wedge \epsilon_I.$$

Now use $\epsilon_i \wedge \epsilon_i = 0$ and $\epsilon_i \wedge \epsilon_j = -\epsilon_j \wedge \epsilon_i$ to obtain that

$$d^2\omega = \sum_{i,j=1}^{n} \frac{\partial^2 f}{\partial x_i \partial x_j} \epsilon_i \wedge (\epsilon_j \wedge \epsilon_I)$$

$$= \sum_{i<j} \left(\frac{\partial^2 f}{\partial x_i \partial x_j} - \frac{\partial^2 f}{\partial x_j \partial x_i}\right) \epsilon_i \wedge \epsilon_j \wedge \epsilon_I = 0. \qquad \square$$

The exterior product in $\mathrm{Alt}^*(\mathbb{R}^n)$ induces an exterior product on $\Omega^*(U)$ upon defining

$$(\omega_1 \wedge \omega_2)(x) = \omega_1(x) \wedge \omega_2(x).$$

The exterior product of a differential p-form and a differential q-form is a differential $(p+q)$-form, so we get a bilinear map

$$\wedge: \Omega^p(U) \times \Omega^q(U) \to \Omega^{p+q}(U).$$

For a smooth function $f \in C^\infty(U, \mathbb{R})$, we have that

$$(f\omega_1) \wedge \omega_2 = f(\omega_1 \wedge \omega_2) = \omega_1 \wedge f\omega_2.$$

This just expresses the bilinearity of the product in $\mathrm{Alt}^*(\mathbb{R}^n)$. Also note that $f \wedge \omega = f\omega$ when $f \in \Omega^0(U)$ and $\omega \in \Omega^p(U)$.

Lemma 3.6 *For $\omega_1 \in \Omega^p(U)$ and $\omega_2 \in \Omega^q(U)$,*

$$d(\omega_1 \wedge \omega_2) = d\omega_1 \wedge \omega_2 + (-1)^p \omega_1 \wedge d\omega_2.$$

Proof. It is sufficient to show the formula when $\omega_1 = f\epsilon_I$ and $\omega_2 = g\epsilon_J$. But then $\omega_1 \wedge \omega_2 = fg \, \epsilon_I \wedge \epsilon_J$, and

$$\begin{aligned}
d(\omega_1 \wedge \omega_2) = d(fg) \wedge \epsilon_I \wedge \epsilon_J &= ((df)g + fdg) \wedge \epsilon_I \wedge \epsilon_J \\
&= df g \wedge \epsilon_I \wedge \epsilon_J + f dg \wedge \epsilon_I \wedge \epsilon_J \\
&= df \wedge \epsilon_I \wedge g\epsilon_J + (-1)^p f\epsilon_I \wedge dg \wedge \epsilon_J \\
&= d\omega_1 \wedge \omega_2 + (-1)^p \omega_1 \wedge d\omega_2.
\end{aligned}$$
$\square$

Summing up, we have introduced an anti-commutative algebra $\Omega^*(U)$ with a *differential*,

$$d: \Omega^*(U) \to \Omega^{*+1}(U), \quad d \circ d = 0$$

and d is a *derivation* (satisfies Lemma 3.6): $(\Omega^*(U), d)$ is a commutative DGA (differential graded algebra). It is called the *de Rham complex* of U.

Theorem 3.7 *There is precisely one linear operator $d: \Omega^p(U) \to \Omega^{p+1}(U)$, $p = 0, 1, \ldots,$ such that*

(i) $f \in \Omega^0(U)$, $df = \frac{\partial f}{\partial x_1}\epsilon_1 + \cdots + \frac{\partial f}{\partial x_n}\epsilon_n$

(ii) $d \circ d = 0$

(iii) $d(\omega_1 \wedge \omega_2) = d\omega_1 \wedge \omega_2 + (-1)^p \omega_1 \wedge d\omega_2$ *if $\omega_1 \in \Omega^p(U)$.*

Proof. We have already defined d with the asserted properties. Conversely assume that d' is a linear operator satisfying (i), (ii) and (iii). We will show that d' is the exterior differential.

The first property tells us that $d = d'$ on $\Omega^0(U)$. In particular $d'x_i = dx_i$ for the i-th projection $x_i: U \to \mathbb{R}$. It follows from Example 3.3 that $d'x_i = \epsilon_i$, the constant function. Since $d' \circ d' = 0$ we have that $d'\epsilon_i = 0$. Then (iii) gives $d'\epsilon_I = 0$. Now let $\omega = f\epsilon_I = f \wedge \epsilon_I$, $f \in C^\infty(U, \mathbb{R})$. Again by using (iii),

$$d'\omega = d'f \wedge \epsilon_I + f \wedge d'\epsilon_I = d'f \wedge \epsilon_I = df \wedge \epsilon_I = d\omega.$$

Since every p-form is the sum of such special p-forms, $d = d'$ on all of $\Omega^p(U)$.□

For an open set U in $\mathbb{R}^3$, $d: \Omega^1(U) \to \Omega^2(U)$ is given as

$$d(f_1\epsilon_1 + f_2\epsilon_2 + f_3\epsilon_3) = df_1 \wedge \epsilon_1 + df_2 \wedge \epsilon_2 + df_3 \wedge \epsilon_3 =$$

$$\left(\frac{\partial f_2}{\partial x_1} - \frac{\partial f_1}{\partial x_2}\right)\epsilon_1 \wedge \epsilon_2 + \left(\frac{\partial f_3}{\partial x_2} - \frac{\partial f_2}{\partial x_3}\right)\epsilon_2 \wedge \epsilon_3 + \left(\frac{\partial f_1}{\partial x_3} - \frac{\partial f_3}{\partial x_1}\right)\epsilon_3 \wedge \epsilon_1.$$

The first equality follows from Theorem 3.7.(iii), as $\epsilon_i: U \to \text{Alt}^1(\mathbb{R}^3)$ is the constant map, and hence $d\epsilon_i = 0$, by (1). Alternatively, we have already noted that the 1-forms ϵ_i and dx_i agree, and hence $d\epsilon_i = d \circ d(x_i) = 0$ by Theorem 3.7.(ii). The second equality comes from the anti-commutativity, $\epsilon_i \wedge \epsilon_j = -\epsilon_j \wedge \epsilon_i$ and Theorem 3.7.(i).

Quite analogously we can calculate that

$$d(g_3\epsilon_1 \wedge \epsilon_2 + g_1\epsilon_2 \wedge \epsilon_3 + g_2\epsilon_3 \wedge \epsilon_1) = \left(\frac{\partial g_1}{\partial x_1} + \frac{\partial g_2}{\partial x_2} + \frac{\partial g_3}{\partial x_3}\right)\epsilon_1 \wedge \epsilon_2 \wedge \epsilon_3.$$

Definition 3.8 The p-th (de Rham) cohomology group is the quotient vector space

$$H^p(U) = \frac{\text{Ker}\big(d: \Omega^p(U) \to \Omega^{p+1}(U)\big)}{\text{Im}\big(d: \Omega^{p-1}(U) \to \Omega^p(U)\big)}.$$

In particular $H^p(U) = 0$ for $p < 0$, and $H^0(U)$ is the kernel of

$$d: C^\infty(U, \mathbb{R}) \to \Omega^1(U),$$

and therefore is the vector space of maps $f \in C^\infty(U, \mathbb{R})$ with vanishing derivatives. This is precisely the space of locally constant maps.

Let $\sim$ be the equivalence relation on the open set U such that $q_1 \sim q_2$ if there exists a continuous curve $\alpha: [a, b] \to U$ with $\alpha(a) = q_1$ and $\alpha(b) = q_2$. The equivalence classes partition U into disjoint open subsets, namely the *connected components* of U. A connected component of U is a maximal non-empty subset W of U that cannot be written as the disjoint union of two non-empty open subsets of W (in the topology induced by $\mathbb{R}^n$). An open set $U \subseteq \mathbb{R}^n$ has at most countably many connected components (in each of them one can choose a point with rational coordinates.)

Lemma 3.9 $H^0(U)$ *is the vector space of maps $U \to \mathbb{R}$ that are constant on each connected component of U.*

Proof. A locally constant function $f: U \to \mathbb{R}$ gives a partition of U into the mutually disjoint open sets $f^{-1}(c)$, $c \in \mathbb{R}$. Consequently $f: U \to \mathbb{R}$ is locally constant precisely when f is constant on each connected component of U. $\square$

It follows that $\dim_{\mathbb{R}} H^0(U)$ (considered as a non-negative integer or ∞) is precisely the number of connected components of U.

The elements in $\Omega^p(U)$ with $d\omega = 0$ are called the *closed* p-forms. The elements of the image $d(\Omega^{p-1}(U)) \subset \Omega^p(U)$ are the *exact* p-forms. The p-th cohomology group thus measures whether every closed p-form is exact. This condition is satisfied precisely when $H^p(U) = 0$. A closed p-form $\omega \in \Omega^p(U)$ gives a cohomology class, denoted by

$$[\omega] = \omega + d\Omega^{p-1}(U) \in H^p(U),$$

and $[\omega] = [\omega']$ if and only if $\omega - \omega'$ is exact. In general the vector space of closed p-forms and the vector space of exact p-forms are infinite-dimensional. In contrast $H^p(U)$ usually has finite dimension.

We can define a bilinear, associative and anti-commutative product

$$(3) \qquad H^p(U) \times H^q(U) \to H^{p+q}(U)$$

by setting $[\omega_1] [\omega_2] = [\omega_1 \wedge \omega_2]$. It is well-defined because

$$(\omega_1 + d\eta_1) \wedge (\omega_2 + d\eta_2) = \omega_1 \wedge \omega_2 + d\eta_1 \wedge \omega_2 + \omega_1 \wedge d\eta_2 + d\eta_1 \wedge d\eta_2$$
$$= \omega_1 \wedge \omega_2 + d\big(\eta_1 \wedge \omega_2 + (-1)^p \omega_1 \wedge \eta_2 + \eta_1 \wedge d\eta_2\big).$$

We want to make $U \to H^p(U)$ into a *contravariant functor*. Thus to a smooth map $\phi: U_1 \to U_2$ between open sets $U_1 \subset \mathbb{R}^n$ and $U_2 \subset \mathbb{R}^m$, we shall define a linear map

$$H^p(\phi): H^p(U_2) \to H^p(U_1),$$

such that:

$$(4) \qquad \begin{aligned} H^p(\phi_2 \circ \phi_1) &= H^p(\phi_1) \circ H^p(\phi_2) \\ H^p(\mathrm{id}) &= \mathrm{id}. \end{aligned}$$

We first make $\Omega^*(-)$ into a contravariant functor.

Definition 3.10 Let $U_1 \subset \mathbb{R}^n$ and $U_2 \subset \mathbb{R}^m$ be open sets and $\phi\colon U_1 \to U_2$ a smooth map. The induced morphism $\Omega^p(\phi)\colon \Omega^p(U_2) \to \Omega^p(U_1)$ is defined by

$$\Omega^p(\phi)(\omega)_x = \mathrm{Alt}^p(D_x\phi) \circ \omega(\phi(x)), \qquad \Omega^0(\phi)(\omega)_x = \omega_{\phi(x)}.$$

Frequently one writes ϕ^* instead of $\Omega^p(\phi)$. We note that the analogue of (4) is satisfied. Indeed,

$$\phi^*(\omega)_x(\xi_1, \ldots, \xi_p) = \omega_{\phi(x)}(D_x\phi(\xi_1), \ldots, D_x\phi(\xi_p)),$$

and using the chain rule $D_x(\psi \circ \phi) = D_{\phi(x)}(\psi) \circ D_x(\phi)$, for $\phi\colon U_1 \to U_2$, $\psi\colon U_2 \to U_3$, it is easy to see that

$$\Omega^p(\psi \circ \phi) = \Omega^p(\phi) \circ \Omega^p(\psi), \qquad \Omega^p(\mathrm{id}_U) = \mathrm{id}_{\Omega^p(U)}.$$

It should be noted that $\Omega^p(i)(\omega) = \omega \circ i$ when $i\colon U_1 \hookrightarrow U_2$ is an inclusion, since then $D_x i = \mathrm{id}$.

Example 3.11 For the constant 1-form $\epsilon_i \in \Omega^1(U_2)$ we have that

$$\phi^*(\epsilon_i) = \sum_{k=1}^{n} \frac{\partial\phi_i}{\partial x_k}\epsilon_k = d\phi_i$$

with ϕ_i the i-th coordinate function. To see this, let $\zeta \in \mathbb{R}^n$. Then

$$\phi^*(\epsilon_i)(\zeta) = \epsilon_i(D_x\phi(\zeta)) = \epsilon_i\left(\sum_{k=1}^{m}\left(\sum_{l=1}^{n}\frac{\partial\phi_k}{\partial x_l}\zeta^l\right)e_k\right)$$

$$= \sum_{l=1}^{n}\frac{\partial\phi_i}{\partial x_l}\zeta^l = \sum_{l=1}^{n}\frac{\partial\phi_i}{\partial x_l}\epsilon_l(\zeta) = d\phi_i(\zeta). \qquad \square$$

Theorem 3.12 *With Definition 3.10 we have the relations*

(i) $\phi^*(\omega \wedge \tau) = \phi^*(\omega) \wedge \phi^*(\tau)$
(ii) $\phi^*(f) = f \circ \phi$ if $f \in \Omega^0(U_2)$
(iii) $d\phi^*(\omega) = \phi^*(d\omega)$.

Conversely, if $\phi'\colon \Omega^*(U_2) \to \Omega^*(U_1)$ *is a linear operator that satisfies the three conditions, then* $\phi' = \phi^*$.

Proof. Let $x \in U_1$ and let $\xi_1, \ldots, \xi_{p+q}$ be vectors in $\mathbb{R}^n$. Then

$$\phi^*(\omega \wedge \tau)_x(\xi_1, \ldots, \xi_{p+q})$$
$$= (\omega \wedge \tau)_{\phi(x)}(D_x\phi(\xi_1), \ldots, D_x\phi(\xi_{p+q}))$$
$$= \sum \mathrm{sign}(\sigma)\Big[\omega_{\phi(x)}(D_x\phi(\xi_{\sigma(1)}), \ldots, D_x\phi(\xi_{\sigma(p)}))$$
$$\tau_{\phi(x)}(D_x\phi(\xi_{\sigma(p+1)}), \ldots, D_x\phi(\xi_{\sigma(p+q)}))\Big]$$
$$= \sum \mathrm{sign}(\sigma)\phi^*(\omega)_x(\xi_{\sigma(1)}, \ldots, \xi_{\sigma(p)})\phi^*(\tau)_x(\xi_{\sigma(p+1)}, \ldots, \xi_{\sigma(p+q)})$$
$$= (\phi^*(\omega)_x \wedge \phi^*(\tau)_x)(\xi_1, \ldots, \xi_{p+q}).$$

This shows (i) when $p > 0$ and $q > 0$. If $p = 0$ or $q = 0$ the proof is quite analogous, but easier. Property (ii) is contained in the definition of ϕ^* for degree 0. So we are left with (iii). We shall first show that $d\phi^*(f) = \phi^*(df)$ when $f \in \Omega^0(U_2)$. We have that

$$df = \sum_{k=1}^{m} \frac{\partial f}{\partial x_k} \epsilon_k = \sum_{k=1}^{m} \frac{\partial f}{\partial x_k} \wedge \epsilon_k,$$

when ϵ_k is considered as the element in $\Omega^1(U_2)$ with constant value ϵ_k. From (i) and (ii) we obtain

$$\phi^*(df) = \sum_{k=1}^{m} \phi^* \left(\frac{\partial f}{\partial x_k} \right) \wedge \phi^*(\epsilon_k) = \sum_{k=1}^{m} \left(\frac{\partial f}{\partial x_k} \circ \phi \right) \wedge \left(\sum_{l=1}^{n} \frac{\partial \phi_k}{\partial x_l} \epsilon_l \right)$$

$$= \sum_{k=1}^{m} \sum_{l=1}^{n} \left(\frac{\partial f}{\partial x_k} \circ \phi \right) \left(\frac{\partial \phi_k}{\partial x_l} \right) \epsilon_l = \sum_{l=1}^{n} \left(\sum_{k=1}^{m} \left(\frac{\partial f}{\partial x_k} \circ \phi \right) \frac{\partial \phi_k}{\partial x_l} \right) \epsilon_l$$

$$= \sum_{l=1}^{n} \frac{\partial (f \circ \phi)}{\partial x_l} \epsilon_l = d(f \circ \phi) = d(\phi^*(f)).$$

In the more general case $\omega = f\epsilon_I = f \wedge \epsilon_I$, Lemma 3.6 gives $d\omega = df \wedge \epsilon_I$, because $d\epsilon_I = 0$. Hence

$$\phi^*(d\omega) = \phi^*(df) \wedge \phi^*(\epsilon_I) = d(\phi^* f) \wedge \phi^*(\epsilon_I)$$
$$= d(\phi^*(f) \wedge \phi^*(\epsilon_I)) = d(\phi^*\omega).$$

The second last equality uses Lemma 3.6 and the fact that $d\phi^*(\epsilon_I) = 0$:

$$d\phi^*(\epsilon_I) = d\left(\phi^*(\epsilon_{i_1}) \wedge \ldots \wedge \phi^*(\epsilon_{i_p}) \right)$$
$$= \sum (-1)^{k-1} \phi^*(\epsilon_{i_1}) \wedge \ldots \wedge d\phi^*(\epsilon_{i_k}) \wedge \ldots \wedge \phi^*(\epsilon_{i_p}) = 0$$

since $d\phi^*(\epsilon_{i_k}) = 0$ by Example 3.11 and Lemma 3.5. $\square$

In the following it will be convenient to use the notation of Example 3.3 and write

$$dx_I = dx_{i_1} \wedge \ldots \wedge dx_{i_p}$$

instead of the (constant) p-form $\epsilon_I = \epsilon_{i_1} \wedge \ldots \wedge \epsilon_{i_p}$. An arbitrary p-form can then be written as

$$\omega(x) = \sum \omega_I(x) dx_I$$

and Example 3.11 becomes $\phi^*(dy_i) = d\phi_i$ when $y_i \colon U_2 \to \mathbb{R}$ is the i-th coordinate function and $\phi_i = y_i \circ \phi$ the i-th coordinate of ϕ; cf. Theorem 3.12.(ii),(iii).

Example 3.13

(i) Let $\gamma\colon (a,b) \to U$ be a smooth curve in U, $\gamma = (\gamma_1, \dots, \gamma_n)$, and let

$$\omega = f_1\, dx_1 + \cdots + f_n dx_n$$

be a 1-form on U. Then we have that

$$\begin{aligned}
\gamma^*(\omega) &= \gamma^*(f_1) \wedge \gamma^*(dx_1) + \cdots + \gamma^*(f_n) \wedge \gamma^*(dx_n) \\
&= \gamma^*(f_1)d(\gamma^*(x_1)) + \cdots + \gamma^*(f_n)d(\gamma^*(x_n)) \\
&= (f_1 \circ \gamma)d\gamma_1 + \cdots + (f_n \circ \gamma)d\gamma_n \\
&= \big[(f_1 \circ \gamma)\gamma_1' + \cdots + (f_n \circ \gamma)\gamma_n'\big]dt = \big\langle f(\gamma(t)), \gamma'(t) \big\rangle\, dt.
\end{aligned}$$

Here $\langle\,,\,\rangle$ is the usual inner product. Compare Example 1.8.

(ii) Let $\phi\colon U_1 \to U_2$ be a smooth map between open sets in $\mathbb{R}^n$. Then

$$\phi^*(dx_1 \wedge \dots \wedge dx_n) = \det(D_x\phi)dx_1 \wedge \dots \wedge dx_n.$$

Indeed, from Theorem 3.12,

$$\begin{aligned}
\phi^*(dx_1 \wedge \dots \wedge dx_n) &= \phi^*(dx_1) \wedge \dots \wedge \phi^*(dx_n) = d\phi^*(x_1) \wedge \dots \wedge d\phi^*(x_n) \\
&= d\phi_1 \wedge \dots \wedge d\phi_n = \det(D_x\phi)dx_1 \wedge \dots \wedge dx_n.
\end{aligned}$$

The last equality is a consequence of Lemma 2.13.

Example 3.14
If $\phi\colon \mathbb{R}^n \times \mathbb{R} \to \mathbb{R}^n$ is given by $\phi(x,t) = \psi(t)x$, where $\psi(t)$ is a smooth real valued function. Then

$$\phi^*(dx_i) = x_i\psi'(t)dt + \psi(t)dx_i.$$

To a smooth map $\phi\colon U_1 \to U_2$ we can now associate a linear map

$$H^p(\phi) : H^p(U_2) \to H^p(U_1)$$

by setting $H^p(\phi)[\omega] = [\Omega^p(\phi)(\omega)] \ (= [\phi^*(\omega)])$. The definition is independent of the choice of representative, since $\phi^*(\omega + dv) = \phi^*(\omega) + \phi^*(dv) = \phi^*(\omega) + d\phi^*(v)$. Furthermore,

$$H^{p+q}(\phi)([\omega_1][\omega_2]) = (H^p(\phi)[\omega_1])(H^q(\phi)[\omega_2])$$

such that $H^*(\phi)\colon H^*(U_2) \to H^*(U_1)$ is a homomorphism of graded algebras.

Theorem 3.15 (Poincaré's lemma) *If U is a star-shaped open set then $H^p(U) = 0$ for $p > 0$, and $H^0(U) = \mathbb{R}$.*

Proof. We may assume U to be star-shaped with respect to the origin $0 \in \mathbb{R}^n$, and wish to construct a linear operator

$$S_p: \Omega^p(U) \to \Omega^{p-1}(U)$$

such that $dS_p + S_{p+1}d = \mathrm{id}$ when $p > 0$ and $S_1 d = \mathrm{id} - e$, where $e(\omega) = \omega(0)$ for $\omega \in \Omega^0(U)$. Such an operator immediately implies our theorem, since $dS_p(\omega) = \omega$ for a closed p-form, $p > 0$, and hence $[\omega] = 0$. If $p = 0$ we have $\omega - \omega(0) = S_1 d\omega = 0$, and ω must be constant.

First we construct

$$\hat{S}_p: \Omega^p(U \times \mathbb{R}) \to \Omega^{p-1}(U).$$

Every $\omega \in \Omega^p(U \times \mathbb{R})$ can be written in the form

$$\omega = \sum f_I(x,t)dx_I + \sum g_J(x,t)dt \wedge dx_J$$

where $I = (i_1, \ldots, i_p)$ and $J = (j_1, \ldots, j_{p-1})$. We define

$$\hat{S}_p(\omega) = \sum \left(\int_0^1 g_J(x,t)dt \right) dx_J.$$

Then we have that

$$
\begin{aligned}
d\hat{S}_p(\omega) + \hat{S}_{p+1}d(\omega) &= \sum_{J,i} \left(\int_0^1 \frac{\partial g_J(x,t)}{\partial x_i}dt \right) dx_i \wedge dx_J \\
&+ \sum_I \left(\int_0^1 \frac{\partial f_I(x,t)}{\partial t}dt \right) dx_I - \sum_{J,i} \left(\int_0^1 \frac{\partial g_J}{\partial x_i}dt \right) dx_i \wedge dx_J \\
&= \sum \left(\int_0^1 \frac{\partial f_I(x,t)}{\partial t}dt \right) dx_I \\
&= \sum f_I(x,1)dx_I - \sum f_I(x,0)dx_I.
\end{aligned}
$$

We apply this result to $\phi^*(\omega)$, where

$$\phi: U \times \mathbb{R} \to U, \ \phi(x,t) = \psi(t)x$$

and $\psi(t)$ is a smooth function for which

$$
\begin{cases}
\psi(t) = 0 & \text{if } t \leq 0 \\
\psi(t) = 1 & \text{if } t \geq 1 \\
0 \leq \psi(t) \leq 1 & \text{otherwise.}
\end{cases}
$$

Define $S_p(\omega) = \hat{S}_p(\phi^*(\omega))$ with $\hat{S}_p \colon \Omega^p(U \times \mathbb{R}) \to \Omega^{p-1}(U)$ as above. Assume that $\omega = \sum h_I(x)dx_I$. From Example 3.14 we have

$$\phi^*(\omega) = \sum h_I(\psi(t)x)(d\psi(t)x_{i_1} + \psi(t)dx_{i_1}) \wedge \ldots \wedge \left(d\psi(t)x_{i_p} + \psi(t)dx_{i_p}\right)$$

In the notation used above we then get that

$$\sum f_I(x,t)dx_I = \sum h_I(\psi(t)x)\psi(t)^p dx_I.$$

This implies that

$$dS_p(\omega) + S_{p+1}d(\omega) = \begin{cases} \sum_I h_I(x)dx_I = \omega & p > 0 \\ \omega(x) - \omega(0) & p = 0. \end{cases}$$

$\square$

4. CHAIN COMPLEXES AND THEIR HOMOLOGY

In this chapter we present some general algebraic definitions and viewpoints, which should illuminate some of the constructions of Chapter 3. The algebraic results will be applied later to de Rham cohomology in Chapters 5 and 6.

A sequence of vector spaces and linear maps

(1) $$A \xrightarrow{f} B \xrightarrow{g} C$$

is said to be *exact* when $\operatorname{Im} f = \operatorname{Ker} g$, where as above

$$\operatorname{Ker} g = \{b \in B | g(b) = 0\} \quad \text{(the kernel of } g\text{)}$$
$$\operatorname{Im} f = \{f(a) | a \in A\} \quad \text{(the image of } f\text{)}.$$

Note that $A \xrightarrow{f} B \to 0$ is exact precisely when f is surjective and that $0 \to B \xrightarrow{g} C$ is exact precisely when g is injective. A sequence $A^* = \{A^i, d^i\}$,

(2) $$\cdots \to A^{i-1} \xrightarrow{d^{i-1}} A^i \xrightarrow{d^i} A^{i+1} \xrightarrow{d^{i+1}} A^{i+2} \to \cdots$$

of vector spaces and linear maps is called a *chain complex* provided $d^{i+1} \circ d^i = 0$ for all i. It is exact if

$$\operatorname{Ker} d^i = \operatorname{Im} d^{i-1}$$

for all i. An exact sequence of the form

(3) $$0 \to A \xrightarrow{f} B \xrightarrow{g} C \to 0$$

is called *short exact*. This is equivalent to requiring that

$$f \text{ is injective,} \quad g \text{ is surjective} \quad \text{and} \quad \operatorname{Im} f = \operatorname{Ker} g.$$

The *cokernel* of a linear map $f: A \to B$ is

$$\operatorname{Cok}(f) = B/\operatorname{Im} f.$$

For a short exact sequence, g induces an isomorphism

$$g: \operatorname{Cok}(f) \xrightarrow{\cong} C.$$

Every (long) exact sequence, as in (2), induces short exact sequences (which can be used to calculate A^i)

$$0 \to \operatorname{Im} d^{i-1} \to A^i \to \operatorname{Im} d^i \to 0.$$

Furthermore the isomorphisms

$$A^{i-1}/\text{Im}\, d^{i-2} \cong A^{i-1}/\text{Ker}\, d^{i-1} \overset{d^{i-1}}{\underset{\cong}{\to}} \text{Im}\, d^{i-1}$$

are frequently applied in concrete calculations.

The *direct sum* of vector spaces A and B is the vector space

$$A \oplus B = \{(a, b) | a \in A, b \in B\}$$
$$\lambda(a, b) = (\lambda a, \lambda b), \quad \lambda \in \mathbb{R}$$
$$(a_1, b_1) + (a_2, b_2) = (a_1 + a_2, b_1 + b_2).$$

If $\{a_i\}$ and $\{b_j\}$ are bases of A and B, respectively, then $\{(a_i, 0), (0, b_j)\}$ is a basis of $A \oplus B$. In particular

$$\dim(A \oplus B) = \dim A + \dim B.$$

Lemma 4.1 *Suppose* $0 \to A \overset{f}{\to} B \overset{g}{\to} C \to 0$ *is a short exact sequence of vector spaces. Then B is finite-dimensional if both A and C are, and $B \cong A \oplus C$.*

Proof. Choose a basis $\{a_i\}$ of A and $\{c_j\}$ of C. Since g is surjective there exist $b_j \in B$ with $g(b_j) = c_j$. Then $\{f(a_i), b_j\}$ is a basis of B: For $b \in B$ we have $g(b) = \sum \lambda_j c_j$. Hence $b - \sum \lambda_j b_j \in \text{Ker}\, g$. Since $\text{Ker}\, g = \text{Im}\, f$, $b - \sum \lambda_j b_j = f(a)$, so

$$b - \sum \lambda_j b_j = f\left(\sum \mu_i a_i\right) = \sum \mu_i f(a_i).$$

This shows that b can be written as a linear combination of $\{b_j\}$ and $\{f(a_i)\}$. It is left to the reader to show that $\{b_j, f(a_i)\}$ are linearly independent. $\square$

Definition 4.2 For a chain complex $A^* = \{\cdots \to A^{p-1} \overset{d^{p-1}}{\to} A^p \overset{d^p}{\to} A^{p+1} \to \cdots\}$ we define the p-th cohomology vector space to be

$$H^p(A^*) = \text{Ker}\, d^p/\text{Im}\, d^{p-1}.$$

The elements of $\text{Ker}\, d^p$ are called *p-cycles* (or are said to be *closed*) and the elements of $\text{Im}\, d^{p-1}$ are called *p-boundaries* (or said to be *exact*). The elements of $H^p(A^*)$ are called *cohomology classes*.

A *chain map* $f: A^* \to B^*$ between chain complexes consists of a family $f^p: A^p \to B^p$ of linear maps, satisfying $d_B^p \circ f^p = f^{p+1} \circ d_A^p$. A chain map is illustrated as the commutative diagram

$$
\begin{array}{ccccccc}
\cdots \longrightarrow & A^{p-1} & \overset{d^{p-1}}{\longrightarrow} & A^p & \overset{d^p}{\longrightarrow} & A^{p+1} & \longrightarrow \cdots \\
& \downarrow{f^{p-1}} & & \downarrow{f^p} & & \downarrow{f^{p+1}} & \\
\cdots \longrightarrow & B^{p-1} & \overset{d^{p-1}}{\longrightarrow} & B^p & \overset{d^p}{\longrightarrow} & B^{p+1} & \longrightarrow \cdots
\end{array}
$$

Lemma 4.3 *A chain map $f: A^* \to B^*$ induces a linear map*

$$f^* = H^*(f): H^p(A^*) \to H^p(B^*), \quad \text{for all } p.$$

Proof. Let $a \in A^p$ be a cycle ($d^p a = 0$) and $[a] = a + \operatorname{Im} d^{p-1}$ its corresponding cohomology class in $H^p(A^*)$. We define $f^*([a]) = [f^p(a)]$. Two remarks are needed. First, we have $d_B^p f^p(a) = f^{p+1} d_A^p(a) = f^{p+1}(0) = 0$. Hence $f^p(a)$ is a cycle. Second, $[f^p(a)]$ is independent of which cycle a we choose in the class $[a]$. If $[a_1] = [a_2]$ then $a_1 - a_2 \in \operatorname{Im} d_A^{p-1}$, and $f^p(a_1 - a_2) = f^p d_A^{p-1}(x) = d_B^{p-1} f^{p-1}(x)$. Hence $f^p(a_1) - f^p(a_2) \in \operatorname{Im} d_B^{p-1}$, and $f^p(a_1), f^p(a_2)$ define the same cohomology class. $\qquad\square$

A *category* $\mathcal{C}$ consists of "objects" and "morphisms" between them, such that "composition" is defined. If $f: C_1 \to C_2$ and $g: C_2 \to C_3$ are morphisms, then there exists a morphism $g \circ f: C_1 \to C_3$. Furthermore it is to be assumed that $\operatorname{id}_C: C \to C$ is a morphism for every object C of $\mathcal{C}$. The concept is best illustrated by examples:

> The category of open sets in Euclidean spaces, where the morphisms are the smooth maps.

> The category of vector spaces, where the morphisms are the linear maps.

> The category of abelian groups, where the morphisms are homomorphisms.

> The category of chain complexes, where the morphisms are the chain maps.

> A category with just one object is the same as a semigroup, namely the semigroup of morphisms of the object.

> Every partially ordered set is a category with one morphism from c to d, when $c \leq d$.

A *contravariant functor* $F: \mathcal{C} \to \mathcal{V}$ between two categories maps every object $C \in \operatorname{ob} \mathcal{C}$ to an object $F(C) \in \operatorname{ob} \mathcal{V}$, and every morphism $f: C_1 \to C_2$ in $\mathcal{C}$ to a morphism $F(f): F(C_2) \to F(C_1)$ in $\mathcal{V}$, such that

$$F(g \circ f) = F(f) \circ F(g), \quad F(\operatorname{id}_C) = \operatorname{id}_{F(C)}.$$

A *covariant functor* $F: \mathcal{C} \to \mathcal{V}$ is an assignment in which $F(f): F(C_1) \to F(C_2)$, and

$$F(g \circ f) = F(g) \circ F(f), \quad F(\operatorname{id}_C) = \operatorname{id}_{F(C)}.$$

Functors thus are the "structure-preserving" assignments between categories. The contravariant ones change the direction of the arrows, the covariant ones preserve directions. We give a few examples:

> Let A be a vector space and $F(C) = \text{Hom}(C, A)$, the linear maps from C to A. For $\phi: C_1 \to C_2$, $\text{Hom}(\phi, A): \text{Hom}(C_2, A) \to \text{Hom}(C_1, A)$ is given by $\text{Hom}(\phi, A)(\psi) = \psi \circ \phi$. This is a contravariant functor from the category of vector spaces to itself.

> $F(C) = \text{Hom}(A, C)$, $\quad F(\phi): \psi \mapsto \phi \circ \psi$. This is a covariant functor from the category of vector spaces to itself.

> Let $\mathcal{U}$ be the category of open sets in Euclidean spaces and smooth maps, and Vect the category of vector spaces. The vector space of differential p-forms on $U \in \mathcal{U}$ defines a contravariant functor

$$\Omega^p: \mathcal{U} \to \text{Vect}.$$

> Let Vect* be the category of chain complexes. The de Rham complex defines a contravariant functor $\Omega^*: \mathcal{U} \to \text{Vect}^*$.

> For every p the homology $H^p: \text{Vect}^* \to \text{Vect}$ is a covariant functor.

> The composition of the two functors above is exactly the de Rham cohomology functor $H^p: \mathcal{U} \to \text{Vect}$. It is contravariant.

A short exact sequence of chain complexes

$$0 \to A^* \xrightarrow{f} B^* \xrightarrow{g} C^* \to 0$$

consists of chain maps f and g such that $0 \to A^p \xrightarrow{f} B^p \xrightarrow{g} C^p \to 0$ is exact for every p.

Lemma 4.4 *For a short exact sequence of chain complexes the sequence*

$$H^p(A^*) \xrightarrow{f^*} H^p(B^*) \xrightarrow{g^*} H^p(C^*)$$

is exact.

Proof. Since $g^p \circ f^p = 0$ we have

$$g^* \circ f^*([a]) = g^*([f^p(a)]) = [g^p(f^p(a))] = 0$$

for every cohomology class $[a] \in H^p(A^*)$. Conversely, assume for $[b] \in H^p(B)$ that $g^*[b] = 0$. Then $g^p(b) = d_C^{p-1}(c)$. Since g^{p-1} is surjective, there exists

$b_1 \in B^{p-1}$ with $g^{p-1}(b_1) = c$. It follows that $g^p\big(b - d_B^{p-1}(b_1)\big) = 0$. Hence there exist $a \in A^p$ with $f^p(a) = b - d_B^{p-1}(b_1)$. We will show that a is a p-cycle. Since f^{p+1} is injective, it is sufficient to note that $f^{p+1}\big(d_A^p(a)\big) = 0$. But

$$f^{p+1}\big(d_A^p(a)\big) = d_B^p\big(f^p(a)\big) = d_B^p\big(b - d_B^{p-1}(b_1)\big) = 0$$

since b is a p-cycle and $d^p \circ d^{p-1} = 0$. We have thus found a cohomology class $[a] \in H^p(A)$, and $f^*[a] = [b - d_B^{p-1}(b_1)] = [b]$. □

One might expect that the sequence of Lemma 4.4 could be extended to a short exact sequence, but this is not so. The problem is that, even though $g^p \colon B^p \to C^p$ is surjective, the pre-image $(g^p)^{-1}(c)$ of a p-cycle with $c \in C^p$ need not contain a cycle. We shall measure when this is the case by introducing

Definition 4.5 For a short exact sequence of chain complexes $0 \to A^* \xrightarrow{f} B^* \xrightarrow{g} C^* \to 0$ we define

$$\partial^* \colon H^p(C^*) \to H^{p+1}(A^*)$$

to be the linear map given by

$$\partial^*([c]) = \left[(f^{p+1})^{-1}\Big(d_B^p\big((g^p)^{-1}(c)\big)\Big)\right].$$

There are several things to be noted. The definition expresses that for every $b \in (g^p)^{-1}(c)$ we have $d_B^p(b) \in \mathrm{Im}\big(f^{p+1}\big)$, and that the uniquely determined $a \in A^{p+1}$ with $f^{p+1}(a) = d_B^p(b)$ is a $(p+1)$-cycle. Finally it is postulated that $[a] \in H^{p+1}(A^*)$ is independent of the choice of $b \in (g^p)^{-1}(c)$.

In order to prove these assertions it is convenient to write the given short exact sequence in a diagram:

The slanted arrow indicates the definition of ∂^*. We shall now prove the necessary assertions which, when combined, make ∂^* well-defined. Namely:

(i)   If $g^p(b) = c$ and $d_C^p(c) = 0$ then $d_B^p(b) \in \mathrm{Im} f^{p+1}$.

(ii) If $f^{p+1}(a) = d_B^p(b)$ then $d_A^{p+1}(a) = 0$.

(iii) If $g^p(b_1) = g^p(b_2) = c$ and $f^{p+1}(a_i) = d_B^p(b_i)$ then $[a_1] = [a_2] \in H^{p+1}(A^*)$.

The first assertion follows, because $g^{p+1}d_B^p(b) = d_C^p(c) = 0$, and $\operatorname{Ker} g^{p+1} = \operatorname{Im} f^{p+1}$; (ii) uses the injectivity of f^{p+2} and that $f^{p+2}d_A^{p+1}(a) = d_B^{p+1}f^{p+1}(a) = d_B^{p+1}d_B^p(b) = 0$; (iii) follows since $b_1 - b_2 = f^p(a)$ so that $d_B^p(b_1) - d_B^p(b_2) = d_B^pf^p(a) = f^{p+1}d_A^p(a)$, and therefore $(f^{p+1})^{-1}(d_B^p(b_1)) = (f^{p+1})^{-1}(d_B^p(b_2)) + d_A^p(a)$.

Example 4.6 Here is a short exact sequence of chain complexes (the dots indicate that the chain groups are zero) with $\partial^* \neq 0$:

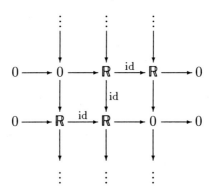

One can easily verify that $\partial^* \colon \mathbb{R} \to \mathbb{R}$ is an isomorphism.

Lemma 4.7 *The sequence* $H^p(B^*) \xrightarrow{g^*} H^p(C^*) \xrightarrow{\partial^*} H^{p+1}(A^*)$ *is exact.*

Proof. We have $\partial^*g^*([b]) = \partial^*[g^p(b)] = [(f^{p+1})^{-1}(d_B(b))] = 0$. Conversely assume that $\partial^*[c] = 0$. Choose $b \in B^p$ with $g^p(b) = c$ and $a \in A^p$, such that

$$d_B^p(b) = f^{p+1}(d_A^p a).$$

Now we have $d_B^p(b - f^p(a)) = 0$ and $g^p(b - f^p(a)) = c$. Hence $g^*[b - f^p(a)] = [c]$. □

Lemma 4.8 *The sequence* $H^p(C^*) \xrightarrow{\partial^*} H^{p+1}(A^*) \xrightarrow{f^*} H^{p+1}(B^*)$ *is exact.*

Proof. We have $f^*\partial^*([c]) = [d_B^p(b)] = 0$, where $g^p(b) = c$. Conversely assume that $f^*[a] = 0$, i.e. $f^{p+1}(a) = d_B^p(b)$. Then $d_C^p(g^p(b)) = g^{p+1}f^{p+1}(a) = 0$, and $\partial^*[g^p(b)] = [a]$. □

We can sum up Lemmas 4.4, 4.7 and 4.8 in the important

Theorem 4.9 (Long exact homology sequence). *Let* $0 \to A^* \xrightarrow{f} B^* \xrightarrow{g} C^* \to 0$ *be a short exact sequence of chain complexes. Then the sequence*

$$\cdots \to H^p(A^*) \xrightarrow{f^*} H^p(B^*) \xrightarrow{g^*} H^p(C^*) \xrightarrow{\partial^*} H^{p+1}(A^*) \xrightarrow{f^*} H^{p+1}(B^*) \to \cdots$$

is exact. □

Definition 4.10 Two chain maps $f, g: A^* \to B^*$ are said to be *chain-homotopic*, if there exist linear maps $s: A^p \to B^{p-1}$ satisfying

$$d_B s + s d_A = f - g: A^p \to B^p$$

for every p.

In the form of a diagram, a chain homotopy is given by the slanted arrows.

The name *chain homotopy* will be explained in Chapter 6.

Lemma 4.11 *For two chain-homotopic chain maps $f, g: A^* \to B^*$ we have that*

$$f^* = g^*: H^p(A^*) \to H^p(B^*).$$

Proof. If $[a] \in H^p(A^*)$ then

$$(f^* - g^*)[a] = [f^p(a) - g^p(a)] = \left[d_B^{p-1} s(a) + s d_A^p(a) \right] = \left[d_B^{p-1} s(a) \right] = 0. \quad \square$$

Remark 4.12 In the proof of the Poincaré lemma in Chapter 3 we constructed linear maps

$$S^p: \Omega^p(U) \to \Omega^{p-1}(U)$$

with $d^{p-1} S^p + S^{p+1} d^p = \mathrm{id}$ for $p > 0$. This is a chain homotopy between id and 0 (for $p > 0$), such that

$$\mathrm{id}^* = 0^*: H^p(U) \to H^p(U), \quad p > 0.$$

However $\mathrm{id}^* = \mathrm{id}$ and $0^* = 0$. Hence $\mathrm{id} = 0$ on $H^p(U)$, and $H^p(U) = 0$ when $p > 0$.

Lemma 4.13 *If A^* and B^* are chain complexes then*

$$H^p(A^* \oplus B^*) = H^p(A^*) \oplus H^p(B^*).$$

Proof. It is obvious that

$$\mathrm{Ker}\left(d_{A \oplus B}^p \right) = \mathrm{Ker}\, d_A^p \oplus \mathrm{Ker}\, d_B^p$$
$$\mathrm{Im}\left(d_{A \oplus B}^{p-1} \right) = \mathrm{Im}\, d_A^{p-1} \oplus \mathrm{Im}\, d_B^{p-1}$$

and the lemma follows. $\square$

5. THE MAYER-VIETORIS SEQUENCE

This chapter introduces a fundamental calculational technique for de Rham cohomology, namely the so-called Mayer-Vietoris sequence, which calculates $H^*(U_1 \cup U_2)$ as a "function" of $H^*(U_1)$, $H^*(U_2)$ and $H^*(U_1 \cap U_2)$. Here U_1 and U_2 are open sets in $\mathbb{R}^n$. By iteration we get a calculation of $H^*(U_1 \cup \cdots \cup U_n)$ as a "function" of $H^*(U_\alpha)$, where α runs over the subsets of $\{1, \dots, n\}$ and $U_\alpha = U_{i_1} \cap \cdots \cap U_{i_r}$ when $\alpha = \{i_1, \dots, i_r\}$. Combined with the Poincaré lemma, this yields a *principal* calculation of $H^*(U)$ for quite general open sets in $\mathbb{R}^n$. If, for instance, U can be covered by a finite number of convex open sets U_i, then every U_α will also be convex and $H^*(U_\alpha)$ thus known from the Poincaré lemma.

Theorem 5.1 *Let U_1 and U_2 be open sets of $\mathbb{R}^n$ with union $U = U_1 \cup U_2$. For $\nu = 1, 2$, let $i_\nu \colon U_\nu \to U$ and $j_\nu \colon U_1 \cap U_2 \to U_\nu$ be the corresponding inclusions. Then the sequence*

$$0 \to \Omega^p(U) \xrightarrow{I^p} \Omega^p(U_1) \oplus \Omega^p(U_2) \xrightarrow{J^p} \Omega^p(U_1 \cap U_2) \to 0$$

is exact, where $I^p(\omega) = (i_1^*(\omega), i_2^*(\omega))$, $J^p(\omega_1, \omega_2) = j_1^*(\omega_1) - j_2^*(\omega_2)$.

Proof. For a smooth map $\phi \colon V \to W$ and a p-form $\omega = \Sigma f_I dx_I \in \Omega^p(W)$,

$$\Omega^p(\phi)(\omega) = \phi^*(\omega) = \Sigma(f_I \circ \phi) d\phi_{i_1} \wedge \dots \wedge d\phi_{i_p}.$$

In particular, if ϕ is an inclusion of open sets in $\mathbb{R}^n$, i.e. $\phi_i(x) = x_i$, then

$$d\phi_{i_1} \wedge \dots \wedge d\phi_{i_p} = dx_{i_1} \wedge \dots \wedge dx_{i_p}.$$

Hence

(1) $$\phi^*(\omega) = \Sigma(f_I \circ \phi) dx_I.$$

This will be used for $\phi = i_\nu, j_\nu$, $\nu = 1, 2$. It follows from (1) that I^p is injective. If namely $I^p(\omega) = 0$ then $i_1^*(\omega) = 0 = i_2^*(\omega)$, and

$$i_\nu^*(\omega) = \Sigma(f_I \circ i_\nu) dx_I = 0$$

if and only if $f_I \circ i_\nu = 0$ for all I. However $f_I \circ i_1 = 0$ and $f_I \circ i_2 = 0$ imply that $f_I = 0$ on all of U, since U_1 and U_2 cover U. Similarly we show that $\operatorname{Ker} J^p = \operatorname{Im} I^p$. First,

$$J^p \circ I^p(\omega) = j_2^* i_2^*(\omega) - j_1^* i_1^*(\omega) = j^*(\omega) - j^*(\omega) = 0,$$

where $j: U_1 \cap U_2 \to U$ is the inclusion. Hence $\operatorname{Im} I^p \subseteq \operatorname{Ker} J^p$. To show the converse inclusion we start with two p-forms $\omega_\nu \in \Omega^p(U_\nu)$,

$$\omega_1 = \Sigma f_I dx_I, \qquad \omega_2 = \Sigma g_I dx_I.$$

Since $J^p(\omega_1, \omega_2) = 0$ we have that $j_1^*(\omega_1) = j_2^*(\omega_2)$, which by (1) translates into $f_I \circ j_1 = g_I \circ j_2$ or $f_I(x) = g_I(x)$ for $x \in U_1 \cap U_2$. We define a smooth function $h_I: U \to \mathbb{R}^n$ by

$$h_I(x) = \begin{cases} f_I(x), & x \in U_1 \\ g_I(x), & x \in U_2. \end{cases}$$

Then $I^p(\sum h_I dx_I) = (\omega_1, \omega_2)$. Finally we show that J^p is surjective. To this end we use a partition of unity $\{p_1, p_2\}$ with support in $\{U_1, U_2\}$, i.e. smooth functions

$$p_\nu: U \to [0, 1], \quad \nu = 1, 2$$

for which $\operatorname{supp}_U(p_\nu) \subset U_\nu$, and such that $p_1(x) + p_2(x) = 1$ for $x \in U$ (cf. Appendix A).

Let $f: U_1 \cap U_2 \to \mathbb{R}$ be a smooth function. We use $\{p_1, p_2\}$ to extend f to U_1 and U_2. Since $\operatorname{supp}_U(p_1) \cap U_2 \subset U_1 \cap U_2$ we can define a smooth function by

$$f_2(x) = \begin{cases} -f(x)p_1(x) & \text{if } x \in U_1 \cap U_2 \\ 0 & \text{if } x \in U_2 - \operatorname{supp}_U(p_1). \end{cases}$$

Analogously we define

$$f_1(x) = \begin{cases} f(x)p_2(x) & \text{if } x \in U_1 \cap U_2 \\ 0 & \text{if } x \in U_1 - \operatorname{supp}_U(p_2). \end{cases}$$

Note that $f_1(x) - f_2(x) = f(x)$ when $x \in U_1 \cap U_2$, because $p_1(x) + p_2(x) = 1$. For a differential form $\omega \in \Omega^p(U_1 \cap U_2)$, $\omega = \sum f_I dx_I$, we can apply the above to each of the functions $f_I: U_1 \cap U_2 \to \mathbb{R}$. This yields the functions $f_{I,\nu}: U_\nu \to \mathbb{R}$, and thus the differential forms $\omega_\nu = \sum f_{I,\nu} dx_I \in \Omega^p(U_\nu)$. With this choice $J^p(\omega_1, \omega_2) = \omega$. $\square$

It is clear that

$$I: \Omega^*(U) \to \Omega^*(U_1) \oplus \Omega^*(U_2)$$
$$J: \Omega^*(U_1) \oplus \Omega^*(U_2) \to \Omega^*(U_1 \cap U_2)$$

are chain maps, so that Theorem 5.1 yields a short exact sequence of chain complexes. From Theorem 4.9 one thus obtains a long exact sequence of cohomology vector spaces. Finally Lemma 4.13 tells us that

$$H^p(\Omega^*(U_1) \oplus \Omega^*(U_2)) = H^p(\Omega^*(U_1)) \oplus H^p(\Omega^*(U_2)).$$

We have proved:

Theorem 5.2 (Mayer-Vietoris) *Let U_1 and U_2 be open sets in $\mathbb{R}^n$ and $U = U_1 \cup U_2$. There exists an exact sequence of cohomology vector spaces*

$$\cdots \to H^p(U) \xrightarrow{I^*} H^p(U_1) \oplus H^p(U_2) \xrightarrow{J^*} H^p(U_1 \cap U_2) \xrightarrow{\partial^*} H^{p+1}(U) \to \cdots$$

Here $I^*([\omega]) = (i_1^*[\omega], i_2^*[\omega])$ and $J^*([\omega_1], [\omega_2]) = j_1^*[\omega_1] - j_2^*[\omega_2]$ in the notation of Theorem 5.1. □

Corollary 5.3 *If U_1 and U_2 are disjoint open sets in $\mathbb{R}^n$ then*

$$I^*: H^p(U_1 \cup U_2) \to H^p(U_1) \oplus H^p(U_2)$$

is an isomorphism.

Proof. It follows from Theorem 5.1 that

$$I^p: \Omega^p(U_1 \cup U_2) \to \Omega^p(U_1) \oplus \Omega^p(U_2)$$

is an isomorphism, and Lemma 4.13 gives that the corresponding map on cohomology is also an isomorphism. □

Example 5.4 We use Theorem 5.2 to calculate the de Rham cohomology vector spaces of the punctured plane $\mathbb{R}^2 - \{0\}$. Let

$$U_1 = \mathbb{R}^2 - \{(x_1, x_2) \mid x_1 \geq 0, \ x_2 = 0\}$$
$$U_2 = \mathbb{R}^2 - \{(x_1, x_2) \mid x_1 \leq 0, \ x_2 = 0\}.$$

These are star-shaped open sets, such that $H^p(U_1) = H^p(U_2) = 0$ for $p > 0$ and $H^0(U_1) = H^0(U_2) = \mathbb{R}$. Their intersection

$$U_1 \cap U_2 = \mathbb{R}^2 - \mathbb{R} = \mathbb{R}_+^2 \cup \mathbb{R}_-^2$$

is the disjoint union of the open half-planes $x_2 > 0$ and $x_2 < 0$. Hence

$$(2) \qquad H^p(U_1 \cap U_2) = \begin{cases} 0 & \text{if } p > 0 \\ \mathbb{R} \oplus \mathbb{R} & \text{if } p = 0 \end{cases}$$

by the Poincaré lemma and Corollary 5.3. From the Mayer-Vietoris sequence we have

$$\cdots \to H^p(U_1) \oplus H^p(U_2) \xrightarrow{J^*} H^p(U_1 \cap U_2) \xrightarrow{\partial^*}$$
$$H^{p+1}(\mathbb{R}^2 - \{0\}) \xrightarrow{I^*} H^{p+1}(U_1) \oplus H^{p+1}(U_2) \to \cdots$$

For $p > 0$,

$$0 \to H^p(U_1 \cap U_2) \xrightarrow{\partial^*} H^{p+1}(\mathbb{R}^2 - \{0\}) \to 0$$

is exact, i.e. ∂^* is an isomorphism, and $H^q(\mathbb{R}^2 - \{0\}) = 0$ for $q \geq 2$ according to (2).

If $p = 0$, one gets the exact sequence

$$(3) \qquad \begin{array}{c} H^{-1}(U_1 \cap U_2) \to H^0(\mathbb{R}^2 - \{0\}) \xrightarrow{I^0} H^0(U_1) \oplus H^0(U_2) \xrightarrow{J^0} \\ H^0(U_1 \cap U_2) \xrightarrow{\partial^*} H^1(\mathbb{R}^2 - \{0\}) \xrightarrow{I^1} H^1(U_1) \oplus H^1(U_2). \end{array}$$

Since $H^{-1}(U) = 0$ for all open sets, and in particular $H^{-1}(U_\nu) = 0$, I^0 is injective. Since $H^1(U_\nu) = 0$, ∂^* is surjective, and the sequence (3) reduces to the exact sequence

$$\begin{array}{ccccc} & \mathbb{R} \oplus \mathbb{R} & & \mathbb{R} \oplus \mathbb{R} & \\ & \| & & \| & \\ 0 \to H^0(\mathbb{R}^2 - \{0\}) \xrightarrow{I^0} H^0(U_1) \oplus H^0(U_2) \xrightarrow{J^0} H^0(U_1 \cap U_2) \xrightarrow{\partial^*} H^1(\mathbb{R}^2 - \{0\}) \to 0. \end{array}$$

However, $\mathbb{R}^2 - \{0\}$ is connected. Hence $H^0(\mathbb{R}^2 - \{0\}) \cong \mathbb{R}$, and since I^0 is injective we must have that $\operatorname{Im} I^0 \cong \mathbb{R}$. Exactness gives $\operatorname{Ker} J^0 \cong \mathbb{R}$, so that J^0 has rank 1. Therefore $\operatorname{Im} J^0 \cong \mathbb{R}$ and, once again, by exactness

$$\partial^* : H^0(U_1 \cap U_2) \, / \, \operatorname{Im} J^0 \xrightarrow{\cong} H^1(\mathbb{R}^2 - \{0\}).$$

Since $H^0(U_1 \cap U_2) \, / \, \operatorname{Im} J^0 \cong \mathbb{R}$, we have shown

$$H^p(\mathbb{R} - \{0\}) = \begin{cases} 0 & \text{if } p \geq 2 \\ \mathbb{R} & \text{if } p = 1 \\ \mathbb{R} & \text{if } p = 0. \end{cases}$$

In the proof above we could alternatively have calculated

$$J^0 : H^0(U_1) \oplus H^0(U_2) \to H^0(U_1 \cap U_2)$$

by using Lemma 3.9: $H^0(U)$ consists of locally constant functions. If f_i is a constant function on U_i, then

$$J^0(f_1) = f_{1|U_1 \cap U_2} \quad \text{and} \quad J^0(f_2) = -f_{2|U_1 \cap U_2}$$

so that $J^0(a, b) = a - b$.

Theorem 5.5 *Assume that the open set U is covered by convex open sets $U_1, \ldots, U_r$. Then $H^p(U)$ is finitely generated.*

Proof. We use induction on the number of open sets. If $r = 1$ the assertion follows from the Poincaré lemma. Assume the assertion is proved for $r - 1$ and let $V = U_1 \cup \cdots \cup U_{r-1}$, such that $U = V \cup U_r$. From Theorem 5.2 we have the exact sequence

$$H^{p-1}(V \cap U_r) \xrightarrow{\partial^*} H^p(U) \xrightarrow{I^*} H^p(V) \oplus H^p(U_r)$$

which by Lemma 4.1 yields

$$H^p(U) \simeq \mathrm{Im}\partial^* \oplus \mathrm{Im}I^*.$$

Now both V and $V \cap U_r = (U_1 \cap U_r) \cup \cdots \cup (U_{r-1} \cap U_r)$ are unions of $(r-1)$ convex open sets. Therefore Theorem 5.5 holds for $H^*(V \cap U_r), H^*(V)$ and $H^*(U_r)$, and hence also for $H^*(U)$. $\qquad\square$

6. HOMOTOPY

In this chapter we show that de Rham cohomology is functorial on the category of continuous maps between open sets in Euclidean spaces and calculate $H^*(\mathbb{R}^n - \{0\})$.

Definition 6.1 Two continuous maps $f_\nu: X \to Y$, $\nu = 0, 1$ between topological spaces are said to be *homotopic*, if there exists a continuous map

$$F: X \times [0, 1] \to Y$$

such that $F(x, \nu) = f_\nu(x)$ for $\nu = 0, 1$ and all $x \in X$.

This is denoted by $f_0 \simeq f_1$, and F is called a *homotopy* from f_0 to f_1. It is convenient to think of F as a family of continuous maps $f_t: X \to Y$ $(0 \leq t \leq 1)$, given by $f_t(x) = F(x, t)$, which deform f_0 to f_1.

Lemma 6.2 *Homotopy is an equivalence relation.*

Proof. If F is a homotopy from f_0 to f_1, a homotopy from f_1 to f_0 is defined by $G(x, t) = F(x, 1 - t)$. If $f_0 \simeq f_1$ via F and $f_1 \simeq f_2$ via G, then $f_0 \simeq f_2$ via

$$H(x, t) = \begin{cases} F(x, 2t) & 0 \leq t \leq \frac{1}{2} \\ G(x, 2t - 1) & \frac{1}{2} \leq t \leq 1. \end{cases}$$

Finally we have that $f \simeq f$ via $F(x, t) = f(x)$. $\qquad\square$

Lemma 6.3 *Let X, Y and Z be topological spaces and let $f_\nu: X \to Y$ and $g_\nu: Y \to Z$ be continuous maps for $\nu = 0, 1$. If $f_0 \simeq f_1$ and $g_0 \simeq g_1$ then $g_0 \circ f_0 \simeq g_1 \circ f_1$.*

Proof. Given homotopies F from f_0 to f_1 and G from g_0 to g_1, the homotopy H from $g_0 \circ f_0$ to $g_1 \circ f_1$ can be defined by $H(x, t) = G(F(x, t), t)$. $\qquad\square$

Definition 6.4 A continuous map $f: X \to Y$ is called a *homotopy equivalence*, if there exists a continuous map $g: Y \to X$, such that $g \circ f \simeq \mathrm{id}_X$ and $f \circ g \simeq \mathrm{id}_Y$. Such a map g is said to be a *homotopy inverse* to f.

Two topological spaces X and Y are called *homotopy equivalent* if there exists a homotopy equivalence between them. We say that X is *contractible*, when X is homotopy equivalent to a single-point space. This is the same as saying that id_X is homotopic to a constant map. The equivalence classes of topological spaces defined by the relation homotopy equivalence are called *homotopy types*.

Example 6.5 Let $Y \subseteq \mathbb{R}^m$ have the topology induced by $\mathbb{R}^m$. If, for the continuous maps $f_\nu: X \to Y$, $\nu = 0, 1$, the line segment in $\mathbb{R}^m$ from $f_0(x)$ to $f_1(x)$ is contained in Y for all $x \in X$, we can define a homotopy $F: X \times [0, 1] \to Y$ from f_0 to f_1 by

$$F(x, t) = (1 - t)f_0(x) + t\, f_1(x).$$

In particular this shows that a star-shaped set in $\mathbb{R}^m$ is contractible.

Lemma 6.6 *If U, V are open sets in Euclidean spaces, then*

(i) *Every continuous map $h: U \to V$ is homotopic to a smooth map.*
(ii) *If two smooth maps $f_\nu: U \to V$, $\nu = 0, 1$ are homotopic, then there exists a smooth map $F: U \times \mathbb{R} \to V$ with $F(x, \nu) = f_\nu(x)$ for $\nu = 0, 1$ and all $x \in U$ (F is called a smooth homotopy from f_0 to f_1).*

Proof. We use Lemma A.9 to approximate h by a smooth map $f: U \to V$. We can choose f such that V contains the line segment from $h(x)$ to $f(x)$ for every $x \in U$. Then $h \simeq f$ by Example 6.5.

Let G be a homotopy from f_0 to f_1. Use a continuous function $\psi: \mathbb{R} \to [0, 1]$ with $\psi(t) = 0$ for $t \leq \frac{1}{3}$ and $\psi(t) = 1$ for $t \geq \frac{2}{3}$ to construct

$$H: U \times \mathbb{R} \to V; \quad H(x, t) = G(x, \psi(t)).$$

Since $H(x, t) = f_0(x)$ for $t \leq \frac{1}{3}$ and $H(x, t) = f_1(x)$ for $t \geq \frac{2}{3}$, H is smooth on $U \times (-\infty, \frac{1}{3}) \cup U \times (\frac{2}{3}, \infty)$. Lemma A.9 allows us to approximate H by a smooth map $F: U \times \mathbb{R} \to V$ such that F and H have the same restriction on $U \times \{0, 1\}$. For $\nu = 0, 1$ and $x \in U$ we have that $F(x, \nu) = H(x, \nu) = f_\nu(x)$. $\square$

Theorem 6.7 *If $f, g: U \to V$ are smooth maps and $f \simeq g$ then the induced chain maps*

$$f^*, g^*: \Omega^*(V) \to \Omega^*(U)$$

are chain-homotopic (see Definition 4.10).

Proof. Recall, from the proof of Theorem 3.15, that every p-form ω on $U \times \mathbb{R}$ can be written as

$$\omega = \sum f_I(x, t)dx_I + \sum g_J(x, t)dt \wedge dx_J.$$

If $\phi: U \to U \times \mathbb{R}$ is the inclusion map $\phi(x) = \phi_0(x) = (x, 0)$, then

$$\phi^*(\omega) = \sum f_I(x, 0)d\phi_I = \sum f_I(x, 0)dx_I.$$

Indeed, $\phi^*(dt \wedge dx_J) = 0$ since the last component (the t-component) of ϕ is constant; see Example 3.11. Analogously, for $\phi_1(x) = (x, 1)$, we have that

$$\phi_1^*(\omega) = \sum f_I(x, 1)dx_I.$$

In the proof of Theorem 3.15 we constructed

$$\hat{S}_p \colon \Omega^p(U \times \mathbb{R}) \to \Omega^{p-1}(U)$$

such that

(1) $$(d\hat{S}_p + \hat{S}_{p+1}d)(\omega) = \phi_1^*(\omega) - \phi_0^*(\omega).$$

Consider the composition $U \xrightarrow{\phi_\nu} U \times \mathbb{R} \xrightarrow{F} V$, where F is a smooth homotopy between f and g. Then we have that $F \circ \phi_0 = f$ and $F \circ \phi_1 = g$. We define

$$S_p \colon \Omega^p(V) \to \Omega^{p-1}(U)$$

to be $S_p = \hat{S}_p \circ F^*$, and assert that

$$dS_p + S_{p+1}d = g^* - f^*.$$

This follows from (1) applied to $F^*(\omega)$, because

$$d\hat{S}_p(F^*(\omega)) + \hat{S}_{p+1}dF^*(\omega) = \phi_1^*F^*(\omega) - \phi_0^*F^*(\omega)$$
$$= (F \circ \phi_1)^*(\omega) - (F \circ \phi_0)^*(\omega) = g^*(\omega) - f^*(\omega).$$

Furthermore $\hat{S}_{p+1}dF^*(\omega) = \hat{S}_{p+1}F^*d(\omega) = S_{p+1}d(\omega)$, since F^* is a chain map.

$\square$

In the situation of Theorem 6.7, Lemma 4.11 states that $f^* = g^* \colon H^p(V) \to H^p(U)$. For a continuous map $\phi \colon U \to V$ we can find a smooth map $f \colon U \to V$ with $\phi \simeq f$ by (i) of Lemma 6.6, and by Lemma 6.2 and the result above we see that $f^* \colon H^p(V) \to H^p(U)$ is independent of the choice of f. Hence we can define

$$\phi^* = H^p(\phi) \colon H^p(V) \to H^p(U)$$

by setting $\phi^* = f^*$, where $f \colon U \to V$ is a smooth map homotopic to ϕ.

Theorem 6.8 *For $p \in \mathbb{Z}$ and open sets U, V, W in Euclidean spaces we have*

(i) *If $\phi_0, \phi_1 \colon U \to V$ are homotopic continuous maps, then*

$$\phi_0^* = \phi_1^* \colon H^p(V) \to H^p(U).$$

(ii) *If $\phi \colon U \to V$ and $\psi \colon V \to W$ are continuous, then $(\psi \circ \phi)^* = \phi^* \circ \psi^* \colon H^p(W) \to H^p(U)$.*

(iii) *If the continuous map $\phi \colon U \to V$ is a homotopy equivalence, then*

$$\phi^* \colon H^p(V) \to H^p(U)$$

is an isomorphism.

Proof. Choose a smooth map $f: U \to V$ with $\phi_0 \simeq f$. Lemma 6.2 gives that $\phi_1 \simeq f$ and (i) immediately follows. Part (ii), with smooth ϕ and ψ, follows from the formula

$$\Omega^p(\psi \circ \phi) = \Omega^p(\phi) \circ \Omega^p(\psi).$$

In the general case, choose smooth maps $f: U \to V$ and $g: V \to W$ with $\phi \simeq f$ and $\psi \simeq g$. Lemma 6.3 shows that $\psi \circ \phi \simeq g \circ f$, and we get

$$(\psi \circ \phi)^* = (g \circ f)^* = f^* \circ g^* = \phi^* \circ \psi^*.$$

If $\psi: V \to U$ is a homotopy inverse to ϕ, i.e.

$$\psi \circ \phi \simeq \mathrm{id}_U \quad \text{and} \quad \phi \circ \psi \simeq \mathrm{id}_V,$$

then it follows from (ii) that $\psi^*: H^p(U) \to H^p(V)$ is inverse to ϕ^*. $\square$

This result shows that $H^p(U)$ depends only on the homotopy type of U. In particular we have:

Corollary 6.9   (Topological invariance) *A homeomorphism* $h: U \to V$ *between open sets in Euclidean spaces induces isomorphisms* $h^*: H^p(V) \to H^p(U)$ *for all* p.

Proof. The corollary follows from Theorem 6.8.(iii), as $h^{-1}: V \to U$ is a homotopy inverse to h. $\square$

Corollary 6.10 *If* $U \subseteq \mathbb{R}^n$ *is an open contractible set, then* $H^p(U) = 0$ *when* $p > 0$ *and* $H^0(U) = \mathbb{R}$.

Proof. Let $F: U \times [0,1] \to U$ be a homotopy from $f_0 = \mathrm{id}_U$ to a constant map f_1 with value $x_0 \in U$. For $x \in U$, $F(x,t)$ defines a continuous curve in U, which connects x to x_0. Hence U is connected and $H^0(U) = \mathbb{R}$ by Lemma 3.9. If $p > 0$ then $\Omega^p(f_1): \Omega^p(U) \to \Omega^p(U)$ is the zero map. Hence by Theorem 6.8.(i) we get that

$$\mathrm{id}_{H^p(U)} = f_0^* = f_1^* = 0$$

and thus $H^p(U) = 0$. $\square$

In the proposition below, $\mathbb{R}^n$ is identified with the subspace $\mathbb{R}^n \times \{0\}$ of $\mathbb{R}^{n+1}$ and $\mathbb{R} \cdot 1$ denotes the 1-dimensional subspace consisting of constant functions.

Proposition 6.11 *For an arbitrary closed subset A of $\mathbb{R}$ with $A \neq \mathbb{R}^n$ we have isomorphisms*

$$H^{p+1}\left(\mathbb{R}^{n+1} - A\right) \cong H^p(\mathbb{R}^n - A) \qquad \text{for } p \geq 1$$
$$H^1\left(\mathbb{R}^{n+1} - A\right) \cong H^0(\mathbb{R}^n - A)/\mathbb{R} \cdot 1$$
$$H^0\left(\mathbb{R}^{n+1} - A\right) \cong \mathbb{R}.$$

Proof. Define open subsets of $\mathbb{R}^{n+1} = \mathbb{R}^n \times \mathbb{R}$,

$$U_1 = \mathbb{R}^n \times (0, \infty) \cup (\mathbb{R}^n - A) \times (-1, \infty)$$
$$U_2 = \mathbb{R}^n \times (-\infty, 0) \cup (\mathbb{R}^n - A) \times (-\infty, 1).$$

Then $U_1 \cup U_2 = \mathbb{R}^{n+1} - A$ and $U_1 \cap U_2 = (\mathbb{R}^n - A) \times (-1, 1)$. Let $\phi: U_1 \to U_1$ be given by adding 1 to the $(n+1)$-st coordinate. For $x \in U_1$, U_1 contains the line segments from x to $\phi(x)$ and from $\phi(x)$ to a fixed point in $\mathbb{R}^n \times (0, \infty)$. As in Example 6.5 we get homotopies from id_{U_1} to ϕ and from ϕ to a constant map. It follows that U_1 is contractible. Analogously U_2 is contractible, and $H^p(U_\nu)$ is described in Corollary 6.10.

Let pr be the projection of $U_1 \cap U_2 = (\mathbb{R}^n - A) \times (-1, 1)$ on $\mathbb{R}^n - A$. Define $i : \mathbb{R}^n - A \to U_1 \cap U_2$ by $i(y) = (y, 0)$. We have $\mathrm{pr} \circ i = \mathrm{id}_{\mathbb{R}^n - A}$ and $i \circ \mathrm{pr} \simeq \mathrm{id}_{U_1 \cap U_2}$. From Theorem 6.8.(iii) we conclude that

$$\mathrm{pr}^*: H^p(\mathbb{R}^n - A) \to H^p(U_1 \cap U_2)$$

is an isomorphism for every p. Theorem 5.2 gives isomorphisms

$$\partial^* : H^p(U_1 \cap U_2) \to H^{p+1}\left(\mathbb{R}^{n+1} - A\right)$$

for $p \geq 1$. By composition with pr^* one obtains the first part of Proposition 6.11. Consider the exact sequence

$$0 \to H^0\left(\mathbb{R}^{n+1} - A\right) \xrightarrow{I^*} H^0(U_1) \oplus H^0(U_2)$$
$$\xrightarrow{J^*} H^0(U_1 \cap U_2) \xrightarrow{\partial^*} H^1\left(\mathbb{R}^{n+1} - A\right) \to 0.$$

An element of $H^0(U_1) \oplus H^0(U_2)$ is given by a pair of constant functions on U_1 and U_2 with values a_1 and a_2. Their image under J^* is by Theorem 5.2 the constant function on $U_1 \cap U_2$ with the value $a_1 - a_2$. This shows that

$$\mathrm{Ker}\, \partial^* = \mathrm{Im}\, J^* = \mathbb{R} \cdot 1 \, ,$$

and we obtain the isomorphisms

$$H^1\left(\mathbb{R}^{n+1} - A\right) \cong H^0(U_1 \cap U_2)/_{\mathbb{R} \cdot 1} \cong H^0(\mathbb{R}^n - A)/\mathbb{R} \cdot 1.$$

We also have that $\dim(\mathrm{Im}\,(I^*)) = \dim(\mathrm{Ker}\,(J^*)) = 1$, so $H^0\left(\mathbb{R}^{n+1} - A\right) \cong \mathbb{R}.\;\square$

Addendum 6.12 *In the situation of Proposition* 6.11 *we have a diffeomorphism*

$$R: \mathbb{R}^{n+1} - A \to \mathbb{R}^{n+1} - A$$

defined by $R(x_1, \ldots, x_n, x_{n+1}) = (x_1, \ldots, x_n, -x_{n+1})$. *The induced linear map*

$$R^*: H^{p+1}\left(\mathbb{R}^{n+1} - A\right) \to H^{p+1}\left(\mathbb{R}^{n+1} - A\right)$$

is multiplication by -1 *for* $p \geq 0$.

Proof. In the notation of the proof above we have commutative diagrams, in which the horizontal diffeomorphisms are restrictions of R:

$$
\begin{array}{ccc}
\mathbb{R}^{n+1} - A \xrightarrow{R} \mathbb{R}^{n+1} - A & \qquad & \mathbb{R}^{n+1} - A \xrightarrow{R} \mathbb{R}^{n+1} - A \\
\uparrow{i_1} \qquad\qquad \uparrow{i_2} & & \uparrow{i_2} \qquad\qquad \uparrow{i_1} \\
U_1 \xrightarrow{R_1} U_2 & & U_2 \xrightarrow{R_2} U_1 \\
\uparrow{j_1} \qquad\qquad \uparrow{j_2} & & \uparrow{j_2} \qquad\qquad \uparrow{j_1} \\
U_1 \cap U_2 \xrightarrow{R_0} U_1 \cap U_2 & & U_1 \cap U_2 \xrightarrow{R_0} U_1 \cap U_2
\end{array}
$$

In the proof of Proposition 6.11 we saw that

$$\partial^*: H^p(U_1 \cap U_2) \to H^{p+1}\left(\mathbb{R}^{n+1} - A\right)$$

is surjective. Therefore it is sufficient to show that $R^* \circ \partial^*([\omega]) = -\partial^*([\omega])$ for an arbitrary closed p-form ω on $U_1 \cap U_2$.

Using Theorem 5.1 we can find $\omega_\nu \in \Omega^p(U_\nu)$, $\nu = 0, 1$, with $\omega = j_1^*(\omega_1) - j_2^*(\omega_2)$. The definition of ∂^* (see Definition 4.5) shows that $\partial^*([\omega]) = [\tau]$ where $\tau \in \Omega^{p+1}\left(\mathbb{R}^{n+1} - A\right)$ is determined by $i_\nu^*(\tau) = d\omega_\nu$ for $\nu = 1, 2$. Furthermore we get

$$-R_0^*\omega = R_0^* \circ j_2^*(\omega_2) - R_0^* \circ j_1^*(\omega_1) = j_1^*(R_1^*\omega_2) - j_2^*(R_2^*\omega_1)$$
$$i_1^*(R^*\tau) = R_1^*(i_2^*\tau) = R_1^*(d\omega_2) = d(R_1^*\omega_2)$$
$$i_2^*(R^*\tau) = R_2^*(i_1^*\tau) = R_2^*(d\omega_1) = d(R_2^*\omega_1).$$

These equations and the definition of ∂^* give $\partial^*(-[R_0^*\omega]) = [R^*\tau]$. Hence

(2) $$\partial^* \circ R_0^*([\omega]) = -R_0^* \circ \partial^*([\omega]).$$

For the projection $\mathrm{pr}: U_1 \cap U_2 \to \mathbb{R}^n - A$ we have that $\mathrm{pr} \circ R_0 = \mathrm{pr}$ and therefore the composition

$$H^p(\mathbb{R}^n - A) \xrightarrow{\mathrm{pr}^*} H^p(U_1 \cap U_2) \xrightarrow{R_0^*} H^p(U_1 \cap U_2)$$

is identical with pr^*. Since pr^* is an isomorphism, R_0^* is forced to be the identity map on $H^p(U_1 \cap U_2)$, and the left-hand side in (2) is $\partial^*[\omega]$. This completes the proof. $\square$

Theorem 6.13 *For $n \geq 2$ we have the isomorphisms*

$$H^p(\mathbb{R}^n - \{0\}) \cong \begin{cases} \mathbb{R} & \text{if } p = 0, \ n-1 \\ 0 & \text{otherwise.} \end{cases}$$

Proof. The case $n = 2$ was shown in Example 5.4. The general case follows from induction on n, via Proposition 6.11.   □

An invertible real $n \times n$ matrix A defines a linear isomorphism $\mathbb{R}^n \to \mathbb{R}^n$, and a diffeomorphism

$$f_A: \mathbb{R}^n - \{0\} \to \mathbb{R}^n - \{0\}.$$

Lemma 6.14   *For each $n \geq 2$, the induced map $f_A^*: H^{n-1}(\mathbb{R}^n - \{0\}) \to H^{n-1}(\mathbb{R}^n - \{0\})$ operates by multiplication by $\det A / |\det A| \in \{\pm 1\}$.*

Proof. Let B be obtained from A by replacing the r-th row by the sum of the r-th row and c times the s-th row, where $r \neq s$ and $c \in \mathbb{R}$,

$$B = (I + cE_{r,s})A,$$

where I is the identity matrix and $E_{r,s}$ is the matrix with entry 1 in its r-th row and s-th column and zeros elsewhere. A homotopy between f_A and f_B is defined by the matrices

$$(I + tcE_{r,s})\, A, \quad 0 \leq t \leq 1.$$

From Theorem 6.8 it follows that $f_A^* = f_B^*$. Furthermore $\det A = \det B$. By a sequence of elementary operations of this kind, A can be changed to $\mathrm{diag}\,(1, \ldots, 1, d)$, where $d = \det A$. Hence it suffices to prove the assertion for diagonal matrices. The matrices

$$\mathrm{diag}\left(1, \ldots, 1, \frac{|d|^t d}{|d|}\right), \quad 0 \leq t \leq 1$$

yield a homotopy, which reduces the problem to the two cases $A = \mathrm{diag}(1, \ldots, 1, \pm 1)$, so f_A is either the identity or the map R from Addendum 6.12. This proves the assertion.   □

From topological invariance (see Corollary 6.9) and the calculation in Theorem 6.13, supplemented with

$$H^p(\mathbb{R}^1 - \{0\}) \cong \begin{cases} \mathbb{R} \oplus \mathbb{R} & \text{if } p = 0 \\ 0 & \text{if } p \neq 0 \end{cases}$$

we get

Proposition 6.15 *If $n \neq m$ then $\mathbb{R}^n$ and $\mathbb{R}^m$ are not homeomorphic.*

Proof. A possible homeomorphism $\mathbb{R}^n \to \mathbb{R}^m$ may be assumed to map 0 to 0, and would induce a homeomorphism between $\mathbb{R}^n - \{0\}$ and $\mathbb{R}^m - \{0\}$. Hence

$$H^p(\mathbb{R}^n - \{0\}) \cong H^p(\mathbb{R}^m - \{0\})$$

for all p, in conflict with our calculations. $\square$

Remark 6.16 We offer the following more conceptual proof of Addendum 6.12. Let

$$
\begin{array}{ccccccccc}
0 & \longrightarrow & A^* & \xrightarrow{f^*} & B^* & \xrightarrow{g^*} & C^* & \longrightarrow & 0 \\
& & \downarrow{\alpha^*} & & \downarrow{\beta^*} & & \downarrow{\gamma^*} & & \\
0 & \longrightarrow & A_1^* & \xrightarrow{f_1^*} & B_1^* & \xrightarrow{g_1^*} & C_1^* & \longrightarrow & 0
\end{array}
$$

be a commutative diagram of chain complexes with exact rows. It is not hard to prove that the diagram

$$
\begin{array}{ccc}
H^p(C^*) & \xrightarrow{\partial^*} & H^{p+1}(A^*) \\
\downarrow{\gamma^*} & & \downarrow{\alpha^*} \\
H^p(C_1^*) & \xrightarrow{\partial_1^*} & H^{p+1}(A_1^*)
\end{array}
$$

is commutative. In the situation of Addendum 6.12 consider the diagram

$$
\begin{array}{ccccccccc}
0 & \longrightarrow & \Omega^*(U) & \xrightarrow{I^*} & \Omega^*(U_1) \oplus \Omega^*(U_2) & \longrightarrow & \Omega^*(U_1 \cap U_2) & \longrightarrow & 0 \\
& & \downarrow{R^*} & & \downarrow{R} & & \downarrow{-R_0^*} & & \\
0 & \longrightarrow & \Omega^*(U) & \xrightarrow{I^*} & \Omega^*(U_1) \oplus \Omega^*(U_2) & \longrightarrow & \Omega^*(U_1 \cap U_2) & \longrightarrow & 0
\end{array}
$$

with $R(\omega_1, \omega_2) = (R_1^* \omega_2, R_2^* \omega_1)$. This gives equation (2) of the proof of the addendum.

7. APPLICATIONS OF DE RHAM COHOMOLOGY

Let us introduce the standard notation

$$D^n = \{x \in \mathbb{R}^n \mid \|x\| \leq 1\} \qquad \text{(the } n\text{-ball)}$$
$$S^{n-1} = \{x \in \mathbb{R}^n \mid \|x\| = 1\} \qquad \text{(the } (n-1)\text{-sphere)}$$

A fixed point for a map $f: X \to X$ is a point $x \in X$, such that $f(x) = x$.

Theorem 7.1 (Brouwer's fixed point theorem, 1912) *Every continuous map* $f: D^n \to D^n$ *has a fixed point.*

Proof. Assume that $f(x) \neq x$ for all $x \in D^n$. For every $x \in D^n$ we can define the point $g(x) \in S^{n-1}$ as the point of intersection between S^{n-1} and the half-line from $f(x)$ through x.

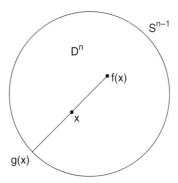

We have that $g(x) = x + tu$, where $u = \frac{x - f(x)}{\|x - f(x)\|}$, and

$$t = -x \cdot u + \sqrt{1 - \|x\|^2 + (x \cdot u)^2}.$$

Here $x \cdot u$ denotes the usual inner product. The expression for $g(x)$ is obtained by solving the equation $(x + tu) \cdot (x + tu) = 1$. There are two solutions since the line determined by $f(x)$ and x intersects S^{n-1} in two points. We are interested in the solution with $t \geq 0$. Since g is continuous with $g_{|S^{n-1}} = \mathrm{id}_{S^{n-1}}$, the theorem follows from the lemma below. $\qquad\square$

Lemma 7.2 *There is no continuous map* $g: D^n \to S^{n-1}$, *with* $g_{|S^{n-1}} = \mathrm{id}_{S^{n-1}}$.

Proof. We may assume that $n \geq 2$. For the map $r: \mathbb{R}^n - \{0\} \to \mathbb{R}^n - \{0\}, r(x) = x/\|x\|$, we get that $\mathrm{id}_{\mathbb{R}^n - \{0\}} \simeq r$, because $\mathbb{R}^n - \{0\}$ always contains the line segment between x and $r(x)$ (see Example 6.5). If g is of the indicated type, then $g(t \cdot r(x))$, $0 \leq t \leq 1$ defines a homotopy from a constant map to r. This shows that $\mathbb{R}^n - \{0\}$ is contractible. Corollary 6.10 asserts that $H^{n-1}(\mathbb{R}^n - \{0\}) = 0$, which contradicts Theorem 6.13. $\square$

The tangent space of S^n in the point $x \in S^n$ is $T_x S^n = \{x\}^\perp$, the orthogonal complement in $\mathbb{R}^{n+1}$ to the position vector. A tangent vector field on S^n is a continuous map $v \colon S^n \to \mathbb{R}^{n+1}$ such that $v(x) \in T_x S^n$ for every $x \in S^n$.

Theorem 7.3 *The sphere S^n has a tangent vector field v with $v(x) \neq 0$ for all $x \in S^n$ if and only if n is odd.*

Proof. Such a vector field v can be extended to a vector field w on $\mathbb{R}^n - \{0\}$ by setting

$$w(x) = v\left(\frac{x}{\|x\|}\right).$$

We have that $w(x) \neq 0$ and $w(x) \cdot x = 0$. The expression

$$F(x, t) = (\cos \pi t)x + (\sin \pi t)w(x)$$

defines a homotopy from $f_0 = \mathrm{id}_{(\mathbb{R}^{n+1} - \{0\})}$ to the antipodal map f_1, $f_1(x) = -x$. Theorem 6.8.(i) shows that f_1^* is the identity on $H^n(\mathbb{R}^{n+1} - \{0\})$, which by Theorem 6.13 is 1-dimensional. On the other hand Lemma 6.14 evaluates f_1^* to be multiplication with $(-1)^{n+1}$. Hence n is odd.

Conversely, for $n = 2m - 1$, we can define a vector field v with

$$v(x_1, x_2, \ldots, x_{2m}) = (-x_2, x_1, -x_4, x_3, \ldots, -x_{2m}, x_{2m-1}). \qquad \square$$

In 1962 J. F. Adams solved the so-called "vector field problem": find the maximal number of linearly independent tangent vector fields one may have on S^n. (Tangent vector fields $v_1, \ldots, v_d$ on S^n are called linearly independent if for every $x \in S^n$ the vectors $v_1(x), \ldots, v_d(x)$ are linearly independent.)

Adams' theorem *For $n = 2m - 1$, let $2m = (2c + 1)2^{4a+b}$, where $0 \leq b \leq 3$. The maximal number of linearly independent tangent vector fields on S^n is equal to $2^b + 8a - 1$.*

Lemma 7.4 (Urysohn–Tietze) *If $A \subseteq \mathbb{R}^n$ is closed and $f \colon A \to \mathbb{R}^m$ continuous, then there exists a continuous map $g \colon \mathbb{R}^n \to \mathbb{R}^m$ with $g_{|A} = f$.*

Proof. We denote Euclidean distance in $\mathbb{R}^n$ by $d(x, y)$ and for $x \in \mathbb{R}^n$ we define

$$d(x, A) = \inf_{y \in A} d(x, y).$$

For $p \in \mathbb{R}^n - A$ we have an open neighborhood $U_p \subseteq \mathbb{R}^n - A$ of p given by

$$U_p = \left\{ x \in \mathbb{R}^n \;\middle|\; d(x,p) < \frac{1}{2} d(p,A) \right\}.$$

These sets cover $\mathbb{R}^n - A$ and we can use Theorem A.1 to find a subordinate partition of unity ϕ_p. We define g by

$$g(x) = \begin{cases} f(x) & \text{if } x \in A \\ \displaystyle\sum_{p \in \mathbb{R}^n - A} \phi_p(x) f(a(p)) & \text{if } x \in \mathbb{R}^n - A \end{cases}$$

where for $p \in \mathbb{R}^n - A$, $a(p) \in A$ is chosen such that

$$d(p, a(p)) < 2d(p, A).$$

Since the sum is locally finite on $\mathbb{R}^n - A$, g is smooth on $\mathbb{R}^n - A$.

The only remaining problem is the continuity of g at a point x_0 on the boundary of A. If $x \in U_p$ then

$$d(x_0, p) \leq d(x_0, x) + d(x, p) < d(x_0, x) + \frac{1}{2} d(p, A) \leq d(x_0, x) + \frac{1}{2} d(p, x_0).$$

Hence $d(x_0, p) < 2d(x_0, x)$ for $x \in U_p$. Since $d(p, a(p)) < 2d(p, A) \leq 2d(x_0, p)$ we get for $x \in U_p$ that $d(x_0, a(p)) \leq d(x_0, p) + d(p, a(p)) < 3d(x_0, p) < 6d(x_0, x)$. For $x \in \mathbb{R}^n - A$ we have

$$g(x) - g(x_0) = \sum_{p \in \mathbb{R}^n - A} \phi_p(x)(f(a(p)) - f(x_0))$$

and

(1) $$\|g(x) - g(x_0)\| \leq \sum_{p} \phi_p(x) \|f(a(p)) - f(x_0)\|,$$

where we sum over the points p with $x \in U_p$.

For an arbitrary $\epsilon > 0$ choose $\delta > 0$ such that $\|f(y) - f(x_0)\| < \epsilon$ for every $y \in A$ with $d(x_0, y) < 6\delta$. If $x \in \mathbb{R}^n - A$ and $d(x, x_0) < \delta$ then, for p with $x \in U_p$, we have that $d(x_0, a(p)) < 6\delta$ and $\|f(a(p)) - f(x_0)\| < \epsilon$. Then (1) yields

$$\|g(x) - g(x_0)\| \leq \sum_{p} \phi_p(x) \cdot \epsilon = \epsilon.$$

Continuity of g at x_0 follows. $\qquad\square$

Remark 7.5 The proof above still holds, with marginal changes, when $\mathbb{R}^n$ is replaced by a metric space and $\mathbb{R}^m$ by a locally convex topological vector space.

Lemma 7.6 *Let $A \subseteq \mathbb{R}^n$ and $B \subseteq \mathbb{R}^m$ be closed sets and let $\phi: A \to B$ be a homeomorphism. There is a homeomorphism h of $\mathbb{R}^{n+m}$ to itself, such that*

$$h(x, 0_m) = (0_n, \phi(x))$$

for all $x \in A$.

Proof. By Lemma 7.4 we can extend ϕ to a continuous map $f_1: \mathbb{R}^n \to \mathbb{R}^m$. A homeomorphism $h_1: \mathbb{R}^n \times \mathbb{R}^m \to \mathbb{R}^n \times \mathbb{R}^m$ is defined by

$$h_1(x, y) = (x, y + f_1(x)).$$

The inverse to h_1 is obtained by subtracting $f_1(x)$ instead. Analogously we can extend ϕ^{-1} to a continuous map $f_2: \mathbb{R}^m \to \mathbb{R}^n$ and define a homeomorphism $h_2: \mathbb{R}^n \times \mathbb{R}^m \to \mathbb{R}^n \times \mathbb{R}^m$ by

$$h_2(x, y) = (x + f_2(y), y).$$

If h is defined to be $h = h_2^{-1} \circ h_1$, then we have for $x \in A$ that

$$h(x, 0_m) = h_2^{-1}(x, f_1(x)) = h_2^{-1}(x, \phi(x)) = (x - f_2(\phi(x)), \phi(x)) = (0_n, \phi(x)).$$

$\square$

We identify $\mathbb{R}^n$ with the subspace of $\mathbb{R}^{n+m}$ consisting of vectors of the form $(x_1, \ldots, x_n, 0, \ldots, 0)$.

Corollary 7.7 *If $\phi: A \to B$ is a homeomorphism between closed subsets A and B of $\mathbb{R}^n$, then ϕ can be extended to a homeomorphism $\tilde{\phi}: \mathbb{R}^{2n} \to \mathbb{R}^{2n}$.*

Proof. We merely have to compose the homeomorphism h from Lemma 7.6 with the homeomorphism of $\mathbb{R}^{2n} = \mathbb{R}^n \times \mathbb{R}^n$ to itself that switches the two factors. $\square$

Note that $\tilde{\phi}$ by restriction gives a homeomorphism between $\mathbb{R}^{2n} - A$ and $\mathbb{R}^{2n} - B$. In contrast it can occur that $\mathbb{R}^n - A$ is not homeomorphic to $\mathbb{R}^n - B$. A well-known example is Alexander's "horned sphere" Σ in $\mathbb{R}^3$: Σ is homeomorphic to S^2, but $\mathbb{R}^3 - \Sigma$ is not homeomorphic to $\mathbb{R}^3 - S^2$. This and numerous other examples are treated in [Rushing].

Theorem 7.8 *Assume that $A \neq \mathbb{R}^n$ and $B \neq \mathbb{R}^n$ are closed subsets of $\mathbb{R}^n$. If A and B are homeomorphic, then*

$$H^p(\mathbb{R}^n - A) \cong H^p(\mathbb{R}^n - B).$$

Proof. By induction on m Proposition 6.11 yields isomorphisms

$$H^{p+m}(\mathbb{R}^{n+m} - A) \cong H^p(\mathbb{R}^n - A) \qquad \text{(for } p > 0)$$

$$H^m(\mathbb{R}^{n+m} - A) \cong H^0(\mathbb{R}^n - A)/\mathbb{R}\cdot 1$$

for all $m \geq 1$. Analogously for B. From Corollary 7.7 we know that $\mathbb{R}^{2n} - A$ and $\mathbb{R}^{2n} - B$ are homeomorphic. Topological invariance (Corollary 6.9) shows that they have isomorphic de Rham cohomologies. We thus have the isomorphisms

$$H^p(\mathbb{R}^n - A) \cong H^{p+n}(\mathbb{R}^{2n} - A) \cong H^{p+n}(\mathbb{R}^{2n} - B) \cong H^p(\mathbb{R}^n - B)$$

for $p > 0$ and

$$H^0(\mathbb{R}^n - A)/\mathbb{R} \cdot 1 \cong H^n(\mathbb{R}^{2n} - A) \cong H^n(\mathbb{R}^{2n} - B) \cong H^0(\mathbb{R}^n - B)/\mathbb{R} \cdot 1. \quad \square$$

For a closed set $A \subseteq \mathbb{R}^n$ the open complement $U = \mathbb{R}^n - A$ will always be a disjoint union of at most countably many connected components, which all are open. If there are infinitely many, then $H^0(U)$ will have infinite dimension. Otherwise the number of connected components is equal to $\dim H^0(U)$.

Corollary 7.9 *If A and B are two homeomorphic closed subsets of $\mathbb{R}^n$, then $\mathbb{R}^n - A$ and $\mathbb{R}^n - B$ have the same number of connected components.*

Proof. If $A \neq \mathbb{R}^n$ and $B \neq \mathbb{R}^n$ the assertion follows from Theorem 7.8 and the remarks above. If $A = \mathbb{R}^n$ and $B \neq \mathbb{R}^n$ then $\mathbb{R}^{n+1} - A$ has precisely 2 connected components (the open half-spaces), while $\mathbb{R}^{n+1} - B$ is connected. Hence this case cannot occur. $\qquad \square$

Theorem 7.10 (Jordan–Brouwer separation theorem) *If $\Sigma \subseteq \mathbb{R}^n$ $(n \geq 2)$ is homeomorphic to S^{n-1} then*

(i) *$\mathbb{R}^n - \Sigma$ has precisely 2 connected components U_1 and U_2, where U_1 is bounded and U_2 is unbounded.*

(ii) *Σ is the set of boundary points for both U_1 and U_2.*

We say U_1 is the domain inside Σ and U_2 the domain outside Σ.

Proof. Since Σ is compact, Σ is closed in $\mathbb{R}^n$. To show (i), it suffices, by Corollary 7.9, to verify it for $S^{n-1} \subseteq \mathbb{R}^n$. The two connected components of $\mathbb{R}^n - S^{n-1}$ are

$$\mathring{D}^n = \{x \in \mathbb{R}^n \mid \|x\| < 1\} \quad \text{and} \quad W = \{x \in \mathbb{R}^n \mid \|x\| > 1\}$$

By choosing $r = \max_{x \in \Sigma} \|x\|$, the connected set

$$rW = \{x \in \mathbb{R}^n \mid \|x\| > r\}$$

will be contained in one of the two components in $\mathbb{R}^n - \Sigma$, and the other component must be bounded. This completes the proof of (i).

Let $p \in \Sigma$ be given and consider an open neighbourhood V of p in $\mathbb{R}^n$. The set $A = \Sigma - (\Sigma \cap V)$ is closed and homeomorphic to a corresponding proper closed subset B of S^{n-1}. It is obvious that $\mathbb{R}^n - B$ is connected, so by Corollary 7.9 the same is the case for $\mathbb{R}^n - A$. For $p_1 \in U_1$ and $p_2 \in U_2$, we can find a continuous curve $\gamma : [a, b] \to \mathbb{R}^n - A$ with $\gamma(a) = p_1$ and $\gamma(b) = p_2$. By (i) the curve must intersect Σ, i.e. $\gamma^{-1}(\Sigma)$ is non-empty. The closed set $\gamma^{-1}(\Sigma) \subseteq [a, b]$ has a first element c_1 and a last element c_2, which both belong to (a, b). Hence $\gamma(c_1) \in \Sigma \cap V$ and $\gamma(c_2) \in \Sigma \cap V$ are points of contact for $\gamma([a, c_1)) \subseteq U_1$ and $\gamma((c_2, b]) \subseteq U_2$ respectively. Therefore we can find $t_1 \in [a, c_1)$ and $t_2 \in (c_2, b]$, such that $\gamma(t_1) \in U_1 \cap V$ and $\gamma(t_2) \in U_2 \cap V$. This shows that p is a boundary point for both U_1 and U_2, and proves (ii). $\square$

Theorem 7.11 *If $A \subseteq \mathbb{R}^n$ is homeomorphic to D^k, with $k \leq n$, then $\mathbb{R}^n - A$ is connected.*

Proof. Since A is compact, A is closed. By Corollary 7.9 it is sufficient to prove the assertion for $D^k \subseteq \mathbb{R}^k \subseteq \mathbb{R}^n$. This is left to the reader. $\square$

Theorem 7.12 (Brouwer) *Let $U \subseteq \mathbb{R}^n$ be an arbitrary open set and $f : U \to \mathbb{R}^n$ an injective continuous map. The image $f(U)$ is open in $\mathbb{R}^n$, and f maps U homeomorphically to $f(U)$.*

Proof. It is sufficient to prove that $f(U)$ is open; the same will then hold for $f(W)$, where $W \subseteq U$ is an arbitrary open subset. This proves continuity of the inverse function from $f(U)$ to U. Consider a closed sphere.

$$D = \{x \in \mathbb{R}^n \mid \|x - x_0\| \leq \delta\}$$

contained in U with boundary S and interior $\mathring{D} = D - S$. It is sufficient to show that $f(\mathring{D})$ is open. The case $n = 1$ follows from elementary theorems about continuous functions of one variable, so we assume $n \geq 2$.

Both S and $\Sigma = f(S)$ are homeomorphic to S^{n-1}. Let U_1 and U_2 be the two connected components of $\mathbb{R}^n - \Sigma$ from Theorem 7.10. They are open; U_1 is bounded and U_2 is unbounded. By Theorem 7.11, $\mathbb{R}^n - f(D)$ is connected. Since this set is disjoint from Σ, it must be contained in U_1 or U_2. As $f(D)$ is compact, $\mathbb{R}^n - f(D)$ is unbounded. We must have $\mathbb{R}^n - f(D) \subseteq U_2$. It follows that $\Sigma \cup U_1 = \mathbb{R}^n - U_2 \subseteq f(D)$. Hence

$$U_1 \subseteq f(\mathring{D}).$$

Since $\mathring{D}$ is connected, $f(\mathring{D})$ will also be connected (even though it is not known whether or not $f(\mathring{D})$ is open). Since $f(\mathring{D}) \subseteq U_1 \cup U_2$, we must have that $U_1 = f(\mathring{D})$. This completes the proof. $\square$

Corollary 7.13 (Invariance of domain) *If $V \subseteq \mathbb{R}^n$ has the topology induced by $\mathbb{R}^n$ and is homeomorphic to an open subset of $\mathbb{R}^n$ then V is open in $\mathbb{R}^n$.*

Proof. This follows immediately from Theorem 7.12. □

Corollary 7.14 (Dimension invariance) *Let $U \subseteq \mathbb{R}^n$ and $V \subseteq \mathbb{R}^m$ be non-empty open sets. If U and V are homeomorphic then $n = m$.*

Proof. Assume that $m < n$. From Corollary 7.13 applied to V, considered as a subset of $\mathbb{R}^n$ via the inclusion $\mathbb{R}^m \subseteq \mathbb{R}^n$, it follows that V is open in $\mathbb{R}^n$. This contradicts that V is contained in a proper subspace. □

Example 7.15 A *knot* in $\mathbb{R}^3$ is a subset $\Sigma \subseteq \mathbb{R}^3$ that is homeomorphic to S^1. The corresponding knot-complement is the open set $U = \mathbb{R}^3 - \Sigma$. We show:
$$H^p(U) \cong \begin{cases} \mathbb{R} & \text{if } 0 \leq p \leq 2 \\ 0 & \text{otherwise.} \end{cases}$$
According to Theorem 7.8, it is sufficient to show this for the "trival knot" $S^1 \subseteq \mathbb{R}^2 \subseteq \mathbb{R}^3$. First we calculate

(2) $$H^p(\mathbb{R}^2 - S^1) = H^p(\mathring{D}^2) \oplus H^p(\mathbb{R}^2 - D^2).$$

Here $\mathring{D}^2$ is star-shaped, while $\mathbb{R}^2 - D^2$ is diffeomorphic to $\mathbb{R}^2 - \{0\}$. Using Theorem 3.15 and Example 5.4 it follows that (2) has dimension 2 for $p = 0$, dimension 1 for $p = 1$, and dimension 0 for $p \geq 2$. Apply Proposition 6.11.

An analogous calculation of $H^*(\mathbb{R}^n - \Sigma)$ can be done for a higher-dimensional knot $\Sigma \subseteq \mathbb{R}^n$, where Σ is homeomorphic to S^k ($1 \leq k \leq n - 2$). See Exercise 7.2.

Proposition 7.16 *Let $\Sigma \subseteq \mathbb{R}^n$ ($n \geq 2$) be homeomorphic to S^{n-1} and let U_1 and U_2 be the interior and exterior domains of Σ. Then*
$$H^p(U_1) \cong \begin{cases} \mathbb{R} & \text{if } p = 0 \\ 0 & \text{otherwise} \end{cases} \quad \text{and} \quad H^p(U_2) \cong \begin{cases} \mathbb{R} & \text{if } p = 0, n - 1 \\ 0 & \text{otherwise.} \end{cases}$$

Proof. The case $p = 0$ follows from Theorem 7.10. Set $W = \mathbb{R}^n - D^n$. For $p > 0$ there are isomorphisms
$$H^p(U_1) \oplus H^p(U_2) \cong H^p(\mathbb{R}^n - \Sigma) \cong H^p(\mathbb{R}^n - S^{n-1})$$
$$\cong H^p(\mathring{D}^n) \oplus H^p(W) \cong H^p(W).$$
The inclusion map $i : W \to \mathbb{R}^n - \{0\}$ is a homotopy equivalence with homotopy inverse defined by
$$g(x) = \frac{\|x\| + 1}{\|x\|} x.$$

The two required homotopies are given by Example 6.5. From Theorem 6.8.(iii) we have that $H^p(i)$ is an isomorphism. The calculation from Theorem 6.13 yields

$$H^p(W) \cong \begin{cases} \mathbb{R} & \text{if } p = 0, n-1 \\ 0 & \text{otherwise.} \end{cases}$$

We now have that $H^p(U_1) = 0$ and $H^p(U_2) = 0$ when $p \notin \{0, n-1\}$. On the other hand the dimensions of $H^{n-1}(U_1)$ and $H^{n-1}(U_2)$ are 0 or 1, so it suffices to show that $H^{n-1}(U_2) \neq 0$.

Without loss of generality we may assume that $0 \in U_1$ and that the bounded set $U_1 \cup \Sigma$ is contained in D^n. We thus have a commutative diagram of inclusion maps

$$\begin{array}{ccc} & & \mathbb{R}^n - \{0\} \\ & \nearrow^{i} & \uparrow \\ W & \longrightarrow & U_2 \end{array}$$

and apply H^{n-1} to get the commutative diagram

$$\begin{array}{ccc} & & H^{n-1}(\mathbb{R}^n - \{0\}) \cong \mathbb{R} \\ \small{H^{n-1}(i)} \nearrow & & \downarrow \\ H^{n-1}(W) & \longrightarrow & H^{n-1}(U_2) \end{array}$$

where $H^{n-1}(i)$ is an isomorphism. It follows that $H^{n-1}(U_2) \neq 0$. $\square$

Remark 7.17 The above result about $H^*(U_1)$ might suggest that U_1 is contractible (cf. Corollary 6.10). In general, however, this is not the case. In *Topological Embeddings*, Rushing discusses several examples for $n = 3$, where U_1 is not simply connected (i.e. there exists a continuous map $S^1 \to U_1$, which is not homotopic to a constant map). Hence U_1 is not contractible either. Corresponding examples can be found for $n > 3$.

If $n = 2$ a theorem by Schoenflies (cf. [Moise]) states that there exists a homeomorphism

$$h: U_1 \cup \Sigma \to D^2.$$

By Theorem 7.12, such a homeomorphism applied to $h_{|U_1}$ and $h^{-1}_{|\overset{\circ}{D}^2}$ will map U_1 homeomorphically to $\overset{\circ}{D}^2$.

A result by M. Brown from 1960 shows that the conclusion in Schoenflies' theorem is also valid if $n > 2$, provided it is additionally assumed that Σ is *flat* in $\mathbb{R}^n$, that is, there exists a $\delta > 0$ and a continuous injective map $\phi: S^{n-1} \times (-\delta, \delta) \to \mathbb{R}^n$ with $\Sigma = \phi(S^{n-1} \times \{0\})$.

Example 7.18 One can also calculate the cohomology of "$\mathbb{R}^n$ with m holes", i.e. the cohomology of

$$V = \mathbb{R}^n - \left(\bigcup_{j=1}^{m} K_j \right).$$

The "holes" K_j in $\mathbb{R}^n$ are disjoint compact sets with boundary Σ_j, homeomorphic to S^{n-1}. Hence the interiors $\overset{\circ}{K}_j = K_j - \Sigma_j$ are exactly the interior domains of Σ_j. One has

$$\text{(3)} \qquad H^p(V) \cong \begin{cases} \mathbb{R} & \text{if } p = 0 \\ \mathbb{R}^m & \text{if } p = n - 1 \\ 0 & \text{otherwise.} \end{cases}$$

We use induction on m. The case $n = 1$ follows from Proposition 7.16. Assume the assertion is true for

$$V_1 = \mathbb{R}^n - \left(\bigcup_{j=1}^{m-1} K_j \right).$$

Let $V_2 = \mathbb{R}^n - K_m$. Then $V_1 \cup V_2 = \mathbb{R}^n$ and $V_1 \cap V_2 = V$. For $p \geq 0$ we have the exact Mayer–Vietoris sequence

$$H^p(\mathbb{R}^n) \overset{I^*}{\to} H^p(V_1) \oplus H^p(V_2) \overset{J^*}{\to} H^p(V) \to 0.$$

If $p = 0$ then $H^0(\mathbb{R}^n) \cong \mathbb{R}$ and I^* is injective. We get $H^0(V_1) \cong \mathbb{R}$ by induction and $H^0(V_2) \cong \mathbb{R}$ from Proposition 7.16. The exact sequence yields $H^0(V) \cong \mathbb{R}$. If $p > 0$ then $H^p(\mathbb{R}^n) = 0$ and the exact sequence gives the isomorphism

$$H^p(V_1) \oplus H^p(V_2) \cong H^p(V).$$

Now (3) follows by induction.

8. SMOOTH MANIFOLDS

A topological space X has a countable topological base, when there exists a countable system of open sets $\mathcal{V} = \{U_i \mid i \in \mathbb{N}\}$, such that every open set can be written in the form $\bigcup_{i \in I} U_i$, where $I \subseteq \mathbb{N}$.

For instance $\mathbb{R}^n$ has a countable basis for the topology given by

$$\mathcal{V} = \{\mathring{D}^n(x; \epsilon) \mid x = (x_1, \ldots, x_n), \ x_i \in \mathbb{Q}; \ \epsilon \in \mathbb{Q}, \ \epsilon > 0\}$$

where $\mathring{D}(x; \epsilon)$ is the open ball with center at x and radius ϵ.

A topological space X is a Hausdorff space when, for arbitrary distinct $x, y \in X$, there exist open neighborhoods U_x and U_y with $U_x \cap U_y = \emptyset$.

Definition 8.1 A topological manifold M is a topological Hausdorff space that has a countable basis for its topology and that is locally homeomorphic to $\mathbb{R}^n$. The number n is called the dimension of M.

Remark 8.2 Every open ball $\mathring{D}^n(0, \epsilon)$ in $\mathbb{R}^n$ is diffeomorphic to $\mathbb{R}^n$ via the map Φ given by

$$\Phi(y) = \begin{cases} \tan(\pi \|y\|/2\epsilon) \cdot y/\|y\| & \text{if } y \neq 0, \|y\| < \epsilon \\ 0 & \text{if } y = 0. \end{cases}$$

(Smoothness of Φ and Φ^{-1} at 0 can be shown by means of the Taylor series at 0 for tan and Arctan.) Thus in Definition 8.1 it does not matter whether we require that M^n is locally homeomorphic to $\mathbb{R}^n$ or to an open set in $\mathbb{R}^n$.

Definition 8.3

    (i) A chart (U, h) on an n-dimensional manifold is a homeomorphism $h: U \rightarrow U'$, where U is an open set in M and U' is an open set in $\mathbb{R}^n$.

   (ii) A system $A = \{h_i: U_i \rightarrow U_i' \mid i \in J\}$ of charts is called an atlas, provided $\{U_i \mid i \in J\}$ covers M.

 (iii) An atlas is smooth when all of the maps

$$h_{ji} = h_j \circ h_i^{-1}: h_i(U_i \cap U_j) \rightarrow h_j(U_i \cap U_j)$$

are smooth. They are called chart transformations (or transition functions) for the given atlas.

Note in Definition 8.3.(iii) that $h_i(U_i \cap U_j)$ is open in $\mathbb{R}^n$.

Two smooth atlases A_1, A_2 are smoothly equivalent if $A_1 \cup A_2$ is a smooth atlas. This defines an equivalence relation on the set of atlases on M. A smooth structure on M is an equivalence class $\mathcal{A}$ of smooth atlases on M.

Definition 8.4 A smooth manifold is a pair $(M, \mathcal{A})$ consisting of a topological manifold M and a smooth structure $\mathcal{A}$ on M.

Usually $\mathcal{A}$ is suppressed from the notation and we write M instead of $(M, \mathcal{A})$.

Example 8.5 The n-dimensional sphere $S^n = \{x \in \mathbb{R}^{n+1} \mid \|x\| = 1\}$ is an n-dimensional smooth manifold. We define an atlas with $2(n+1)$ charts $(U_{\pm i}, h_{\pm i})$ where
$$U_{+i} = \{x \in S^n \mid x_i > 0\}, \qquad U_{-i} = \{x \in S^n \mid x_i < 0\}$$
and $h_{\pm i} : U_{\pm i} \to \mathring{D}^n$ is the map given by $h_{\pm i}(x) = (x_1, \ldots, \hat{x}_i, \ldots, x_{n+1})$. The circumflex over x_i denotes that x_i is omitted. The inverse map is
$$h_{\pm i}^{-1}(u) = \left(u_1, \ldots, u_{i-1}, \pm \sqrt{1 - \|u\|^2}, u_i, \ldots, u_n\right)$$
It is left to the reader to prove that the chart transformations are smooth.

Example 8.6 (The projective space $\mathbb{RP}^n$) On S^n we define an equivalence relation:
$$x \sim y \iff x = y \text{ or } x = -y.$$
The equivalence classes $[x] = \{x, -x\}$ define the set $\mathbb{RP}^n$. Alternatively one can consider $\mathbb{RP}^n$ as all lines in $\mathbb{R}^{n+1}$ through 0. Let π be the canonical projection
$$\pi : S^n \to \mathbb{RP}^n; \quad \pi(x) = [x].$$
We give $\mathbb{RP}^n$ the quotient topology, i.e.
$$U \subseteq \mathbb{RP}^n \text{ open} \iff \pi^{-1}(U) \subseteq S^n \text{ open}.$$
With the conventions of Example 8.5, $\pi(U_{-i}) = \pi(U_{+i})$. We define $U_i = \pi(U_{\pm i}) \subseteq \mathbb{RP}^n$, and note that $\pi^{-1}(U_i) = U_{+i} \cup U_{-i}$ with $U_{+i} \cap U_{-i} = \emptyset$. An equivalence class $[x] \in U_i$ has exactly one representative in U_{+i} and one representative in U_{-i}. Hence $\pi : U_{+i} \to U_i$ is a homeomorphism. We define
$$h_i = h_{+i} \circ \pi^{-1} : U_i \to \mathring{D}^n,$$
$i = 1, \ldots, n$. This gives a smooth atlas on $\mathbb{RP}^n$.

Definition 8.7 Consider smooth manifolds M_1 and M_2 and a continuous map $f : M_1 \to M_2$. The map f is called smooth at $x \in M_1$ if there exist charts $h_1 : U_1 \to U_1'$ and $h_2 : U_2 \to U_2'$ on M_1 and M_2 with $x \in U_1$ and $f(x) \in U_2$, such that
$$h_2 \circ f \circ h_1^{-1} : h_1\left(f^{-1}(U_2)\right) \to U_2'$$
is smooth in a neighborhood of $h_1(x)$. If f is smooth at all points of M_1 then f is said to be smooth.

Since chart transformations (by Definition 8.3.(iii)) are smooth, we have that Definition 8.7 is independent of the choice of charts in the given atlases for M_1 and M_2. A composition of two smooth maps is smooth.

A *diffeomorphism* $f: M_1 \to M_2$ between smooth manifolds is a smooth map that has a smooth inverse. In particular a diffeomorphism is a homeomorphism.

As soon as we have chosen an atlas $\mathcal{A}$ on a manifold M, we know which functions on M are smooth. In particular we know when a homeomorphism $f: V \to V'$ between an open set $V \subset M$ and an open set $V' \subseteq \mathbb{R}^n$ is a diffeomorphism. We can therefore define a new *maximal* atlas, $\mathcal{A}_{\max}$, associated with the given smooth structure:

$$\mathcal{A}_{\max} = \{f: V \to V' \mid V \subseteq M^n \text{ open, } V' \subseteq \mathbb{R}^n \text{ open, } f \text{ diffeomorphism}\}.$$

The inverse diffeomorphisms $f^{-1}: V' \to V$ will be called local parametrizations. From Remark 8.2 it follows that every point in a smooth manifold M^n has an open neighborhood $V \subseteq M^n$ that is diffeomorphic to $\mathbb{R}^n$.

From now on chart *will mean a chart in the maximal atlas.*

Definition 8.8 A subset $N \subset M^n$ of a smooth manifold is said to be a *smooth submanifold* (of dimension k), if the following condition is satisfied: for every $x \in N$ there exists a chart $h: U \to U'$ on M such that

(1) $$x \in U \quad \text{and} \quad h(U \cap N) = U' \cap \mathbb{R}^k,$$

where $\mathbb{R}^k \subseteq \mathbb{R}^n$ is the standard subspace.

It is easy to see that a smooth submanifold N of a smooth manifold M is a smooth manifold again. A smooth atlas on N is given by all $h: U \cap N \to U' \cap \mathbb{R}^k$, where (U, h) are charts on M satisfying (1).

Example 8.9 The n-sphere S^n is a smooth submanifold of $\mathbb{R}^{n+1}$. In fact the charts $(U_{\pm i}, h_{\pm i})$ from Example 8.5 can easily be extended to diffeomorphisms

For every $p \in N^n$ there exists an open neighborhood $V \subseteq \mathbb{R}^{n+k}$, an open set $U' \subseteq \mathbb{R}^n$ and a homeomorphism

$$g: U' \to N \cap V,$$

such that g is smooth (considered as a map from U' to V) and such that $D_x g: \mathbb{R}^n \to \mathbb{R}^{n+k}$ is injective.

This is the usual definition of an embedded manifold (regular surface when $n = 2$). Theorem 8.11 tells us that every smooth manifold is diffeomorphic to an *embedded manifold*. Conversely, if $N \subseteq \mathbb{R}^{n+k}$ satisfies the above condition, then the implicit function theorem shows that it is a submanifold in the sense of Definition 8.8. A theorem by H. Whitney asserts that the codimension k in Theorem 8.11 can always be taken to be less or equal to $n + 1$. On the other hand k cannot be arbitrarily small. $\mathbb{RP}^2$ cannot be embedded in $\mathbb{R}^3$.

Lemma 8.12 *Let M^n be an n-dimensional smooth manifold. For $p \in M$ there exist smooth maps*

$$\phi_p: M \to \mathbb{R}, \qquad f_p: M \to \mathbb{R}^n$$

such that $\phi_p(p) > 0$, and f_p maps the open set $M - \phi_p^{-1}(0)$ diffeomorphically onto an open subset of $\mathbb{R}^n$.

Proof. Choose a chart $h: V \to V'$ with $p \in V$. By Lemma A.7 we can find a function $\psi \in C^\infty(\mathbb{R}^n, \mathbb{R})$ with compact support $\mathrm{supp}_{\mathbb{R}^n}(\psi) \subseteq V'$, such that ψ is constantly equal to 1 on an open neighborhood $U' \subset V'$ of $h(p)$. The smooth map f_p can now be defined by

$$f_p(q) = \begin{cases} \psi(h(q))h(q) & \text{if } q \in V \\ 0 & \text{otherwise.} \end{cases}$$

On the open neighborhood $U = h^{-1}(U')$ the function f_p coincides with h and therefore maps U diffeomorphically onto U'. Choose $\psi_0 \in C^\infty(\mathbb{R}^n, \mathbb{R})$ with compact support $\mathrm{supp}_{\mathbb{R}^n}(\psi_0) \subseteq U'$ and $\psi_0(h(p)) > 0$, and let

$$\phi_p(q) = \begin{cases} \psi_0(h(q)) & \text{if } q \in V \\ 0 & \text{otherwise.} \end{cases}$$

Since $M - \phi_p^{-1}(0) \subseteq U$, the final assertion holds. $\qquad\square$

Proof of Theorem 8.11 (M compact). For every $p \in M$ choose ϕ_p and f_p as in Lemma 8.12. By compactness M can be covered by a finite number of the sets $M - \phi_p^{-1}(0)$. After a change of notation we have smooth functions

$$\phi_j: M \to \mathbb{R}, \qquad f_j: M \to \mathbb{R}^n \qquad (1 \le j \le d)$$

satisfying the following conditions:

   (i) The open sets $U_j = M - \phi_j^{-1}(0)$ cover M.
   (ii) $f_{j|U_j}$ maps U_j diffeomorphically onto an open set $U_j' \subseteq \mathbb{R}^n$.

We define a smooth map $f \colon M \to \mathbb{R}^{nd+d}$ by setting

$$f(q) = (f_1(q), \ldots, f_d(q), \phi_1(q), \ldots, \phi_d(q)).$$

Assuming $f(q_1) = f(q_2)$, we can by (i) choose j such that $q_1 \in U_j$. Then $\phi_j(q_2) = \phi_j(q_1) \neq 0$, $q_2 \in U_j$, and by (ii), $q_1 = q_2$. Hence f is injective. Since M is compact, f is a homeomorphism from M to $f(M)$. Let

$$\pi_1 \colon \mathbb{R}^{nd+d} \to \mathbb{R}^n, \qquad \pi_2 \colon \mathbb{R}^{nd+d} \to \mathbb{R}^{n(d-1)+d}$$

be the projections on the first n coordinates and the last $n(d-1)+d$ coordinates, respectively. By (ii) $\pi_1 \circ f = f_1$ is a diffeomorphism from U_1 to U_1'. In particular π_1 maps $f(U_1)$ bijectively onto U_1'. Hence $f(U_1)$ is the graph of the smooth map

$$g_1 \colon U_1' \to \mathbb{R}^{n(d-1)+d} ; \quad g_1 = \pi_2 \circ f \circ \left(f_{1|U_1} \right)^{-1}.$$

Define a diffeomorphism h_1 from $\pi_1^{-1}(U_1')$ to itself by the formula

$$h_1(x, y) = (x, y - g_1(x)), \quad x \in U_1', \ y \in \mathbb{R}^{n(d-1)+n}.$$

We see that h_1 maps $f(U_1)$ bijectively onto $U_1' \times \{0\}$. Since $f(U_1)$ is open in $f(M)$, $f(U_1) = f(M) \cap W_1$ for an open set $W_1 \subseteq \mathbb{R}^{nd+d}$, which can be chosen to be contained in $\pi_1^{-1}(U_1')$. The restriction $h_{1|W_1}$ is a diffeomorphism from W_1 onto an open set W_1', and it maps $f(M) \cap W_1$ bijectively onto $W_1' \cap \mathbb{R}^n$, as required by Definition 8.8. The remaining $f(U_j)$ are treated analogously. Hence $f(M)$ is a smooth submanifold of $\mathbb{R}^{nd+d}$. Note also that $f_{|U_1} \colon U_1 \to f(U_1)$ is a diffeomorphism, namely the composite of $f_{1|U_1} \colon U_1 \to U_1'$ and the inverse to the diffeomorphism $f(U_1) \to U_1'$ induced by π_1. The remaining U_j are treated analogously. Hence $f \colon M \to f(M)$ is a diffeomorphism. $\qquad \square$

Remark 8.13 The general case of Theorem 8.11 is shown in standard text books on differential topology. To get $k = n + 1$ (Whitney's embedding theorem) one uses Theorem 11.6 below. In the proofs above one can change "s~ manifold" to "topological manifold", "smooth map" to "c~~

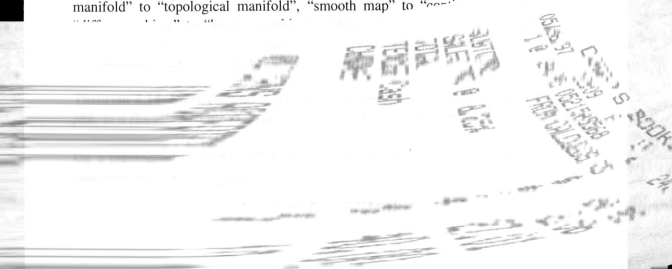

consisting of the maps $M \to \mathbb{R}$, that are smooth in the structure $\mathcal{A}$ on M (and the standard structure on $\mathbb{R}$). Usually $\mathcal{A}$ is suppressed from the notation, and the $\mathbb{R}$-algebra of smooth real-valued functions on M is denoted by $C^\infty(M, \mathbb{R})$. This subalgebra of $C^0(M, \mathbb{R})$ uniquely determines the smooth structure on M. This is a consequence of the following Proposition 8.15 applied to the identity maps id_M

$$(M, \mathcal{A}_1) \overset{\leftarrow}{\to} (M, \mathcal{A}_2).$$

Proposition 8.15 *If $g\colon N \to M$ is a continuous map between smooth manifolds N and M, then g is smooth if and only if the homomorphism*

$$g^*\colon C^0(M, \mathbb{R}) \to C^0(N, \mathbb{R})$$

given by $g^*(\psi) = \psi \circ g$ maps $C^\infty(M, \mathbb{R})$ to $C^\infty(N, \mathbb{R})$.

Proof. "Only if" follows because a composition of two smooth maps is smooth. Conversely if the condition on g^* is satisfied, Lemma 8.12 applied to $p = g(q)$ yields a smooth map $f\colon M^n \to \mathbb{R}^n$ and an open neighborhood V of p in M, such that the restriction $f_{|V}$ is a diffeomorphism of V onto an open subset of $\mathbb{R}^n$. For the j-th coordinate function $f_j \in C^\infty(M, \mathbb{R})$ we have $f_j \circ g = g^*(f_j) \in C^\infty(N, \mathbb{R})$, so that $f \circ g\colon N \to \mathbb{R}^n$ is smooth. Using the chart $f_{|V}$ on M, g is seen to be smooth at q. $\qquad\square$

Remark 8.16 There is a quite elaborate theory which attempts to classify n-dimensional smooth and topological manifolds up to diffeomorphism and homeomorphism. Every connected 1-dimensional smooth or topological manifold is diffeomorphic or homeomorphic to $\mathbb{R}$ or S^1. For $n = 2$ there is a complete classification of the compact connected surfaces. There are two infinite families of them:

Orientable surfaces:

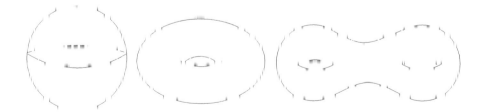

Non-orientable surfaces: $\mathbb{RP}^2$, Klein's bottle, etc. See e.g. [Hirsch] [Massey]

In dimension 3, one meets a famous open problem: the Poincaré conjecture, which asserts that every compact topological 3-manifold that is homotopy equivalent to S^3 is homeomorphic to S^3. It is known that every topological 3-manifold M^3 has a smooth structure $\mathcal{A}$ and that two homeomorphic smooth 3-manifolds also are diffeomorphic. In the mid 1950s J. Milnor discovered smooth 7-manifolds that are homeomorphic to S^7, but not diffeomorphic to S^7. In collaboration with M. Kervaire he classified such exotic n-spheres. For example they showed that there are exactly 28 *oriented* diffeomorphism classes of exotic 7-spheres. In 1960 Kervaire described a topological 10-manifold that has no smooth structure. During the 1960s the so-called "surgery" technique was developed, which in principle classifies all manifolds of a specified homotopy type, but only for $n \geq 5$.

In the early 1980s M. Freedman completely classified the simply-connected compact topological 4-manifolds; see [Freedman–Quinn]. At the same time S. Donaldson proved some very surprising results about smooth compact 4-manifolds, which showed that there is a tremendous difference between smooth and topological 4-manifolds. Donaldson used methods originating in mathematical physics (Yang–Mills theory). This has led to a wealth of new results on smooth 4-manifolds; see [Donaldson–Kronheimer].

One of the most bizarre conclusions of the work of Donaldson and Freedman is that there exists a smooth structure on $\mathbb{R}^4$ such that the resulting smooth 4-manifold "$\mathbb{R}^4$" is not diffeomorphic to the usual $\mathbb{R}^4$. It was proved earlier by S. Smale that every smooth structure on $\mathbb{R}^n$ for $n \neq 4$ is diffeomorphic to the standard $\mathbb{R}^n$.

9. DIFFERENTIAL FORMS ON SMOOTH MANIFOLDS

In this chapter we define the de Rham complex $\Omega^*(M)$ of a smooth manifold M^m and generalize the material of earlier chapters to the manifold case.

For a given point $p \in M^m$ we shall construct an m-dimensional real vector space T_pM called the tangent space at p. Moreover we want a smooth map $f: M \to N$ to induce a linear map $D_pf: T_pM \to T_{f(p)}M$ known as the tangent map of f at p.

Remarks 9.1

(i) In the case $p \in U \subseteq \mathbb{R}^m$, where U is open, one usually identifies the tangent space to U at p with $\mathbb{R}^m$. Better suited for generalization is the following description: Consider the set of smooth parametrized curves $\gamma: I \to U$ with $\gamma(0) = p$, defined on open intervals around 0. An equivalence relation on this set is given by the condition $\gamma_1'(0) = \gamma_2'(0)$. There is a 1-1 correspondence between equivalence classes and $\mathbb{R}^m$, which to the class $[\gamma]$ of γ associates the velocity vector $\gamma'(0) \in \mathbb{R}^m$.

(ii) Consider a further open set $V \subseteq \mathbb{R}^n$ and a smooth map $F: U \to V$. The Jacobi matrix of F evaluated at $p \in U$ defines a linear map $D_pF: \mathbb{R}^m \to \mathbb{R}^n$. For $\gamma: I \to U$, $\gamma(0) = p$ as in (i) the chain rule implies that $D_pF(\gamma'(0)) = (F \circ \gamma)'(0)$. Interpreting tangent spaces as given by equivalence classes of curves we have

$$D_pF([\gamma]) = [F \circ \gamma]. \tag{1}$$

In particular the equivalence class of $F \circ \gamma$ depends only on $[\gamma]$

Let (U, h) be a smooth chart around $p \in M^m$. On the set of smooth curves $\alpha: I \to M$ with $\alpha(0) = p$ defined on open intervals around 0 we have an equivalence relation

$$\alpha_1 \sim \alpha_2 \Leftrightarrow (h \circ \alpha_1)'(0) = (h \circ \alpha_2)'(0). \tag{2}$$

This equivalence relation is independent of the choice of (U, h). In fact, if $(\widetilde{U}, \widetilde{h})$ is another smooth chart around p, one finds that

$$(h \circ \alpha_1)'(0) = (h \circ \alpha_2)'(0) \Leftrightarrow (\widetilde{h} \circ \alpha_1)'(0) = (\widetilde{h} \circ \alpha_2)'(0)$$

by applying the last statement of Remark 9.1.(ii) to the transition diffeomorphism $F = h \circ \widetilde{h}^{-1}$ and its inverse.

Definition 9.2 The *tangent space* $T_p M^m$ is the set of equivalence classes with respect to (2) of smooth curves $\alpha: I \to M$, $\alpha(0) = p$.

We give $T_p M$ the structure of an m-dimensional vector space defined by the following condition: if (U, h) is a smooth chart in M with $p \in U$, then

$$\Phi_h: T_p M \to \mathbb{R}^m, \quad \Phi_h([\alpha]) = (h \circ \alpha)'(0),$$

is a linear isomorphism; here $[\alpha] \in T_p M$ is the equivalence class of α.

By definition Φ_h is a bijection. The linear structure on $T_p M$ is well-defined. This can be seen from the following commutative diagram, where $F = h \circ \tilde{h}^{-1}$, $q = \tilde{h}(p)$

$$
\begin{array}{ccc}
 & & \mathbb{R}^m \\
 & \overset{\Phi_{\tilde{h}}}{\nearrow} & \\
T_p M & & \cong \downarrow D_q F \\
 & \overset{\Phi_h}{\searrow} & \\
 & & \mathbb{R}^m
\end{array}
$$

Lemma 9.3 *Let* $f: M^m \to N^n$ *be a smooth map and* $p \in M$.

(i) There is a linear map $D_p f: T_p M \to T_{f(p)} N$ given in terms of representing curves by

$$D_p f([\alpha]) = [f \circ \alpha].$$

(ii) If (U, h) is a chart around p in M and (V, g) a chart around $f(p)$ in N, then we have the following commutative diagram:

$$
\begin{array}{ccc}
T_p M & \overset{D_p f}{\longrightarrow} & T_{f(p)} M \\
\cong \downarrow \Phi_h & & \cong \downarrow \Phi_g \\
\mathbb{R}^m & \underset{D_{h(p)}(g \circ f \circ h^{-1})}{\longrightarrow} & \mathbb{R}^n
\end{array}
$$

Proof. Remark 9.1.(ii) applied to $F = g \circ f \circ h^{-1}$, defined on the open set $h(U \cap f^{-1}(V))$, shows that the bottom map in the diagram is linear and given on representing curves as stated there. Since Φ_h and Φ_g are linear isomorphisms, there exists a linear map $D_p f$ making the diagram commutative. The formula in (i) follows by chasing around the diagram. $\square$

Note that the linear isomorphism Φ_h in Definition 9.2 can be identified with $D_p h$ through the identification discussed in Remark 9.1.(ii). From now on we write $D_p h: T_p M \to \mathbb{R}^m$ for this linear isomorphism, and similarly $D_{h(p)} h^{-1}: \mathbb{R}^m \to T_p M$ for its inverse.

Suppose $M^m \subseteq \mathbb{R}^l$ is a smooth submanifold with inclusion map $i: M \to \mathbb{R}^l$. Definition 8.8 implies that $D_p i: T_p M^m \to T_p \mathbb{R}^l \cong \mathbb{R}^l$ is injective. In this case we usually identify $T_p M$ with the image in $\mathbb{R}^l$, consisting of all vectors $\alpha'(0)$ where $\alpha: I \to M \subseteq \mathbb{R}^l$ is a smooth parametrized curve with $\alpha(0) = p$.

For a composite $\varphi \circ f$ of smooth maps $f: M \to N$, $\varphi: N \to P$ and a point $p \in M$ we have the *chain rule*, immediately from Lemma 9.3(i),

$$D_p(\varphi \circ f) = D_{f(p)}\varphi \circ D_p\varphi. \tag{3}$$

Remark 9.4 Given a smooth chart (U, h) around the point $p \in M^m$ we obtain a basis for $T_p M$

$$\left(\frac{\partial}{\partial x_1}\right)_p, \ldots, \left(\frac{\partial}{\partial x_m}\right)_p,$$

where $\left(\frac{\partial}{\partial x_i}\right)_p$ is the image under $D_{h(p)} h^{-1}: \mathbb{R}^m \to T_p M$ of the i-th standard basis vector $e_i = (0, \ldots 1, \ldots, 0) \in \mathbb{R}^m$. A tangent vector $X_p \in T_p M$ can be written uniquely as

$$X_p = \sum_{i=1}^{m} a_i \left(\frac{\partial}{\partial x_i}\right)_p \tag{4}$$

where $\mathbf{a} = (a_1, \ldots, a_m) \in \mathbb{R}^m$. If $X_p = [\alpha]$, where $\alpha: I \to U$ with $\alpha(0) = p$ is a representing smooth curve, we have

$$\mathbf{a} = (h \circ \alpha)'(0).$$

Given $f \in C^\infty(M, \mathbb{R})$ we have the tangent map

$$D_p f: T_p M \to T_{f(p)} \mathbb{R} \cong \mathbb{R}. \tag{5}$$

The *directional derivative* $X_p f \in \mathbb{R}$ is defined to be the image in $\mathbb{R}$ of X_p under (5), i.e. $X_p f = (f \circ \alpha)'(0)$. In terms of $f \circ h^{-1}$ we have by the chain rule

$$X_p f = \frac{d}{dt}\left(f h^{-1} \circ h\alpha(t)\right)_{|t=0} = \sum_{i=1}^{m} \frac{\partial f h^{-1}}{\partial x_i}(h(p)) a_i.$$

In particular

$$\left(\frac{\partial}{\partial x_j}\right)_p f = \frac{\partial f h^{-1}}{\partial x_j}(h(p)). \tag{6}$$

Under the assumptions of Lemma 9.3.(ii) there is a similar basis $\left(\partial / \partial y_j\right)_{f(p)}$, $1 \leq j \leq n$, for $T_{f(p)} N$ and the matrix of $D_p f$ with respect to our bases for $T_p M$

and $T_{f(p)}N$ is the Jacobian matrix at $h(p)$ of $g \circ f \circ h^{-1}$. Specializing this to the case $N = M$, $f = \mathrm{id}_M$ we find that

$$(7) \qquad \left(\frac{\partial}{\partial x_i}\right)_p = \sum_{j=1}^{m} \frac{\partial \varphi_j}{\partial x_i}(h(p)) \left(\frac{\partial}{\partial y_j}\right)_p.$$

This holds when (U, h) and (V, g) are two charts around p with transition function

$$(\varphi_1, \ldots, \varphi_m) = \varphi = g \circ h^{-1}$$

expressing the y_j-coordinates in terms of the x_i-coordinates.

Suppose X is a function that to each $p \in M$ assigns a tangent vector $X_p \in T_pM$. Given the chart (U, h), formula (4) holds for $p \in U$ with certain coefficient functions $a_i: U \to \mathbb{R}$. If these are smooth in a neighborhood of $p \in U$, X is said to be smooth at p. From (7), this condition is independent of the choice of smooth chart around p. If X is smooth at every point $p \in M$, X is a a *smooth vector field* on M.

Let us next consider families $\omega = \{\omega_p\}_{p \in M}$ of alternating k-forms on T_pM, where $\omega_p \in \mathrm{Alt}^k(T_pM)$. We need a notion of ω being smooth as function of p. Let $g: W \to M$ be a local parametrization, i.e. the inverse of a smooth chart, where W is an open set in $\mathbb{R}^m$. For $x \in W$,

$$D_x g: \mathbb{R}^m \to T_{g(x)}M$$

is an isomorphism, and induces an isomorphism

$$\mathrm{Alt}^k(D_x g): \mathrm{Alt}^k(T_{g(x)}M) \to \mathrm{Alt}^k(\mathbb{R}^m).$$

We define $g^*(\omega): W \to \mathrm{Alt}^k(\mathbb{R}^m)$ to be the function whose value at x is

$$g^*(\omega)_x = \mathrm{Alt}^k(D_x g)(\omega_{g(x)}) \quad (g^*(\omega)_x = \omega_{g(x)} \text{ for } k = 0).$$

Definition 9.5 A family $\omega = \{\omega_p\}_{p \in M}$ of alternating k-forms on T_pM is said to be smooth if $g^*(\omega)$ is a smooth function for every local parametrization. The set of such smooth families is a vector space $\Omega^k(M)$. In particular, $\Omega^0(M) = C^\infty(M, \mathbb{R})$.

Lemma 9.6 *Let $g_i: W_i \to N$ be a family of local parametrizations with $N = \bigcup g_i(W_i)$. If $g_i^*(\omega)$ is smooth for all i, then ω is smooth.*

Proof. Let $g: W \to N$ be any local parametrization and $z \in W$. We show that $g^*(\omega)$ is smooth close to z. Choose an index i with $g(z) \in g_i(W_i)$. Close to z

we can write $g = g_i \circ g_i^{-1} \circ g = g_i \circ h$, where $h = g_i^{-1} \circ g \colon g^{-1}(g_i(W_i)) \to W_i$ is a smooth map between open sets in Euclidean space. Thus

$$g^*(\omega) = (g_i \circ h)^*(\omega) = h^*(g_i^*(\omega))$$

in a neighborhood of z, and the right-hand side is a smooth k-form by assumption.

$\square$

The exterior differential

$$d \colon \Omega^k(M) \to \Omega^{k+1}(M)$$

can be defined via local parametrizations $g \colon W \to M$ as follows. If $\omega = \{\omega_p\}_{p \in M}$ is a smooth k-form on M then

$$(8) \qquad d_p\omega = \mathrm{Alt}^{k+1}\big((D_x g)^{-1}\big) \circ d_x(g^*\omega), \quad p = g(x).$$

It is not immediately obvious that $d_p\omega$ is independent of the choice of local parametrization, but this is indeed the case: Given a local parametrization g, then any other locally has the form $g \circ \phi$, with $\phi \colon U \to W$ a diffeomorphism. Let $\xi_1, \ldots, \xi_{k+1} \in T_p M$. We choose $v_1, \ldots, v_{k+1} \in \mathbb{R}^n$ so that $D_x(g \circ \phi)(v_i) = \xi_i$. We must show that

$$d_y g^*(\omega)(w_1, \ldots, w_{k+1}) = d_x(g \circ \phi)^*(\omega)(v_1, \ldots, v_{k+1})$$

where $\phi(x) = y$ and $D_x\phi(v_i) = w_i$. This follows from the equations

$$(g \circ \phi)^*(\omega) = \phi^*(g^*(\omega))$$
$$d\phi^*(\tau) = \phi^* d(\tau),$$

where $\tau = g^*(\omega)$; see Theorem 3.12. It is obvious that $d \circ d = 0$. Hence we have defined a chain complex

$$\cdots \to \Omega^{k-1}(M) \xrightarrow{d} \Omega^k(M) \xrightarrow{d} \Omega^{k+1}(M) \to \cdots.$$

We have $\Omega^k(M) = 0$ if $k > \dim M$, since $\mathrm{Alt}^k(T_p M) = 0$ when $k > \dim T_p M$. A smooth map $\phi \colon M \to N$ induces a chain map $\phi^* \colon \Omega^*(N) \to \Omega^*(M)$,

$$(9) \quad \phi^*(\tau)_p = \mathrm{Alt}^k(D_p\phi)\big(\tau_{\phi(p)}\big), \quad \tau \in \Omega^k(N); \quad \phi^*(\tau)_p = \tau_{\phi(p)} \quad \text{for } k = 0.$$

One defines a bilinear product $\omega \wedge \tau$ by $(\omega \wedge \tau)_p = \omega_p \wedge \tau_p$,

$$(10) \qquad \wedge \colon \Omega^k(M) \times \Omega^l(M) \to \Omega^{k+l}(M).$$

One shows by choosing local parametrizations that $\phi^*\omega$ and $\omega \wedge \tau$ are smooth. It is equally easy to see that

$$(11) \qquad \begin{aligned} d(\omega \wedge \tau) &= d\omega \wedge \tau + (-1)^k \omega \wedge d\tau \\ \omega \wedge \tau &= (-1)^{kl} \tau \wedge \omega \end{aligned}$$

for $\omega \in \Omega^k(M)$, $\tau \in \Omega^l(M)$.

Definition 9.7 The p-th (de Rham) cohomology of the manifold M, denoted $H^p(M)$, is the p-th cohomology vector space of $\Omega^*(M)$.

The exterior product induces a product $H^p(M) \times H^q(M) \to H^{p+q}(M)$ which makes $H^*(M)$ into a graded algebra. Note that $H^p(M) = 0$ for $p > n = \dim M$ or $p < 0$.

The chain map ϕ^* induced by a smooth map $\phi \colon M \to N$ induces linear maps

$$H^p(\phi) \colon H^p(N) \to H^p(M),$$

and the de Rham cohomology becomes a contravariant functor from the category of smooth manifolds and smooth maps to the category of graded anti-commutative $\mathbb{R}$-algebras.

Definition 9.8

(i) A smooth manifold M^n of dimension n is called *orientable,* if there exists an $\omega \in \Omega^n(M^n)$ with $\omega_p \neq 0$ for all $p \in M$. Such an ω is called an *orientation form* on M.

(ii) Two orientation forms ω, τ on M are equivalent if $\tau = f \cdot \omega$, for some $f \in \Omega^0(M)$ with $f(p) > 0$ for all $p \in M$. An *orientation* of M is an equivalence class of orientation forms on M.

On the Euclidean space $\mathbb{R}^n$ we have the orientation form $dx_1 \wedge \ldots \wedge dx_n$, which represents the *standard orientation* of $\mathbb{R}^n$.

Let M^n be oriented by the orientation form ω. A basis $b_1, \ldots, b_n$ of T_pM is said to be *positively* or *negatively* oriented with respect to ω depending on whether the number

$$\omega_p(b_1, \ldots, b_n) \in \mathbb{R}$$

is positive or negative. (It cannot be 0, because $\omega_p \neq 0$.) The sign depends only on the orientation determined by ω. If ω and τ are two orientation forms on M^n, then $\tau = f \cdot \omega$ for a uniquely determined function $f \in \Omega^0(M)$ with $f(p) \neq 0$ for all $p \in M$. We say that ω and τ determine the same orientation at p, if $f(p) > 0$. Equivalently, ω and τ induce the same positively oriented bases of T_pM. If M is connected, then f has constant sign on M, so we have:

Lemma 9.9 *On a connected orientable smooth manifold there are precisely 2 orientations.* $\square$

If U is an open subset of an oriented manifold M^n, then an orientation of U is induced by using the restriction of an orientation form on M. Conversely we have:

Lemma 9.10 *Let* $V = (V_i)_{i \in I}$ *be an open cover of the smooth submanifold* M^n. *Suppose that all* V_i *have orientations and that the restrictions of the orientations from* V_i *and* V_j *to* $V_i \cap V_j$ *coincide for all* $i \neq j$. *Then* M *has a uniquely determined orientation with the given restriction to* V_i *for all* $i \in I$.

The proof is a typical application of a smooth partition of unity in the following form:

Theorem 9.11 *Let* $V = (V_i)_{i \in I}$ *be an open cover of the smooth manifold* $M^n \subseteq \mathbb{R}^l$. *Then there exist smooth functions* $\phi_i \colon M \to [0,1]$ $(i \in I)$ *that satisfy*

(i) $\mathrm{Supp}_M(\phi_i) \subseteq V_i$ *for all* $i \in I$.
(ii) *Every* $p \in M$ *has an open neighborhood where only finitely many of the functions* ϕ_i $(i \in I)$ *do not vanish.*
(iii) *For every* $p \in M$ *we have* $\sum_{i \in I} \phi_i(p) = 1$.

Proof. Since M has the topology induced by $\mathbb{R}^l$, we can choose an open set $U_i \subseteq \mathbb{R}^l$ with $U_i \cap M = V_i$ for each $i \in I$. By applying Theorem A.1 to $U = \bigcup_{i \in I} U_i$ we get smooth functions $\psi_i \colon U \to [0,1]$ with

(i) $\mathrm{Supp}_U(\psi_i) \subseteq U_i$.
(ii) Local finiteness.
(iii) For every $x \in U$, $\sum_{i \in I} \psi_i(x) = 1$.

Let $\phi_i \colon M \to [0,1]$ be the restriction of ψ_i; conditions (i), (ii) and (iii) of the theorem follow immediately. $\qquad\square$

Proof of Lemma 9.10. Let the orientation of V_i be given by the orientation form $\omega_i \in \Omega^n(V_i)$, and choose smooth functions $\phi_i \colon M \to [0,1]$ as in Theorem 9.11. We can define $\omega \in \Omega^n(M)$ by

$$\omega = \sum_{i \in I} \phi_i \, \omega_i,$$

where $\phi_i \omega_i$ is extended to an n-form on all of M by letting it vanish on $M - \mathrm{Supp}_M(\phi_i)$. This is an orientation form, because if $p \in V_i \subseteq M$ and $b_1, \dots, b_n$ is a basis of T_pM, with $\omega_{i,p}(b_1, \dots, b_n) > 0$, then $\omega_{i',p}(b_1, \dots, b_n) > 0$ for every other i' with $p \in V_{i'}$ and in the formula

$$\omega_p(b_1, \dots, b_n) = \sum_i \phi_i(p) \, \omega_{i,p}(b_1, \dots, b_n),$$

all terms are positive (or zero). Thus ω is an orientation form on M, and $b_1, \dots, b_n$ are positively oriented with respect to ω. The orientation of M determined by ω has the desired property.

If $\tau \in \Omega^n(M^n)$ is another orientation form that gives the orientation of the required type, then $\tau = f \cdot \omega$ and $\tau_p(b_1, \ldots, b_n) = f(p)\omega_p(b_1, \ldots, b_n)$. Since both τ_p and ω_p are positive on $b_1, \ldots, b_n$ we have $f(p) > 0$. Hence ω and τ determine the same orientation. $\qquad\square$

Definition 9.12 Let $\phi: M_1^n \to M_2^n$ be a diffeomorphism between manifolds that are oriented by the orientation forms $\omega_j \in \Omega^n(M_j^n)$. Then $\phi^*(\omega_2)$ is an orientation form on M_1^n. We say ϕ is *orientation-preserving* (resp. orientation-reversing), when $\phi^*(\omega_2)$ determines the same orientation of M_1^n as ω_1 (resp. $-\omega_1$).

Example 9.13 Consider a diffeomorphism $\phi: U_1 \to U_2$ between open subsets U_1, U_2 of $\mathbb{R}^n$, both equipped with the standard orientation of $\mathbb{R}^n$. It follows from Example 3.13.(ii) that ϕ is orientation-preserving if and only if $\det(D_x\phi) > 0$ for all $x \in U_1$. Analogously ϕ is orientation-reversing if and only if all Jacobi determinants are negative.

Around any point on an oriented smooth manifold M^n we can find a chart $h: U \to U'$ such that h is an orientation-preserving diffeomorphism when U is given the orientation of M and U' the orientation of $\mathbb{R}^n$. We call h an *oriented chart* of M. The transition function associated with two oriented charts of M is an orientation-preserving diffeomorphism. For any atlas of M consisting of oriented charts, all Jacobi determinants of the transition functions will be positive. Such an atlas is called *positive*.

Proposition 9.14 *If* $\{h_i: U_i \to U_i' \mid i \in I\}$ *is a positive atlas on* M^n, *then* M^n *has a uniquely determined orientation, so all* h_i *are oriented charts.*

Proof. For $i \in I$ we orient U_i so that h_i is an orientation-preserving diffeomorphism. By Example 9.13, the two orientations on $U_i \cap U_j$ defined by the restriction from U_i and U_j coincide. The assertion follows from Lemma 9.10. $\qquad\square$

Definition 9.15 A *Riemannian structure* (or Riemannian metric) on a smooth manifold M^n is a family of inner products $\langle\,,\,\rangle_p$ on T_pM, for all $p \in M$, that satisfy the following condition: for any local parametrization $f: W \to M$ and any pair $v_1, v_2 \in \mathbb{R}^n$,

$$x \to \langle D_x f(v_1), D_x f(v_2) \rangle_{f(x)}$$

is a smooth function on W.

It is sufficient to have the smoothness condition satisfied for the functions

$$g_{ij}(x) = \langle D_x f(e_i), D_x f(e_j) \rangle_{f(x)}, \quad 1 \le i, j \le n,$$

where $e_1, \ldots, e_n$ is the standard basis of $\mathbb{R}^n$. These functions are called the coefficients of the *first fundamental form*. For $x \in W$ the $n \times n$ matrix $(g_{ij}(x))$ is symmetric and positive definite.

A smooth manifold equipped with a Riemannian structure is called a *Riemannian manifold*. A smooth submanifold $M^n \subseteq \mathbb{R}^l$ has a Riemannian structure defined by letting $\langle\,,\,\rangle_p$ be the restriction to the subspace $T_pM \subseteq \mathbb{R}^l$ of the usual inner product on $\mathbb{R}^l$.

Proposition 9.16 *If M^n is an oriented Riemannian manifold then M^n has a uniquely determined orientation form* vol_M *with*

$$\mathrm{vol}_M(b_1, \ldots, b_n) = 1$$

for every positively oriented orthonormal basis of a tangent space T_pM. We call vol_M *the volume form on M.*

Proof. Let the orientation be given by the orientation form $\omega \in \Omega^n(M^n)$. Consider two positively oriented orthonormal bases $b_1, \ldots, b_n$ and $b'_1, \ldots, b'_n$ in the same tangent space T_pM. There exists an orthogonal $n \times n$ matrix $C = (c_{ij})$ such that

$$b'_i = \sum_{j=1}^n c_{ij}\, b_j,$$

and $\omega_p \in \mathrm{Alt}^n T_pM$ satisfies

(12) $$\omega_p(b'_1, \ldots, b'_n) = (\det C)\omega_p(b_1, \ldots, b_n).$$

Positivity ensures that $\det C > 0$; but then $\det C = 1$. Hence there exists a function $\rho\colon M \to (0, \infty)$ such that $\rho(p) = \omega_p(b_1, \ldots, b_n)$ for every positively oriented orthonormal basis $b_1, \ldots, b_n$ of T_pM. We must show that ρ is smooth; then $\mathrm{vol}_M = \rho^{-1}\omega$ will be the volume form.

Consider an orientation-preserving local parametrization $f\colon W \to M^n$ and set

$$X_j(q) = \left(\frac{\partial}{\partial x_j}\right)_q = D_q f(e_j) \in T_{f(q)}M \quad \text{for } 1 \le j \le n \text{ and } q \in W.$$

These form a positively oriented basis of $T_{f(q)}M$. An application of the Gram–Schmidt orthonormalization process gives an upper triangular matrix $A(q) = (a_{ij}(q))$ of smooth functions on W with $a_{ii}(q) > 0$, such that

(13) $$b_i(q) = \sum_{j=1}^n a_{ij}(q)X_j(q), \quad i = 1, \ldots, n$$

is a positively oriented orthonormal basis of $T_{f(q)}M$. Then

(14) $$\begin{aligned} \rho \circ f(q) &= \omega_{f(q)}(b_1(q), \ldots, b_n(q)) = (\det A(q))\, \omega_{f(q)}(X_1(q), \ldots, X_n(q)) \\ &= (\det A(q))(f^*\omega)_q(e_1, \ldots, e_n). \end{aligned}$$

This shows that ρ is smooth. $\qquad\square$

Addendum 9.17 There is the following formula for vol_M in local coordinates:

$$(15) \qquad f^*(\mathrm{vol}_M) = \sqrt{\det(g_{ij}(x))}\, dx_1 \wedge \ldots \wedge dx_n,$$

where $f: W \to M^n$ is an orientation-preserving local parametrization and

$$g_{ij}(x) = \langle D_x f(e_i),\, D_x f(e_j) \rangle_{f(x)}.$$

Proof. Repeat the proof above starting with $\omega = \mathrm{vol}_M$, so that $\rho(p) = 1$ for all $p \in M$. Formula (14) becomes

$$(16) \qquad f^*(\mathrm{vol}_M) = (\det A(x))^{-1} dx_1 \wedge \ldots \wedge dx_n.$$

The inner product of (13) with the corresponding formula for $b_k(q)$ yields (with a Kronecker delta notation)

$$\delta_{ik} = \langle b_i(q), b_k(q) \rangle_{f(q)} = \sum_{j=1}^n \sum_{l=1}^n a_{ij}(q) g_{jl}(q)\, a_{kl}(q).$$

This is the matrix identity, $I = A(q)\, G(q)\, A(q)^t$, where $G(q) = (g_{jl}(q))$. In particular $(\det A(q))^2 \det G(q) = 1$. Since $\det A(q) = \prod_i a_{ii}(q) > 0$, we obtain

$$(\det A(q))^{-1} = \sqrt{\det G(q)}. \qquad \square$$

Example 9.18 Define an $(n-1)$-form $\omega_0 \in \Omega^{n-1}(\mathbb{R}^n)$ by

$$(17) \qquad \omega_{0x}(w_1, \ldots, w_{n-1}) = \det(x, w_1, \ldots, w_{n-1}) \in \mathrm{Alt}^{n-1}(\mathbb{R}^n),$$

for $x \in \mathbb{R}$. Since $\omega_{0x}(e_1, \ldots, \hat{e}_i, \ldots, e_n) = (-1)^{i-1} x_i$, we have

$$(18) \qquad \omega_0 = \sum_{i=1}^n (-1)^{i-1} x_i dx_1 \wedge \ldots \wedge \widehat{dx_i} \wedge \ldots \wedge dx_n.$$

If $x \in S^{n-1}$ and $w_1, \ldots, w_{n-1}$ is a basis of $T_x S^{n-1}$ then $x, w_1, \ldots, w_{n-1}$ becomes a basis for $\mathbb{R}^n$ and (17) shows that $\omega_{0x} \neq 0$. Hence $\omega_{0|S^{n-1}} = i^*(\omega_0)$ is an orientation form on S^{n-1}. For the orientation of S^{n-1} given by ω_0, the basis $w_1, \ldots, w_{n-1}$ of $T_x S^{n-1}$ is positively oriented if and only if the basis $x, w_1, \ldots, w_{n-1}$ for $\mathbb{R}^n$ is positively oriented.

We give S^{n-1} the Riemannian structure induced by $\mathbb{R}^n$. Then (17) implies that $\mathrm{vol}_{S^{n-1}} = \omega_{0|S^{n-1}}$.

We may construct a closed $(n-1)$-form on $\mathbb{R}^n - \{0\}$ with $\omega_{|S^{n-1}} = \mathrm{vol}_{S^{n-1}}$ by setting $\omega = r^*(\mathrm{vol}_{S^{n-1}})$, where $r: \mathbb{R}^n - \{0\} \to S^{n-1}$ is the map $r(x) = x/\|x\|$. For $x \in \mathbb{R}^n - \{0\}$, $\omega_x \in \mathrm{Alt}^{n-1}(\mathbb{R}^n)$ is given by

$$\omega_x(v_1, \ldots, v_{n-1}) = \omega_{0r(x)}(D_x r(v_1), \ldots, D_x r(v_{n-1}))$$
$$= \|x\|^{-1} \det(x, D_x r(v_1), \ldots, D_x r(v_{n-1})).$$

Now we have

$$D_x r(v) = \begin{cases} 0 & \text{if } v \in \mathbb{R}x \\ \|x\|^{-1} & \text{if } v \in (\mathbb{R}x)^{\perp} \end{cases}$$

so that $D_x r(v) = \|x\|^{-1} w$, where w is the orthogonal projection of v on $(\mathbb{R}x)^{\perp}$. Letting w_i be the orthogonal projection of v_i on $(\mathbb{R}x)^{\perp}$ we have

$$\omega_x(v_1, \ldots, v_{n-1}) = \|x\|^{-n} \det(x, w_1, \ldots, w_{n-1}) = \|x\|^{-n} \det(x, v_1, \ldots, v_{n-1})$$
$$= \|x\|^{-n} \omega_{0x}(v_1, \ldots, v_{n-1}).$$

Hence the closed form ω is given by

$$(19) \qquad \omega = \frac{1}{\|x\|^n} \sum_{i=1}^{n} (-1)^{i-1} x_i \, dx_1 \wedge \ldots \wedge \widehat{dx_i} \wedge \ldots \wedge dx_n.$$

Example 9.19 For the antipodal map

$$A \colon S^{n-1} \to S^{n-1}; \quad Ax = -x$$

we have

$$A^*(\mathrm{vol}_{S^{n-1}}) = (-1)^n \mathrm{vol}_{S^{n-1}},$$

and A is orientation-preserving if and only if n is even. In this case we get an orientation form τ on $\mathbb{RP}^{n-1}$ such that $\pi^*(\tau) = \mathrm{vol}_{S^{n-1}}$, where π is the canonical map $\pi \colon S^{n-1} \to \mathbb{RP}^{n-1}$. For $x \in S^{n-1}$,

$$T_x S^{n-1} \overset{D_x A}{\to} T_{Ax} S^{n-1}$$

is a linear isometry. Hence there exists a Riemannian structure on $\mathbb{RP}^{n-1}$ characterized by the requirement that the isomorphism

$$T_x S^{n-1} \overset{D_x \pi}{\to} T_{\pi(x)} \mathbb{RP}^{n-1}$$

is an isometry for every $x \in S^{n-1}$. If n is even and $\mathbb{RP}^{n-1}$ is oriented as before, one gets $\pi^*(\mathrm{vol}_{\mathbb{RP}^{n-1}}) = \mathrm{vol}_{S^{n-1}}$. Conversely suppose that $\mathbb{RP}^{n-1}$ is orientable, $n \geq 2$. Choose an orientation and let $\mathrm{vol}_{\mathbb{RP}^{n-1}}$ be the resulting volume form. Since $D_x \pi$ is an isometry, $\pi^*(\mathrm{vol}_{\mathbb{RP}^{n-1}})$ must coincide with $\pm \mathrm{vol}_{S^{n-1}}$ in all points, and by continuity the sign is locally constant. Since S^{n-1} is connected the sign is constant on all of S^{n-1}. We thus have that

$$\pi^*(\mathrm{vol}_{\mathbb{RP}^{n-1}}) = \delta \mathrm{vol}_{S^{n-1}},$$

where $\delta = \pm 1$. We can apply A^* and use the equation $\pi \circ A = \pi$ to get

$$(-1)^n \delta \, \mathrm{vol}_{S^{n-1}} = \delta A^*(\mathrm{vol}_{S^{n-1}}) = A^* \pi^*(\mathrm{vol}_{\mathbb{RP}^{n-1}})$$
$$= (\pi \circ A)^*(\mathrm{vol}_{\mathbb{RP}^{n-1}}) = \pi^*(\mathrm{vol}_{\mathbb{RP}^{n-1}}) = \delta \, \mathrm{vol}_{S^{n-1}}.$$

This requires that n is even and implies that $\mathbb{RP}^{n-1}$ is orientable if and only if n is even.

Remark 9.20 For two smooth manifolds M^m and N^n the Cartesian product $M^m \times N^n$ is a smooth manifold of dimension $m+n$. For a pair of charts $h\colon U \to U'$ and $k\colon V \to V'$ of M and N, respectively, we can use $h \times k\colon U \times V \to U' \times V'$ as a chart of $M \times N$. These product charts form a smooth atlas on $M \times N$. For $p \in M$ and $q \in N$ there is a natural isomorphism

$$T_{(p,q)}(M \times N) \cong T_p M \oplus T_q N.$$

If M and N are oriented, one can use oriented charts (U, h) and (V, k). The transition diffeomorphisms between the charts $(U \times V, h \times k)$ satisfy the condition of Proposition 9.14. Hence we obtain a *product orientation* of $M \times N$. If the orientations are specified by orientation forms $\omega \in \Omega^m(M)$ and $\sigma \in \Omega^n(N)$, the product orientation is given by the orientation form $\mathrm{pr}_M^*(\omega) \wedge \mathrm{pr}_N^*(\sigma)$, where pr_M and pr_N are the projections of $M \times N$ on M and N.

In the following we shall consider a smooth submanifold $M^n \subseteq \mathbb{R}^{n+k}$ of dimension n. At every point $p \in M$ we have a normal vector space $T_p M^\perp$ of dimension k. A smooth *normal vector field* Y on an open set $W \subseteq M$ is a smooth map $Y\colon W \to \mathbb{R}^{n+k}$ with $Y(p) \in T_p M^\perp$ for every $p \in W$. In the case $k = 1$, Y is called a *Gauss map* on W when all $Y(p)$ have length 1. Such a map always exists locally since we have the following:

Lemma 9.21 *For every $p_0 \in M^n \subseteq \mathbb{R}^{n+k}$ there exists an open neighborhood W of p_0 on M and smooth normal vector fields $Y_j (1 \le j \le k)$ on W such that $Y_1(p), \ldots, Y_k(p)$ form an orthonormal basis of $T_p M^\perp$ for every $p \in W$.*

Proof. On a coordinate patch around $p_0 \in M$, there exist smooth tangent vector fields $X_1, \ldots, X_n$, which at every point p yield a basis of $T_p M$, cf. Remark 9.4. Choose a basis $V_1, \ldots, V_k$ of $T_{p_0} M^\perp$. Since the $(n + k) \times (n + k)$ determinant

$$\det(X_1(p), \ldots, X_n(p), V_1, \ldots, V_k)$$

is non-zero at p_0, it also non-zero for all p in some open neighborhood W of p_0 on M. Gram–Schmidt orthonormalization applied to the basis

$$X_1(p), \ldots, X_n(p), \quad V_1, \ldots, V_k \quad (p \in W)$$

of $\mathbb{R}^{n+k}$ gives an orthonormal basis

$$\tilde{X}_1(p), \ldots, \tilde{X}_n(p), \ Y_1(p), \ldots, Y_k(p),$$

where the first n vectors span $T_p M$. The formulas of the Gram–Schmidt orthonormalization show that all $\tilde{X}_i$ and Y_j are smooth on W, so that $Y_1, \ldots, Y_k$ have the desired properties. $\qquad\square$

Proposition 9.22 *Let $M^n \subseteq \mathbb{R}^{n+1}$ be a smooth submanifold of codimension* 1.

(i) *There is a 1-1 correspondence between smooth normal vector fields Y on M and n-forms in $\Omega^n(M)$. It associates to Y the n-form $\omega = \omega_Y$ given by*
$$\omega_p(W_1, \ldots, W_n) = \det(Y(p), W_1, \ldots, W_n)$$
for $p \in M$, $W_i \in T_pM$.

(ii) *This induces a 1-1 correspondence between Gauss maps $Y: M \to S^n$ and orientations of M.*

Proof. If $p \in M$ then $Y(p) = 0$ if and only if $\omega_p = 0$. Since ω_Y depends linearly on Y, the map $Y \to \omega_Y$ must be injective. If Y is a Gauss map, then ω_Y is an orientation form and it can be seen that ω_Y is exactly the volume form associated to the orientation determined by ω_Y and the Riemannian structure on M induced from $\mathbb{R}^{n+1}$. If M has a Gauss map Y then (i) follows, since every element in $\Omega^n(M)$ has the form $f \cdot \omega_Y = \omega_{fY}$ for some $f \in C^\infty(M, \mathbb{R})$. Now M can be covered by open sets, for which there exist Gauss maps. For each of these (i) holds, but then the global case of (i) automatically follows.

An orientation of M determines a volume form vol_M, and from (i) one gets a Y with $\omega_Y = \mathrm{vol}_M$. This Y is a Gauss map. $\qquad\square$

Theorem 9.23 (Tubular neighborhoods) *Let $M^n \subseteq \mathbb{R}^{n+k}$ be a smooth submanifold. There exists an open set $V \subseteq \mathbb{R}^{n+k}$ with $M \subseteq V$ and an extension of id_M to a smooth map $r: V \to M$, such that*

(i) *For $x \in V$ and $y \in M$, $\|x - r(x)\| \leq \|x - y\|$, with equality if and only if $y = r(x)$.*

(ii) *For every $p \in M$ the fiber $r^{-1}(p)$ is an open ball in the affine subspace $p + T_pM^\perp$ with center at p and radius $\rho(p)$, where ρ is a positive smooth function on M. If M is compact then ρ can be taken to be constant.*

(iii) *If $\epsilon: M \to \mathbb{R}$ is smooth and $0 < \epsilon(p) < \rho(p)$ for all $p \in M$ then*
$$S_\epsilon = \{x \in V \mid \|x - r(x)\| = \epsilon(r(x))\}$$
is a smooth submanifold of codimension 1 in $\mathbb{R}^{n+k}$.

We call $V(= V_\rho)$ the open tubular neighborhood of M of radius ρ

Proof. We first give a local construction around a point $p_0 \in M$. Choose normal vector fields $Y_1, \ldots, Y_k$ as in Lemma 9.21, defined on an open neighborhood W of p_0 in M for which we have a diffeomorphism $f: \mathbb{R}^n \to W$ with $f(0) = p_0$. Let us define $\Phi: \mathbb{R}^{n+k} \to \mathbb{R}^{n+k}$ by
$$\Phi(x, t) = f(x) + \sum_{j=1}^{k} t_j Y_j(f(x)) \qquad (x \in \mathbb{R}^n,\ t \in \mathbb{R}^k).$$

The Jacobi matrix of Φ at 0 has the columns

$$\frac{\partial f}{\partial x_1}(0), \ldots, \frac{\partial f}{\partial x_n}(0), \ Y_1(p_0), \ldots, Y_k(p_0).$$

The first n form a basis of $T_{p_0} M$ and the last k a basis of $T_{p_0} M^\perp$. By the inverse function theorem, Φ is a local diffeomorphism around 0. There exists a (possibly) smaller open neighbourhood W_0 of p_0 in M and an $\epsilon_0 > 0$, such that

$$\Phi_0(p, t) = p + \sum_{j=1}^{k} t_j\, Y_j(p)$$

defines a diffeomorphism from $W_0 \times \epsilon_0 \mathring{D}^k$ to an open set $V_0 \subseteq \mathbb{R}^{n+k}$. The map $r_0 = \mathrm{pr}_{W_0} \circ \Phi_0^{-1}$ defines a smooth map $r_0 \colon V_0 \to W_0$, which extends id_{W_0}, so that the fiber $r_0^{-1}(p)$ is the open ball in $p + T_p M^\perp$ with center at p and radius ϵ_0 for every $p \in W_0$. By shrinking ϵ_0 and cutting W_0 down we can arrange that the following condition holds:

(20)   For $x \in V_0$ and $y \in M$ we have $\|x - r_0(x)\| \le \|x - y\|$

with equality if and only if $y = r_0(x)$. This can be done as follows.

By Definition 8.8 there exists an open neighborhood $\tilde{W}$ of p_0 in $\mathbb{R}^{n+k}$ such that $M \cap \tilde{W}$ is closed in $\tilde{W}$. In the above we can ensure that $V_0 \subseteq \tilde{W}$ where $M \cap V_0$ remains closed in V_0. Choose compact subsets $K_1 \subseteq K_2$ of W_0 so that (in the induced topology on M) $p_0 \in \mathrm{int} K_1 \subseteq K_1 \subseteq \mathrm{int} K_2$, where $\mathrm{int} K_i$ denotes the interior of K_i. The set

$$B = (\mathbb{R}^{n+k} - V_0) \cup (M \cap V_0 - \mathrm{int} K_2)$$

is closed in $\mathbb{R}^{n+k}$ and disjoint from K_1. There exists an $\epsilon \in (0, \epsilon_0]$ such that $\|b - y\| \ge 2\epsilon$ for all $b \in B$, $y \in K_1$. If we introduce the open set

$$V_0' = \{x \in V_0 \mid r_0(x) \in \mathrm{int} K_1 \text{ and } \|x - r_0(x)\| < \epsilon\},$$

we get for $x \in V_0'$ and $b \in M - K_2 \subseteq B$ that

$$\|x - b\| \ge \|b - r_0(x)\| - \|x - r_0(x)\| > \epsilon.$$

Since $\|x - r_0(x)\| < \epsilon$, the function $y \to \|x - y\|$, defined on M, attains a minimum less than ϵ somewhere on the compact set K_2. Consider such a $y_0 \in K_2$ with

$$\|x - y_0\| = \min_{y \in M} \|x - y\| \le \|x - r_0(x)\| < \epsilon \le \epsilon_0.$$

Hence $x - y_0$ is a normal vector to M at y_0 (see Exercise 9.1), but then $x \in V_0$ and $y_0 = r_0(x)$. This shows that Condition (20) can be satisfied by replacing (W_0, V_0, ϵ_0) by $(\mathrm{int} K_1, V_0', \epsilon)$.

Now all of M can be covered with open sets of the type V_0 with the associated smooth maps r_0 which satisfy Condition (20). If (V_1, r_1) is a different pair then r_0 and r_1 will coincide on $V_0 \cap V_1$. On the union V' of all such open sets (of the type above) we can now define a smooth map $r: V' \to M$, which extends id_M, such that part (i) of the theorem is satisfied. We have $r_{|V_0} = r_0$ for every local (V_0, r_0). If in the above we always choose $\epsilon_0 \leq 1$, then the fiber $r^{-1}(p)$ over a $p \in M$ will be an open ball in $p + T_p M^\perp$ with radius $\tilde{\rho}(p)$ for which $0 < \tilde{\rho}(p) \leq 1$. Thus we have satisfied (i) and (ii), except that $\tilde{\rho}$ might be discontinuous.

The distance function from M,

$$d_M(x) = \inf_{y \in M} \|x - y\|,$$

is continuous on all of $\mathbb{R}^{n+k}$. For $x \in V'$, (i) shows that

$$d_M(x) = \|x - r(x)\| \quad (x \in V').$$

If $p \in M$ and $x \in p + T_p M^\perp$ has distance $\tilde{\rho}(p)$ from p, we can conclude that $d_M(x) = \tilde{\rho}(p)$. In this case (i) excludes that $x \in V'$, so x lies on the boundary of V'. Hence the distance function

$$d: M \to \mathbb{R}; \quad d(p) = \inf_{z \notin V'} \|p - z\|$$

satisfies $0 < d(p) \leq \tilde{\rho}(p)$ and is continuous.

By Lemma A.9 the function $\frac{1}{2} d \circ r: V' \to \mathbb{R}$ can be approximated by a smooth function $\psi: V' \to \mathbb{R}$ such that

$$\left\| \psi(x) - \tfrac{1}{2} d \circ r(x) \right\| \leq \tfrac{1}{4} d \circ r(x)$$

for all $x \in V'$. In particular

$$\tfrac{1}{4} d(x) \leq \psi(x) \leq \tfrac{3}{4} d(x) \quad \text{when } x \in M.$$

Hence the restriction $\rho = \psi_{|M}: M \to \mathbb{R}$ is a positive smooth function with $\rho(p) < \tilde{\rho}(p)$ for all $p \in M$. When M is compact, the same can be achieved for the constant function which takes the value $\rho = \frac{1}{2} \min_{p \in M} d(p)$. If we define

$$V = \left\{ x \in V' \mid \|x - r(x)\| < \rho(r(x)) \right\}$$

both (i) and (ii) hold for the restriction of r to V.

It remains to prove (iii). It is sufficient to show that $S_\epsilon \cap V_0$ is empty or a smooth submanifold of V_0 of codimension 1. The image under the diffeomorphism $\Phi_0^{-1}: V_0 \to W_0 \times \epsilon_0 \mathring{D}^k$ of $S_\epsilon \cap V_0$ is the set

$$S = \{ (p, t) \in W_0 \times \mathbb{R}^k \mid \|t\| = \epsilon(p) < \epsilon_0 \}.$$

The projection of S on W_0 is the open set

$$U = \{ p \in W_0 \mid \epsilon(p) < \epsilon_0 \} \subseteq M.$$

The diffeomorphism $\phi: U \times \mathbb{R}^k \to U \times \mathbb{R}^k$ given by $\phi(p, t) = (p, \epsilon(p)t)$ maps $U \times S^{k-1}$ to S. This yields (iii). $\qquad \qquad \square$

Remark 9.24 In Chapter 11 we will need additional information in the case where $M^n \subseteq \mathbb{R}^{n+k}$ is compact and $\rho > 0$ is constant. To any number ϵ, $0 < \epsilon < \rho$, we define the *closed tubular neighborhood* of radius ϵ around M by

$$N_\epsilon = \{x \in V \mid \|x - r(x)\| \le \epsilon\}.$$

This set is the disjoint union of the closed balls in $p + T_p M^\perp$ with centers at p and radius ϵ. Note that N_ϵ is compact and that S_ϵ is the set of boundary points of N_ϵ in $\mathbb{R}^{n+k}$. By Theorem 9.23.(i) we see for $p \in M$ that the real-valued function on S_ϵ, $x \to \|x - p\|$, attains its minimum value ϵ at all points $x \in S_\epsilon$ with $r(x) = p$. It follows that

$$(21) \qquad\qquad x - r(x) \in T_x S_\epsilon^\perp, \quad x \in S_\epsilon.$$

We end this chapter with a few applications of the existence of tubular neighborhoods. Let (V, i, r) be a tubular neighborhood of M with $i\colon M \to V$ the inclusion map and $r\colon V \to M$ the smooth retraction map such that $r \circ i = \mathrm{id}_M$. In cohomology this gives

$$H^d(i) \circ H^d(r) = \mathrm{id}_{H^d(M)},$$

so that $H^d(i)\colon H^d(V) \to H^d(M)$ is surjective and $H^d(r)\colon H^d(M) \to H^d(V)$ is injective.

Proposition 9.25 *For any compact differentiable manifold M^n all cohomology spaces $H^d(M)$ are finite-dimensional.*

Proof. We may assume that M^n is a smooth submanifold of $\mathbb{R}^{n+k}$ by Theorem 8.11, and that (V, i, r) is a tubular neighborhood. Since M is compact we can find finitely many open balls $U_1, \dots, U_r$ in $\mathbb{R}^{n+k}$ such that their union $U = U_1 \cup \dots \cup U_r$ satisfies $M \subseteq U \subseteq V$. Now we have a smooth inclusion $i\colon M \to U$ and a smooth map $r_{|U}\colon U \to M$ with $r_{|U} \circ i = \mathrm{id}_M$. The argument above shows that

$$H^d(i)\colon H^d(U) \to H^d(M)$$

is surjective, and the assertion now follows from Theorem 5.5. □

Proposition 9.26 *Let M_1 and M_2 be smooth submanifolds of Euclidean spaces.*

(i) *If $f_0, f_1\colon M_1 \to M_2$ are two homotopic smooth maps, then*

$$H^d(f_0) = H^d(f_1)\colon H^d(M_2) \to H^d(M_1).$$

(ii) *Every continuous map $M_1 \to M_2$ is homotopic to a smooth map.*

Proof. Choose tubular neighborhoods (V_ν, i_ν, r_ν) of M_ν, $\nu = 1, 2$. Lemma 6.3 implies that $i_2 \circ f_0 \circ r_1 \simeq i_2 \circ f_1 \circ r_1$. Hence $H^d(i_2 \circ f_0 \circ r_1) = H^d(i_2 \circ f_1 \circ r_1)$, so that

$$H^d(r_1) \circ H^d(f_0) \circ H^d(i_2) = H^d(r_1) \circ H^d(f_1) \circ H^d(i_2).$$

Since $H^d(r_1)$ is injective and $H^d(i_2)$ is surjective, we conclude that $H^d(f_0) = H^d(f_1)$. If $\phi \colon M_1 \to M_2$ is continuous, we can use Lemma 6.6.(i) to find a smooth map $g \colon V_1 \to V_2$ with $g \simeq i_2 \circ \phi \circ r_1$. For the smooth map $f = r_2 \circ g \circ i_1 \colon M_1 \to M_2$, Lemma 6.3 shows that $f \simeq r_2 \circ (i_2 \circ \phi \circ r_1) \circ i_1 = \phi$. $\qquad\square$

Remark 9.27 As in the discussion preceeding Theorem 6.8, the de Rham cohomology can now be made functorial on the category of smooth submanifolds of Euclidean space and continuous maps. Theorem 6.8 and Corollary 6.9 are valid (with the same proofs) with open sets in Euclidean space replaced by smooth submanifolds. By Theorem 8.11 the same can be done for differentiable manifolds in general.

Corollary 9.28 *If $M^n \subseteq \mathbb{R}^{n+k}$ is a smooth submanifold and (V, i, r) an open tubular neighborhood, then $H^d(i) \colon H^d(V) \to H^d(M)$ is an isomorphism with $H^d(r)$ as its inverse.*

Proof. We have $r \circ i = \mathrm{id}_M$ and $i \circ r \simeq \mathrm{id}_V$, as V contains the line segment between x and $r(x)$ for all $x \in V$. By Proposition 9.26.(i) we can conclude that $H^d(r)$ and $H^d(i)$ are inverses. $\qquad\square$

Example 9.29 For $n \geq 1$ we have

$$H^d(S^n) \cong \begin{cases} \mathbb{R} & \text{if } d = 0, n \\ 0 & \text{otherwise.} \end{cases}$$

Let $i \colon S^n \to \mathbb{R}^{n+1} - \{0\}$ be the inclusion and define $r \colon \mathbb{R}^{n+1} - \{0\} \to S^n$ by $r(x) = \frac{x}{\|x\|}$ Then $r \circ i = \mathrm{id}_{S^n}$, $i \circ r \simeq \mathrm{id}_{\mathbb{R}^{n+1} - \{0\}}$ and $H^d(i)$ is an isomorphism. The result follows from Theorem 6.13.

Remark 9.30 Let U_1 and U_2 be open subsets of a smooth submanifold $M^n \subseteq \mathbb{R}^l$. Using Theorem 9.11, the proof of Theorem 5.1 can be carried through without any significant changes. As in Chapter 5 this gives rise to the Mayer–Vietoris sequence

$$\to H^p(U_1 \cup U_2) \xrightarrow{I^*} H^p(U_1) \oplus H^p(U_2) \xrightarrow{J^*} H^p(U_1 \cap U_2) \xrightarrow{\partial^*} H^{p+1}(U_1 \cup U_2) \to$$

Example 9.31 We shall compute the de Rham cohomology of $\mathbb{RP}^{n-1}$ $(n \geq 2)$. With the notation of Example 9.19 we see that

$$\mathrm{Alt}^p(D_x A) \colon \mathrm{Alt}^p\left(T_{\pi(x)}\mathbb{RP}^{n-1}\right) \to \mathrm{Alt}^p\left(T_x S^{n-1}\right)$$

is an isomorphism for every $x \in S^{n-1}$. Therefore

$$\Omega^p(\pi) \colon \Omega^p\left(\mathbb{RP}^{n-1}\right) \to \Omega^p\left(S^{n-1}\right)$$

is a monomorphism, and we find that the image of $\Omega^p(\pi)$ is equal to the set of p-forms ω on S^{n-1} such that $A^*\omega = \omega$. Since $A^* = \Omega^p(A) \colon \Omega^p\left(S^{n-1}\right) \to \Omega^p\left(S^{n-1}\right)$ has order 2 we can decompose it into (± 1)-eigenspaces

$$\Omega^p\left(S^{n-1}\right) = \Omega^p_+\left(S^{n-1}\right) \oplus \Omega^p_-\left(S^{n-1}\right)$$

where

$$\Omega^p_\pm\left(S^{n-1}\right) = \mathrm{Im}\left(\tfrac{1}{2}(\mathrm{id} \pm \Omega^p(A))\right).$$

This in fact decomposes the de Rham complex of S^{n-1} into a direct sum of two subcomplexes

$$\Omega^*\left(S^{n-1}\right) = \Omega^*_+\left(S^{n-1}\right) \oplus \Omega^*_-\left(S^{n-1}\right). \tag{22}$$

There is an isomorphism of chain complexes

$$\Omega^*\left(\mathbb{RP}^{n-1}\right) \cong \Omega^*_+\left(S^{n-1}\right) \tag{23}$$

induced by $\pi \colon S^{n-1} \to \mathbb{RP}^{n-1}$. From (22) we get isomorphisms

$$H^p\left(S^{n-1}\right) \cong H^p\left(\Omega^*_+\left(S^{n-1}\right)\right) \oplus H^p\left(\Omega^*_-\left(S^{n-1}\right)\right)$$
$$\cong H^p_+\left(S^{n-1}\right) \oplus H^p_-\left(S^{n-1}\right)$$

where $H^p_\pm\left(S^{n-1}\right)$ is the (± 1)-eigenspace of A^* on $H^p\left(S^{n-1}\right)$. Combining with (23) we find that

$$H^p\left(\mathbb{RP}^{n-1}\right) \cong H^p_+\left(S^{n-1}\right). \tag{24}$$

There is a commutative diagram with vertical isomorphisms (See Example 9.29)

$$
\begin{array}{ccc}
H^{n-1}(\mathbb{R}^n - \{0\}) & \longrightarrow & H^{n-1}(\mathbb{R}^n - \{0\}) \\
\cong \downarrow i^* & & \cong \downarrow i^* \\
H^{n-1}(S^{n-1}) & \xrightarrow{\;A^*\;} & H^{n-1}(S^{n-1})
\end{array}
$$

where the top map is induced by the linear map $x \to -x$ of $\mathbb{R}^n$ into itself. Lemma 6.14 shows that the bottom map is multiplication by $(-1)^n$. Using (24) and Example 9.29 we finally get

$$H^p\left(\mathbb{RP}^{n-1}\right) \cong \begin{cases} \mathbb{R} & \text{if } p = 0 \text{ or } p = n-1 \text{ with } n \text{ even} \\ 0 & \text{otherwise.} \end{cases} \tag{25}$$

10. INTEGRATION ON MANIFOLDS

Let M^n be an oriented n-dimensional smooth manifold. We define an integral

$$\int_M : \Omega_c^n(M^n) \to \mathbb{R}$$

on the vector space of differential n-forms with compact support. Next we shall consider integration on subsets of M^n, and Stokes' theorem will be proved. Finally we calculate $H^n(M^n)$ for an arbitrary orientable compact connected smooth manifold M^n.

In the special case where $M^n = \mathbb{R}^n$ (with the standard orientation) we can write $\omega \in \Omega_c^n(\mathbb{R}^n)$ uniquely in the form

$$\omega = f(x)dx_1 \wedge \ldots \wedge dx_n,$$

where $f \in C^\infty(\mathbb{R}^n, \mathbb{R})$ has compact support. We then define

$$\int_{\mathbb{R}^n} f(x)dx_1 \wedge \ldots \wedge dx_n = \int_{\mathbb{R}^n} f(x)d\mu_n,$$

where $d\mu_n$ is the usual Lebesgue measure on $\mathbb{R}^n$. The same definition can be used when $\omega \in \Omega_c^n(V)$ for an open set $V \subseteq \mathbb{R}^n$, since ω and f are smoothly extendable to the whole of $\mathbb{R}^n$ by setting them equal to 0 on $\mathbb{R}^n - \mathrm{supp}_V(\omega)$.

Lemma 10.1 *Let $\phi : V \to W$ be a diffeomorphism between open subsets V and W of $\mathbb{R}^n$, and assume that the Jacobi determinant $\det(D_x\phi)$ is of constant sign $\delta = \pm 1$ for $x \in V$. For $\omega \in \Omega_c^n(W)$ we have that*

$$\int_V \phi^*(\omega) = \delta \cdot \int_W \omega.$$

Proof. If ω is written in the form

$$\omega = f(x)dx_1 \wedge \ldots \wedge dx_n$$

with $f \in C_c^\infty(W, \mathbb{R})$, it follows from Example 3.13.(ii) that

$$\phi^*(\omega) = f(\phi(x))\det(D_x\phi) \, dx_1 \wedge \ldots \wedge dx_n$$
$$= \delta f(\phi(x))|\det(D_x\phi)| \, dx_1 \wedge \ldots \wedge dx_n.$$

The assertion follows from the transformation theorem for integrals which states that

$$\int_W f(x)d\mu_n = \int_V f(\phi(x)) \, |\det(D_x\phi)| \, d\mu_n. \qquad \square$$

Proposition 10.2 *For an arbitrary oriented n-dimensional smooth manifold M^n there exists a unique linear map*

$$\int_M : \Omega_c^n(M^n) \to \mathbb{R}$$

with the following property: If $\omega \in \Omega_c^n(M^n)$ has support contained in U, where (U, h) is a positively oriented C^∞-chart, then

(1)
$$\int_M \omega = \int_{h(U)} \left(h^{-1} \right)^* \omega.$$

Proof. First consider $\omega \in \Omega_c^n(M^n)$ with "small" support, i.e. such that $\operatorname{supp}_M(\omega)$ is contained in a coordinate patch. Then (U, h) can be chosen as above and the integral is determined by (1). We must show that the right-hand side is independent of the choice of chart. Assume that $(\tilde{U}, \tilde{h})$ is another positively oriented C^∞-chart with $\operatorname{supp}_M(\omega) \subseteq \tilde{U}$.

The diffeomorphism $\phi \colon V \to W$ from $V = h(U \cap \tilde{U})$ to $W = \tilde{h}(U \cap \tilde{U})$ given by $\phi = \tilde{h} \circ h^{-1}$ has everywhere positive Jacobi determinant. Since

$$\operatorname{supp}_{h(U)}((h^{-1})^*\omega) \subseteq V, \qquad \operatorname{supp}_{\tilde{h}(\tilde{U})}((\tilde{h}^{-1})^*\omega) \subseteq W$$

and $\phi^*(\tilde{h}^{-1})^*(\omega) = (h^{-1})^*(\omega)$, Lemma 10.1 shows that

$$\int_{h(U)} \left(h^{-1} \right)^* \omega = \int_{\tilde{h}(\tilde{U})} (\tilde{h}^{-1})^* \omega.$$

So for $\omega \in \Omega_c^n(M)$ with "small" support the integral defined by (1) is independent of the chart.

Now choose a smooth partition of unity $(\rho_\alpha)_{\alpha \in A}$ on M subordinate to an oriented C^∞-atlas on M. For $\omega \in \Omega_c^n(M)$ we have that

$$\omega = \sum_{\alpha \in A} \rho_\alpha \omega,$$

where every term $\rho_\alpha \omega \in \Omega_c^n(M)$ has "small" support, and where only finitely many terms are non-zero. We define

$$I(\omega) = \sum_{\alpha \in A} \int_M \rho_\alpha \omega,$$

where the term associated to $\alpha \in A$ is given by (1), applied to a U_α with $\operatorname{supp}_M(\rho_\alpha) \subseteq U_\alpha$. It is obvious that I is a linear operator on $\Omega_c^n(M)$. If, in particular, $\operatorname{supp}_M(\omega) \subseteq U$, where (U, h) is a positively oriented C^∞-chart, the terms of the sum can be calculated by (1), applied to (U, h). This yields

$$I(\omega) = \int_M \omega,$$

which shows that I is a linear operator with the desired properties. Uniqueness follows analogously. $\qquad\square$

Lemma 10.3

(i) $\int_M \omega$ changes sign when the orientation of M^n is reversed.

(ii) If $\omega \in \Omega^n_c(M^n)$ has support contained in an open set $W \subset M^n$, then

$$\int_M \omega = \int_W \omega,$$

when W is given the orientation induced by M.

(iii) If $\phi: N^n \to M^n$ is an orientation-preserving diffeomorphism, then we have that

$$\int_M \omega = \int_N \phi^*(\omega)$$

for $\omega \in \Omega^n_c(M)$.

Proof. By a partition of unity, we can restrict ourselves to the case where $\mathrm{supp}_M(\omega)$ is contained in a coordinate patch. All three properties are now easy consequences of Lemma 10.1 and (1). $\qquad\square$

Remark 10.4 In the above we could have considered integrals of continuous n-forms with compact support on M^n. If the orientation of M is given by the orientation form $\sigma \in \Omega^n(M)$, a continuous n-form can be written uniquely as $f\sigma$, where $f \in C^0(M, \mathbb{R})$. The support of $f\sigma$ is equal to the support of f. The integral of (1) extended to continuous n-forms gives rise to a linear operator

$$I_\sigma: C^0_c(M, \mathbb{R}) \to \mathbb{R}; \quad I_\sigma(f) = \int_M f\sigma.$$

This linear operator is positive, i.e. $I_\sigma(f) \geq 0$ for $f \geq 0$. By a partition of unity it is sufficient to show this when $\mathrm{supp}(f) \subseteq U$, where (U, h) is a positive oriented C^∞-chart. Then we have that

$$I_\sigma(f) = \int_{h(U)} f \circ h^{-1}(x)\phi(x)d\mu_n,$$

where ϕ is determined by $(h^{-1})^*(\sigma) = \phi(x)dx_1 \wedge \ldots \wedge dx_n$. Since ϕ is positive we get $I_\sigma(f) \geq 0$.

According to Riesz's representation theorem (see for instance chapter 2 of [Rudin]) I_σ determines a positive measure μ_σ on M which satisfies

$$\int_M f(x)d\mu_\sigma = \int_M f\sigma, \quad f \in C^0_c(M, \mathbb{R}).$$

The entire Lebesgue integration machinery now becomes available, but we shall use only very little of it in the following.

If M^n is an oriented Riemannian manifold, the volume form vol_M will determine a measure μ_M on M^n analogous to the Lebesgue measure on $\mathbb{R}^n$. For a compact set K the volume of K can be defined by

$$\mathrm{Vol}(K) = \int_K \mathrm{vol}_M \in \mathbb{R}.$$

Definition 10.5 Let M^n be a smooth manifold. A subset $N \subseteq M^n$ is called a *domain with smooth boundary* or a *codimension zero submanifold with boundary*, if for every $p \in M$ there exists a C^∞-chart (U, h) around p, such that

(2) $h(U \cap N) = h(U) \cap \mathbb{R}^n_- ,$

where $\mathbb{R}^n_- = \{(x_1, \ldots, x_n) \in \mathbb{R}^n \mid x_1 \leq 0\}$.

Note that (2) is automatically satisfied when p is an interior or an exterior point of N (one can choose (U, h) with $p \in U$, such that $h(U)$ is contained in an open half-space in $\mathbb{R}^n$ defined by either $x_1 < 0$ or $x_1 > 0$). If p is a boundary point of N then $h(p)$ has first coordinate equal to zero. Let (U, h) and (V, k) be smooth charts around a boundary point $p \in \partial N$. The resulting transition diffeomorphism

$$\phi = k \circ h^{-1} \colon h(U \cap V) \to k(U \cap V)$$

induces a map

$$h(U \cap V) \cap \mathbb{R}^n_- \to k(U \cap V) \cap \mathbb{R}^n_- ,$$

which restricts to a diffeomorphism

$$\Psi \colon h(U \cap V) \cap \partial \mathbb{R}^n_- \to k(U \cap V) \cap \partial \mathbb{R}^n_-.$$

The Jacobi matrix at the point $q = h(p) \in \partial \mathbb{R}^n_-$ for $\phi = (\phi_1, \ldots, \phi_n)$ has the form

$$D_q\phi = \begin{pmatrix} \frac{\partial \phi_1}{\partial x_1}(q) & 0 & \cdots & 0 \\ * & & & \\ \vdots & & D_q\Psi & \\ * & & & \end{pmatrix}.$$

We must have $(\partial \phi_1 / \partial x_1)(q) \neq 0$, as $D_q\phi$ is invertible. Since ϕ maps $\mathbb{R}^n_-$ into $\mathbb{R}^n_-$, we have that $(\partial \phi_1 / \partial x_1)(q) > 0$.

A tangent vector $w \in T_pM$ at a boundary point $p \in \partial N$ is said to be *outward directed*, if there exists a C^∞-chart (U, h) around p with $h(U \cap N) = h(U) \cap \mathbb{R}^n_-$ and such that $D_ph(w) \in \mathbb{R}^n$ has a positive first coordinate. This will then also be the case for any other smooth chart around p.

Lemma 10.6 *Let $N \subseteq M^n$ be a domain with smooth boundary. Then ∂N is an $(n-1)$-dimensional smooth submanifold of M^n.*

Suppose M^n $(n \geq 2)$ is oriented. There is an induced orientation of ∂N with the following property: if $p \in \partial N$ and $v_1 \in T_p M$ is an outward directed tangent vector then a basis $v_2, \ldots, v_n$ for $T_p \partial N$ is positively oriented if and only if the basis $v_1, v_2, \ldots, v_n$ for $T_p M$ is positively oriented.

Proof. Every smooth chart (U, h) in M that satisfies $h(U \cap N) = h(U) \cap \mathbb{R}^n_-$ can be restricted to a chart $(U \cap \partial N, h_|)$ on ∂N:

$$h_|: U \cap \partial N \rightarrow h(U) \cap \left(\{0\} \times \mathbb{R}^{n-1}\right).$$

These charts have mutual smooth overlap according to the above. This yields a smooth atlas on ∂N.

Suppose M^n is oriented. Then, possibly changing the sign of x_2, we can choose a positively oriented chart (U, h) of the considered type around any $p \in \partial N$. The resulting smooth charts $(U \cap \partial N, h_|)$ on ∂N have positively oriented transformation diffeomorphisms, and they determine an orientation of ∂N that satisfies the stated property. □

Remarks 10.7

(i) We want to integrate n-forms $\omega \in \Omega^n_c(M)$ over domains N with smooth boundary. In view of Remark 10.4 we can set

$$\int_N \omega = \int_M 1_N \omega$$

where 1_N is the function with value 1 on N and zero outside N. Alternatively, one can prove an extension of Lemma 10.1 which uses the following version of the transformation theorem: Let $\phi: V \rightarrow W$ be a diffeomorphism of open sets in $\mathbb{R}^n$ that maps $\mathbb{R}^n_- \cap V$ to $\mathbb{R}^n_- \cap W$, and let f be a smooth function on W with compact support. Then

$$\int_{\mathbb{R}^n_- \cap W} f(x) d\mu_n = \int_{\mathbb{R}^n_- \cap V} f\phi(x) |\det D_x \phi| d\mu_n.$$

(One could approximate both integrals by integrals over W and V upon multiplying f by a sequence of smooth functions ψ_i with values in $[0, 1]$ and converging to $1_{\mathbb{R}^n_-}$.)

(ii) In the case $n = 1$, Lemma 10.6 holds in the following modified form. An orientation of ∂N consists of a choice of sign, $+$ or $-$, for every point $p \in \partial N$. Let $v_1 \in T_p M$ be outward directed. Then p is assigned the sign $+$ if v_1 is a positively oriented basis of $T_p M$, otherwise the sign is $-$.

A 0-form on ∂N is a function $f : \partial N \to \mathbb{R}$. When f has compact support we define

$$\int_{\partial N} f = \sum_{p \in \partial N} \operatorname{sgn}(p) \, f(p).$$

These conventions are used in the case $n = 1$ of Stokes' theorem below.

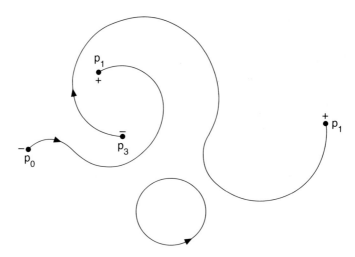

Theorem 10.8 (Stokes' theorem) *Let $N \subseteq M^n$ be a domain with smooth boundary in an oriented smooth manifold. Let ∂N have the induced orientation. For every $\omega \in \Omega^{n-1}(M)$ with $N \cap \operatorname{supp}_M(\omega)$ compact we have*

$$\int_{\partial N} i^*(\omega) = \int_N d\omega,$$

where $i : \partial N \to M$ is the inclusion map.

Proof. We assume that $n \geq 2$ and leave it to the reader to make the necessary changes for the case $n = 1$.

It is clear that $i^*(\omega)$ has compact support. We can choose $f \in \Omega_c^0(M)$ with value constantly equal to 1 on $N \cap \operatorname{supp}_M(\omega)$. Since $f\omega$ coincides with ω on N, both integrals are unchanged when ω is replaced by $f\omega$, so we may assume that ω has compact support.

Choose a smooth atlas on M consisting of charts of the type of Definition 10.5 and a subordinate smooth partition of unity $(\rho_\alpha)_{\alpha \in A}$. The formulas

$$\int_{\partial N} \omega = \sum_\alpha \int_{\partial N} \rho_\alpha \omega, \qquad \int_N d\omega = \sum_\alpha \int_N d(\rho_\alpha \omega)$$

reduce the problem to the case where $\omega \in \Omega_c^{n-1}(M)$, $\operatorname{supp}_M(\omega) \subseteq U$ and (U, h) is a smooth chart with $h(U \cap N) = h(U) \cap \mathbb{R}_-^n$. Furthermore the chart (U, h) is assumed to be positively oriented.

Let $\kappa \in \Omega_c^{n-1}(\mathbb{R}^n)$ be the $(n-1)$-form that is $(h^{-1})^*(\omega)$ on $h(U)$ and 0 on the rest of $\mathbb{R}^n$. By diffeomorphism invariance we then have that

$$\int_{\partial N} \omega = \int_{h(U) \cap \partial \mathbb{R}^n_-} (h^{-1})^*(\omega) = \int_{\partial \mathbb{R}^n_-} \kappa$$

and

$$\int_N d\omega = \int_{h(U) \cap \mathbb{R}^n_-} (h^{-1})^*(d\omega) = \int_{\mathbb{R}^n_-} d\kappa.$$

Hence the proof reduces to the special case where $M = \mathbb{R}^n$, $N = \mathbb{R}^n_-$ and $\omega \in \Omega_c^{n-1}(\mathbb{R}^n)$. This case is treated by direct calculation. We define

$$\omega = \sum_{i=1}^n f_i(x) \, dx_1 \wedge \ldots \wedge \widehat{dx_i} \wedge \ldots \wedge dx_n$$

and choose $b > 0$ such that $\mathrm{supp}_{\mathbb{R}^n} f_i \subseteq [-b, b]^n$, $1 \leq i \leq n$. Using Theorem 3.12,

$$\omega_{|\partial \mathbb{R}^n_-} = f_1(0, x_2, \ldots, x_n) \, dx_2 \wedge \ldots \wedge dx_n.$$

Hence

$$(3) \qquad \int_{\partial \mathbb{R}^n_-} \omega = \int f_1(0, x_2, \ldots, x_n) d\mu_{n-1}.$$

By Theorem 3.7 we have

$$d\omega = \sum_{i=1}^n (-1)^{i-1} \frac{\partial f_i}{\partial x_i} dx_1 \wedge dx_2 \wedge \ldots \wedge dx_n.$$

Hence

$$(4) \qquad \int_{\mathbb{R}^n_-} d\omega = \sum_{i=1}^n (-1)^{i-1} \int_{\mathbb{R}^n_-} \frac{\partial f_i}{\partial x_i} d\mu_n.$$

For $2 \leq i \leq n$ we get

$$\int_{-\infty}^{\infty} \frac{\partial f_i}{\partial x_i} (x_1, \ldots, x_{i-1}, t, x_{i+1}, \ldots, x_n) dt$$
$$= f_i(x_1, \ldots, x_{i-1}, b, x_{i+1}, \ldots, x_n) - f_i(x_1, \ldots, x_{i-1}, -b, x_{i+1}, \ldots, x_n)$$
$$= 0,$$

and then by Fubini's theorem

$$(5) \qquad \int_{\mathbb{R}^n_-} \frac{\partial f_i}{\partial x_i} d\mu_n = 0 \quad (2 \leq i \leq n).$$

When $i = 1$, one gets

$$\int_{-\infty}^{0} \frac{\partial f_1}{\partial x_1}(t, x_2, \ldots, x_n)dt = f_1(0, x_2, \ldots, x_n) - f_1(-b, x_2, \ldots, x_n)$$
$$= f_1(0, x_2, \ldots, x_n),$$

and by Fubini's theorem

(6) $$\int_{\mathbb{R}^n_-} \frac{\partial f_1}{\partial x_1} \, d\mu_n = \int f_1(0, x_2, \ldots, x_n) \, d\mu_{n-1}.$$

By combining Equations (3–6) the desired formula follows. $\qquad\square$

Taking $N = M$ in Theorem 10.8 we have

Corollary 10.9 *If M^n is an oriented smooth manifold and $\omega \in \Omega_c^{n-1}(M)$ then $\int_M d\omega = 0$.*

Remark 10.10 Let ω be a closed d-form on M^n. One way of showing that the cohomology class $[\omega] \in H^d(M)$ is non-zero is to show that

(7) $$\int_Q f^*(\omega) \neq 0$$

for a suitably chosen smooth map $f: Q^d \to M$ from a d-dimensional compact oriented smooth manifold Q^d. If $\omega = d\tau$ for some $\tau \in \Omega^{d-1}(M)$, then Corollary 10.9 yields

$$\int_Q f^*(\omega) = \int_Q d(f^*(\tau)) = 0.$$

This, in essence, was the strategy from Examples 1.2 and 1.7. It can be shown (albeit in a very indirect way via cobordism theory) that $[\omega] = 0$ if and only if all integrals of the form of (7) vanish.

Example 10.11 In Example 9.18 we considered the closed $(n - 1)$-form on $\mathbb{R}^n - \{0\}$,

$$\omega = \frac{1}{\|x\|^n} \sum_{i=1}^{n} (-1)^{i-1} x_i dx_1 \wedge \ldots \wedge \widehat{dx_i} \wedge \ldots \wedge dx_n.$$

Since the pre-image of ω under the inclusion of S^{n-1} is the volume form $\mathrm{vol}_{S^{n-1}}$, which has positive integral over S^{n-1}, we can conclude from Remark 10.10 that $[\omega] \neq 0$ in $H^{n-1}(\mathbb{R}^n - \{0\})$. If $n \geq 2$ then, by Theorem 6.13, $[\omega]$ is a basis of $H^{n-1}(\mathbb{R}^n - \{0\})$. We thus have an isomorphism

$$H^{n-1}(\mathbb{R}^n - \{0\}) \xrightarrow{\cong} \mathbb{R} \quad (n \geq 2)$$

defined by integration over S^{n-1}. The image of $[\omega]$ under this isomorphism is the volume

$$\mathrm{Vol}(S^{n-1}) = \int_{S^{n-1}} \mathrm{vol}_{S^{n-1}}.$$

Example 10.12 The volume of S^{n-1} can be calculated by applying Stokes' theorem to D^n with the standard orientation of $\mathbb{R}^n$ and the $(n-1)$-form on $\mathbb{R}^n$ given by

$$\omega_0 = \sum_{i=1}^{n} (-1)^{i-1} x_i \, dx_1 \wedge \ldots \wedge \widehat{dx_i} \wedge \ldots \wedge dx_n.$$

Since $\omega_{0|S^{n-1}} = \mathrm{vol}_{S^{n-1}}$ and $d\omega_0 = n \, dx_1 \wedge \ldots \wedge dx_n$ we have that

$$\mathrm{Vol}\left(S^{n-1}\right) = \int_{S^{n-1}} \omega_0 = \int_{D^n} d\omega_0 = n\mathrm{Vol}(D^n).$$

By induction on m and Fubini's theorem, it can be shown that

$$\mathrm{Vol}\left(D^{2m}\right) = \frac{\pi^m}{m!}, \quad \mathrm{Vol}\left(D^{2m+1}\right) = \frac{2^{2m+1} m! \pi^m}{(2m+1)!}.$$

This yields

$$\mathrm{Vol}\left(S^{2m-1}\right) = \frac{2\pi^m}{(m-1)!}, \quad \mathrm{Vol}\left(S^{2m}\right) = \frac{2^{2m+1} m! \pi^m}{(2m)!}.$$

We conclude this chapter with a proof of the following:

Theorem 10.13 *If M^n is a connected oriented smooth manifold, then the sequence*

(8)
$$\Omega_c^{n-1}(M) \xrightarrow{d} \Omega_c^n(M) \xrightarrow{\int_M} \mathbb{R} \to 0$$

is exact.

Corollary 10.14 *For a connected compact smooth manifold M^n, integration over M induces an isomorphism*

$$\int_M : H^n(M^n) \xrightarrow{\cong} \mathbb{R}. \qquad \Box$$

In (8) it is obvious that the integral is non-zero and hence surjective. It follows from Corollary 10.9 that the image of d is contained in the kernel of the integral. We show the converse inclusion.

Lemma 10.15 *Theorem* 10.13 *holds for* $M = \mathbb{R}^n$, $n \geq 1$.

Proof. Let $\omega \in \Omega_c^n(\mathbb{R}^n)$ be a differential n-form with $\int_{\mathbb{R}^n} \omega = 0$. We must find $\kappa \in \Omega_c^{n-1}(\mathbb{R}^n)$, such that $d\kappa = \omega$. We can write $\omega = f(\mathbf{x})dx_1 \wedge \ldots \wedge dx_n$, and let

$$\kappa = \sum_{j=1}^n (-1)^{j-1} f_j(\mathbf{x})dx_1 \wedge \ldots \wedge \widehat{dx_j} \wedge \ldots \wedge dx_n.$$

A simple calculation gives

$$d\kappa = \left(\sum_{j=1}^n \frac{\partial f_j}{\partial x_j} \right) dx_1 \wedge \ldots \wedge dx_n.$$

Hence we need to prove the following assertion:

(P_n): Let $f \in C_c^\infty(\mathbb{R}^n)$ be a function with $\int f(x)d\mu_n = 0$. There exist functions $f_1, \ldots, f_n$ in $C_c^\infty(\mathbb{R}^n)$ such that

$$\sum_{j=1}^n \frac{\partial f_j}{\partial x_j} = f.$$

We prove (P_n) by induction. For $n = 1$ we are given a smooth function $f \in C_c^\infty(\mathbb{R})$ with $\int_{-\infty}^\infty f(t)dt = 0$. The problem is solved by setting

$$f_1(x) = \int_{-\infty}^x f(t)dt.$$

Assume (P_{n-1}) for $n \geq 2$, and let $f \in C_c^\infty(\mathbb{R}^n, \mathbb{R})$ be a function with $\int f(x)d\mu_n = 0$. We choose $C > 0$ with $\operatorname{supp}(f) \subseteq [-C, C]^n$ and define

$$(9) \qquad g(x_1, \ldots, x_{n-1}) = \int_{-\infty}^\infty f(x_1, \ldots, x_{n-1}, x_n)dx_n.$$

(The limits can be be replaced by $-C$ and C, respectively). The function g is smooth, since we can differentiate under the integral sign. Furthermore $\operatorname{supp}(g) \subseteq [-C, C]^{n-1}$. Fubini's theorem yields $\int g\, d\mu_{n-1} = \int f d\mu_n = 0$. Using (P_{n-1}) we get functions $g_1, \ldots, g_{n-1}$ in $C_c^\infty(\mathbb{R}^{n-1}, \mathbb{R})$ with

$$(10) \qquad \sum_{j=1}^{n-1} \frac{\partial g_j}{\partial x_j} = g.$$

We choose a function $\rho \in C_c^\infty(\mathbb{R}, \mathbb{R})$ with $\int_{-\infty}^\infty \rho(t)dt = 1$, and define $f_j \in C_c^\infty(\mathbb{R}^n, \mathbb{R})$,

$$(11) \qquad f_j(x_1, \ldots, x_{n-1}, x_n) = g_j(x_1, \ldots, x_{n-1})\rho(x_n), \ 1 \leq j \leq n - 1.$$

Let $h \in C_c^\infty(\mathbb{R}^n)$ be the function

(12)
$$h = f - \sum_{j=1}^{n-1} \frac{\partial f_j}{\partial x_j}.$$

A function $f_n \in C_c^\infty(\mathbb{R}^n)$ with $\partial f_n / \partial x_n = h$ is given by

(13)
$$f_n(x_1, \ldots, x_{n-1}, x_n) = \int_{-\infty}^{x_n} h(x_1, \ldots, x_{n-1}, t)dt.$$

It is obvious that f_n is smooth, but we must show that it has compact support. To this end it is sufficient to show that the integral of (13) vanishes when the upper limit x_n is replaced by ∞. Now (10), (11) and (12) yield that

$$h(x_1, \ldots, x_{n-1}, t) = f(x_1, \ldots, x_{n-1}, t) - \sum_{j=1}^{n-1} \frac{\partial g_j}{\partial x_j}(x_1, \ldots, x_{n-1})\rho(t)$$

$$= f(x_1, \ldots, x_{n-1}, t) - g(x_1, \ldots, x_{n-1})\rho(t).$$

Finally from (9) it follows that

$$\int_{-\infty}^{\infty} h(x_1, \ldots, x_{n-1}, t)dt$$
$$= \int_{-\infty}^{\infty} f(x_1, \ldots, x_{n-1}, t)dt - g(x_1, \ldots, x_{n-1}) \int_{-\infty}^{\infty} \rho(t)dt$$
$$= 0. \qquad \square$$

Lemma 10.16 *Let $(U_\alpha)_{\alpha \in A}$ be an open cover of the connected manifold M, and let $p, q \in M$. There exist indices $\alpha_1, \ldots, \alpha_k$ such that*

(i) $p \in U_{\alpha_1}$ *and* $q \in U_{\alpha_k}$
(ii) $U_{\alpha_i} \cap U_{\alpha_{i+1}} \neq \emptyset$ *when* $1 \leq i \leq k - 1$.

Proof. For a fixed p we define V to be the set of $q \in M$, for which there exists a finite sequence of indices $\alpha_1, \ldots, \alpha_k$ from A, such that (i) and (ii) are satisfied. It is obvious that V is both open and closed in M and that V contains p. Since M is connected, we must have $V = M$. $\qquad \square$

Lemma 10.17 *Let $U \subseteq M$ be an open set diffeomorphic to $\mathbb{R}^n$ and let $W \subseteq U$ be non-empty and open. For every $\omega \in \Omega_c^n(M)$ with $\operatorname{supp} \omega \subseteq U$, there exists a $\kappa \in \Omega_c^{n-1}(M)$ such that $\operatorname{supp} \kappa \subseteq U$ and $\operatorname{supp}(\omega - d\kappa) \subseteq W$.*

Proof. It suffices to prove the lemma when $M = U$, and by diffeomorphism invariance it is enough to consider the case where $M = U = \mathbb{R}^n$.

Choose $\omega_1 \in \Omega_c^n(\mathbf{R}^n)$ with $\operatorname{supp} \omega_1 \subseteq W$ and $\int_{\mathbf{R}^n} \omega_1 = 1$. Then

$$\int_{\mathbf{R}^n} (\omega - a\omega_1) = 0, \quad \text{where } a = \int_{\mathbf{R}^n} \omega.$$

By Lemma 10.15 we can find a $\kappa \in \Omega_c^{n-1}(\mathbf{R}^n)$ with

$$\omega - a w_1 = d\kappa.$$

Hence $\omega - d\kappa = a w_1$ has its support contained in W. $\square$

Lemma 10.18 *Assume that M^n is connected and let $W \subseteq M$ be non-empty and open. For every $\omega \in \Omega_c^n(M)$ there exists a $\kappa \in \Omega_c^{n-1}(M)$ with $\operatorname{supp}(\omega - d\kappa) \subseteq W$.*

Proof. Suppose that $\operatorname{supp} \omega \subseteq U_1$ for some open set $U_1 \subseteq M$ diffeomorphic to $\mathbf{R}^n$. We apply Lemma 10.16 to find open sets $U_2, \ldots, U_k$, diffeomorphic to $\mathbf{R}^n$, such that $U_{i-1} \cap U_i \neq \emptyset$ for $2 \leq i \leq k$ and $U_k \subseteq W$. We use Lemma 10.17 to successively choose $\kappa_1, \kappa_2, \ldots, \kappa_{k-1}$ in $\Omega_c^{n-1}(M)$ such that

$$\operatorname{supp}\left(\omega - \sum_{i=1}^{j} d\kappa_i\right) \subseteq U_j \cap U_{j+1} \quad (1 \leq j \leq k-1).$$

The lemma holds for $\kappa = \sum_{i=1}^{k-1} \kappa_i$.

In the general case we use a partition of unity to write

$$\omega = \sum_{j=1}^{m} \omega_j,$$

where $\omega_j \in \Omega_c^n(M)$ has support contained in an open set diffeomorphic to $\mathbf{R}^n$. The above gives $\tilde{\kappa}_j \in \Omega_c^{n-1}(M)$, $1 \leq j \leq m$, such that $\operatorname{supp}(\omega_j - d\tilde{\kappa}_j) \subseteq W$. For

$$\tilde{\kappa} = \sum_{j=1}^{m} \tilde{\kappa}_j \in \Omega_c^{n-1}(M)$$

we have that

$$\omega - d\tilde{\kappa} = \sum_{j=1}^{m} (\omega_j - d\tilde{\kappa}_j).$$

Hence $\operatorname{supp}(\omega - d\tilde{\kappa}) \subseteq \bigcup_{j=1}^{m} \operatorname{supp}(\omega_j - d\tilde{\kappa}_j) \subseteq W$. $\square$

Proof of Theorem 10.13. Suppose given $\omega \in \Omega_c^n(M)$ with $\int_M \omega = 0$. Choose an open set $W \subseteq M$ diffeomorphic to $\mathbf{R}^n$. By Lemma 10.18 we can find a $\kappa \in \Omega_c^{n-1}(M)$ with $\operatorname{supp}(\omega - d\kappa) \subseteq W$. But then by Corollary 10.9,

$$\int_W (\omega - d\kappa) = \int_M (\omega - d\kappa) = -\int_M d\kappa = 0.$$

Lemma 10.15 implies that Theorem 10.13 holds for W, i.e. there exists a $\tau_0 \in \Omega_c^{n-1}(W)$ that satisfies

$$(\omega - d\kappa)_{|W} = d\tau_0.$$

Let $\tau \in \Omega_c^{n-1}(M)$ be the extension of τ_0 which vanishes outside of $\operatorname{supp}_W(\tau_0)$. Then $\omega - d\kappa = d\tau$, so $\tau + \kappa$ maps to ω under d.   □

11. DEGREE, LINKING NUMBERS AND INDEX OF VECTOR FIELDS

Let $f\colon N^n \to M^n$ be a smooth map between compact connected oriented manifolds of the same dimension n. We have the commutative diagram

(1)
$$
\begin{array}{ccc}
H^n(M) & \xrightarrow{H^n(f)} & H^n(N) \\
\cong \downarrow & & \cong \downarrow \\
\mathbb{R} & \xrightarrow{\deg(f)} & \mathbb{R}
\end{array}
$$

where the vertical isomorphisms are given by integration over M and N respectively; cf. Corollary 10.14. The lower horizontal arrow is multiplication by the real number $\deg(f)$ that makes the diagram commutative. Thus for $\omega \in \Omega^n(M)$,

(2)
$$
\int_N f^*(\omega) = \deg(f) \int_M \omega.
$$

This formulation can be generalized to the case where N is not connected:

Proposition 11.1 *Let $f\colon N^n \to M^n$ be a smooth map between compact n-dimensional oriented manifolds with M connected. There exists a unique $\deg(f) \in \mathbb{R}$ such that (2) holds for all $\omega \in \Omega^n(M)$. We call $\deg(f)$ the degree of f.*

Proof. We write N as a disjoint union of its connected components $N_1, \dots, N_k$ and denote the restriction of f to N_j by f_j. We have already defined $\deg(f_j)$; we set

(3)
$$
\deg(f) = \sum_{j=1}^{k} \deg(f_j).
$$

Thus for $\omega \in \Omega^n(M)$,

$$
\int_N f^*(\omega) = \sum_{j=1}^{k} \int_{N_j} f_j^*(\omega) = \sum_{j=1}^{k} \deg(f_j) \int_M \omega = \deg(f) \int_M \omega. \qquad \square
$$

Corollary 11.2 $\deg(f)$ *depends only on the homotopy class of $f\colon N \to M$.*

Proof. By (3) we can restrict ourselves to the case where N is connected. The assertion then follows from diagram (1), since $H^n(f)$ depends only on the homotopy class of f. $\qquad \square$

Corollary 11.3 *Suppose* $N^n \xrightarrow{f} M^n \xrightarrow{g} P^n$ *are smooth maps between n-dimensional compact oriented manifolds and that M and P are connected. Then*

$$\deg(gf) = \deg(f)\deg(g).$$

Proof. For $\omega \in \Omega^n(P)$,

$$\deg(gf)\int_P \omega = \int_N (gf)^*(\omega) = \int_N f^*(g^*(\omega))$$
$$= \deg(f)\int_M g^*(\omega) = \deg(f)\deg(g)\int_P \omega. \qquad \square$$

Remark 11.4 If $f: M^n \to M^n$ is a smooth map of a connected compact orientable manifold to itself then $\deg(f)$ can be defined by chosing an orientation of M and using it at both the domain and range. Change of orientation leaves $\deg(f)$ unaffected.

We will show that $\deg(f)$ takes only integer values. This follows from an important geometric interpretation of $\deg(f)$ which uses the concept of regular value. In general $p \in M$ is said to be a *regular value* for the smooth map $f: N^n \to M^m$ if

$$D_q f: T_q N \to T_p M$$

is surjective for all $q \in f^{-1}(p)$. In particular, points in the complement of $f(N^n)$ are regular values. Regular values are in rich supply:

Theorem 11.5 (Brown–Sard) *For every smooth map $f: N^n \to M^m$ the set of regular values is dense in M^m.*

When proving Theorem 11.5 one may replace M^m by an open subset $W \subseteq M^m$ diffeomorphic to $\mathbb{R}^n$, and replace N^n by $f^{-1}(W)$. This reduces Theorem 11.5 to the special case where $M^m = \mathbb{R}^m$.

In this case one shows, that *almost all* points in $\mathbb{R}^m$ (in the Lebesgue sense) are regular values. By covering N^n with countably many coordinate patches and using the fact that the union of countably many Lebesgue null-sets is again a null-set, Theorem 11.5 therefore reduces to the following result:

Theorem 11.6 (Sard, 1942) *Let $f: U \to \mathbb{R}^m$ be a smooth map defined on an open set $U \subseteq \mathbb{R}^n$ and let*

$$S = \{x \in U \mid \operatorname{rank} D_x f < m\}.$$

Then $f(S)$ is a Lebesgue null-set in $\mathbb{R}^m$.

Note that $x \in U$ belongs to S if and only if every $m \times m$ submatrix of the Jacobi matrix of f, evaluated at x, has determinant zero. Therefore S is closed in U and we can write S as a union of at most countably many compact subsets $K \subseteq S$. Theorem 11.6 thus follows if $f(K)$ is a Lebesgue null-set for every compact subset K of S. We shall only use and prove these theorems in the case $m = n$, where they follow from

Proposition 11.7 *Let $f: U \to \mathbb{R}^n$ be a C^1-map defined on an open set $U \subseteq \mathbb{R}^n$, and let $K \subseteq U$ be a compact set such that $\det(D_x f) = 0$ for all $x \in K$. Then $f(K)$ is a Lebesgue null-set in $\mathbb{R}^n$.*

Proof. Choose a compact set $L \subseteq U$ which contains K in the interior, $K \subseteq \mathring{L}$. Let $C > 0$ be a constant such that

$$(4) \qquad \sup_{\xi \in L} \|\operatorname{grad}_\xi f_j\| \le C \quad (1 \le j \le n).$$

Here f_j is the j-th coordinate function of f, and $\| \ \|$ denotes the Euclidean norm. Let

$$T = \prod_{i=1}^{n} [t_i, t_i + a]$$

be a cube such that $K \subseteq T$, and let $\epsilon > 0$. Since the functions $\partial f_j / \partial x_i$ are uniformly continuous on L, there exists a $\delta > 0$ such that

$$(5) \quad \|x - y\| \le \delta \Rightarrow \left| \frac{\partial f_j}{\partial x_i}(x) - \frac{\partial f_j}{\partial x_i}(y) \right| \le \epsilon, \quad (1 \le i, j \le n \text{ and } x, y \in L).$$

We subdivide T into a union of N^n closed small cubes T_l with side length $\frac{a}{N}$, and choose N so that

$$(6) \qquad \operatorname{diam}(T_l) = \frac{a\sqrt{n}}{N} \le \delta, \quad T_l \cap K \ne \emptyset \Rightarrow T_l \subseteq L.$$

For a small cube T_l with $T_l \cap K \ne \emptyset$ we pick $x \in T_l \cap K$. If $y \in T_l$ the mean value theorem yields points ξ_j on the line segment between x and y for which

$$(7) \qquad f_j(y) - f_j(x) = \sum_{i=1}^{n} \frac{\partial f_j}{\partial x_i}(\xi_j)(y_i - x_i).$$

Since $\xi_j \in T_l \subseteq L$, the Cauchy–Schwarz inequality and (4) give

$$|f_j(y) - f_j(x)| \le C \|y - x\|$$

and by (6),

$$(8) \qquad \|f(y) - f(x)\| \le C\sqrt{n}\, \mathrm{diam}(T_l) = \frac{anC}{N}.$$

Formula (7) can be rewritten as

$$(9) \qquad f(y) = f(x) + D_x f(y - x) + z,$$

where $z = (z_1, \ldots, z_n)$ is given by

$$z_j = \sum_{i=1}^{n} \left(\frac{\partial f_j}{\partial x_i}(\xi_j) - \frac{\partial f_j}{\partial x_i}(x) \right)(y_i - x_i).$$

By (6), $\|\xi_j - x\| \le \delta$, so that (5) gives $|z_j| \le \epsilon n \frac{a}{N}$. Hence

$$(10) \qquad \|z\| \le \epsilon \frac{a\, n\sqrt{n}}{N}.$$

Since the image of $D_x f$ is a proper subspace of $\mathbb{R}^n$, we may choose an affine hyperplane $H \subseteq \mathbb{R}^n$ with

$$f(x) + \mathrm{Im}(D_x f) \subseteq H.$$

By (9) and (10) the distance from $f(y)$ to H is less than $\epsilon \frac{an\sqrt{n}}{N}$. Then (8) implies that $f(T_l)$ is contained in the set D_l consisting of all points $q \in \mathbb{R}^n$ whose orthogonal projection $\mathrm{pr}(q)$ on H lies in the closed ball in H with radius $\frac{anC}{N}$ and centre $f(x)$ and $\|q - \mathrm{pr}(q)\| \le \epsilon \frac{an\sqrt{n}}{N}$. For the Lebesgue measure μ_n on $\mathbb{R}^n$ we have

$$\mu_n(D_l) = 2\epsilon \frac{an\sqrt{n}}{N} \left(\frac{anC}{N} \right)^{n-1} \mathrm{Vol}(D^{n-1}) = \epsilon \frac{c}{N^n},$$

where $c = 2\, a^n n^{n+\frac{1}{2}} C^{n-1} \mathrm{Vol}(D^{n-1})$. For every small cube T_l with $T_l \cap K \ne \emptyset$ we now have $\mu_n(f(T_l)) \le \epsilon \frac{c}{N^n}$. Since there are at most N^n such small cubes T_l, $\mu_n(f(K)) \le \epsilon c$. This holds for every $\epsilon > 0$ and proves the assertion. $\qquad \square$

Lemma 11.8 *Let $p \in M^n$ be a regular value for the smooth map $f: N^n \to M^n$, with N^n compact. Then $f^{-1}(p)$ consists of finitely many points $q_1, \ldots, q_k$. Moreover, there exist disjoint open neighborhoods V_i of q_i in N^n, and an open neighborhood U of p in M^n, such that*

(i) *$f^{-1}(U) = \bigcup_{i=1}^{k} V_i$*
(ii) *f_i maps V_i diffeomorphically onto U for $1 \le i \le k$.*

Proof. For each $q \in f^{-1}(p)$, $D_q f : T_q N \to T_q M$ is an isomorphism. From the inverse function theorem we know that f is a local diffeomorphism around q. In particular q is an isolated point in $f^{-1}(p)$. Compactness of N implies that $f^{-1}(p)$ consists of finitely many points $q_1, \ldots, q_k$. We can choose mutually disjoint open neighborhoods W_i of q_i in N, such that f maps W_i diffeomorphically onto an open neighborhood $f(W_i)$ of p in M. Let

$$U = \left(\bigcap_{i=1}^k f(W_i) \right) - f \left(N - \bigcup_{i=1}^k W_i \right).$$

Since $N - \bigcup_{i=1}^k W_i$ is closed in N and therefore compact, $f(N - \bigcup_{i=1}^k W_i)$ is also compact. Hence U is an open neighborhood of p in M. We then set $V_i = W_i \cap f^{-1}(U)$. $\qquad\square$

Consider a smooth map $f : N^n \to M^n$ between compact n-dimensional oriented manifolds, with M connected. For a regular value $p \in M$ and $q \in f^{-1}(p)$, define the local index

(11) $\quad \mathrm{Ind}(f; q) = \begin{cases} 1 & \text{if } D_q f : T_q N \to T_p M \text{ preserves orientation} \\ -1 & \text{otherwise.} \end{cases}$

Theorem 11.9 *In the situation above, and for every regular value p,*

$$\deg(f) = \sum_{q \in f^{-1}(p)} \mathrm{Ind}(f; q).$$

In particular $\deg(f)$ is an integer.

Proof. Let q_i, V_i, and U be as in Lemma 11.8. We may assume that U and hence V_i connected. The diffeomorphism $f_{|V_i} : V_i \to U$ is positively or negatively oriented, depending on whether $\mathrm{Ind}(f; q_i)$ is 1 or -1. Let $\omega \in \Omega^n(M)$ be an n-form with

$$\mathrm{supp}_M(\omega) \subseteq U, \quad \int_M \omega = 1.$$

Then $\mathrm{supp}_N(f^*(\omega)) \subseteq f^{-1}(U) = V_1 \cup \ldots \cup V_k$, and we can write

$$f^*(\omega) = \sum_{i=1}^k \omega_i,$$

where $\omega_i \in \Omega^n(N)$ and $\mathrm{supp}(\omega_i) \subseteq V_i$. Here $\omega_{i|V_i} = (f_{|V_i})^*(\omega_{|U})$. The formula is a consequence of the following calculation:

$$\deg(f) = \deg(f) \int_M \omega = \int_N f^*(\omega) = \sum_{i=1}^k \int_N \omega_i = \sum_{i=1}^k \int_{V_i} (f_{|V_i})^*(\omega_{|U})$$

$$- \sum_{i=1}^k \mathrm{Ind}(f; q_i) \int_U \omega_{|U} - \sum_{i=1}^k \mathrm{Ind}(f; q_i) \qquad\sqcap$$

In the special case where $f^{-1}(p) = \emptyset$ the theorem shows that $\deg(f) = 0$ (in the proof above we get $f^*(\omega) = 0$). Thus we have

Corollary 11.10 *If* $\deg(f) \neq 0$, *then* f *is surjective.* $\qquad\square$

Proposition 11.11 *Let* $F: P^{n+1} \to M^n$ *be a smooth map between oriented smooth manifolds, with* M^n *compact and connected. Let* $X \subseteq P$ *be a compact domain with smooth boundary* $N^n = \partial X$, *and suppose* N *is the disjoint union of submanifolds* $N_1^n, \ldots, N_k^n$. *If* $f_i = F_{|N_i}$, *then*

$$\sum_{i=1}^{k} \deg(f_i) = 0.$$

Proof. Let $f = F_{|N}$ so that

$$\deg(f) = \sum_{i=1}^{k} \deg(f_i).$$

On the other hand, if $\omega \in \Omega^n(M)$ has $\int_M \omega = 1$, then

$$\deg(f) = \int_N f^*(\omega) = \int_X dF^*(\omega) = \int_X F^*(d\omega) = 0$$

where the second equation is from Theorem 10.8. $\qquad\square$

We shall give two applications of *degree*. We first consider linking numbers, and then treat indices of vector fields.

Definition 11.12 Let J^d and K^l be two disjoint compact oriented connected smooth submanifolds of $\mathbb{R}^{n+1}$, whose dimensions $d \geq 1$, $l \geq 1$ satisfy $d + l = n$. Their *linking number* is the integer

$$\mathrm{lk}(J, K) = \deg\left(\Psi_{J,K}\right)$$

where

$$\Psi = \Psi_{J,K} : J \times K \to S^n; \quad \Psi(x, y) = \frac{y - x}{\|y - x\|}.$$

Here $J \times K$ is equipped with the product orientation (cf. Remark 9.20) and S^n is oriented as the boundary of D^{n+1} with the standard orientation of $\mathbb{R}^{n+1}$. We note that $\mathrm{lk}(J, K)$ changes sign when the orientation of either J or K is reversed.

Proposition 11.13

(i) $\mathrm{lk}\left(K^l, J^d\right) = (-1)^{(d+1)(l+1)} \mathrm{lk}\left(J^d, K^l\right)$

(ii) *If J and K can be separated by a hyperplane $H \subset \mathbb{R}^{n+1}$ then $\mathrm{lk}(J, K) = 0$.*

(iii) *Let g_t and h_t be homotopies of the inclusions $g_0: J \to \mathbb{R}^{n+1}$ and $h_0: K \to \mathbb{R}^{n+1}$ to smooth embeddings g_1 and h_1, such that $g_t(J) \cap h_t(K) = \emptyset$ for all $t \in [0, 1]$. Then $\mathrm{lk}(J, K) = \mathrm{lk}(g_1(J), h_1(K))$.*

(iv) *Let $\Phi: P^{l+1} \to \mathbb{R}^{n+1} - J$ be a smooth map with P oriented. Given a compact domain $R \subseteq P$ with smooth boundary ∂R, let $Q_1, \ldots, Q_k$ be the connected components of ∂R. Suppose each $\Phi_{|Q_j}$ is a smooth embedding. If $K_i = \Phi(Q_i)$, then*

$$\sum_{i=1}^{k} \mathrm{lk}(J; K_i) = 0.$$

Proof. We look at the commutative diagram

$$
\begin{array}{ccc}
J \times K & \xrightarrow{\Psi_{J,K}} & S^n \\
\downarrow{\scriptstyle T} & & \downarrow{\scriptstyle A} \\
K \times J & \xrightarrow{\Psi_{K,J}} & S^n
\end{array}
$$

where T interchanges factors and A is the antipodal map $Av = -v$. Then (i) follows from Corollary 11.3 upon using that

$$\deg(T) = (-1)^{dl}, \quad \deg(A) = (-1)^{n+1} = (-1)^{d+l+1}.$$

In the situation of (ii) the image of Φ will not contain vectors parallel to H, and the assertion follows from Corollary 11.10.

Assertion (iii) is a consequence of the homotopy property, Corollary 11.2. Indeed, a homotopy $J \times K \times [0, 1] \to S^n$ is given by

$$(h_t(y) - g_t(x)) / \|h_t(y) - g_t(x)\|.$$

Finally (iv) follows from Proposition 11.11 applied to the map $F: J \times P \to S^n$ with

$$F(x, y) = (\Phi(y) - x) / \|\Phi(y) - x\|,$$

and to the domain $X = J \times R$ with boundary components $J \times Q_i$. Indeed, $f_i = F_{|J \times Q_i}$ has degree $\deg(f_i) = \mathrm{lk}(J, K_i)$. $\qquad\square$

Here is a picture to illustrate (iv):

Figure 1

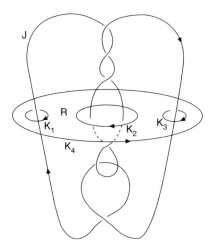

If $\mathrm{lk}(J, K) \neq 0$ then (ii) and (iii) of Proposition 11.13 imply that J and K cannot be deformed to manifolds separated by a hyperplane.

We shall now specialize to the classical case of knots in $\mathbb{R}^3$ where J and K are disjoint oriented submanifolds of $\mathbb{R}^3$ diffeomorphic to S^1. Let us choose smooth regular parametrizations

$$\alpha \colon \mathbb{R} \to J, \qquad \beta \colon \mathbb{R} \to K$$

with periods a and b, respectively, corresponding to a single traversing of J and K, respectively, agreeing with the orientation. For $p \in S^2$, consider the set

$$I(p) = \{(q_1, q_2) \in J \times K \mid q_2 - q_1 = \lambda p, \lambda > 0\}.$$

Let $v(q_1)$ and $w(q_2)$ denote the positively oriented unit tangent vectors to J and K in q_1 and q_2, respectively.

Theorem 11.14 *With the notation above we have:*

(i) *(Gauss)*

$$\mathrm{lk}(J, K) = \frac{1}{4\pi} \int_0^a \int_0^b \frac{\det\left(\alpha(u) - \beta(v), \alpha'(u), \beta'(v)\right)}{\|\alpha(u) - \beta(v)\|^3} \, du \, dv.$$

(ii) *There exists a dense set of points $p \in S^2$ such that*

$$\det\left(q_1 - q_2, v(q_1), w(q_2)\right) \neq 0 \text{ for } (q_1, q_2) \in I(p).$$

(iii) *For such points p, $\mathrm{lk}(J, K) = \sum_{(q_1, q_2) \in I(p)} \delta(q_1, q_2)$, where $\delta(q_1, q_2)$ is the sign of the determinant in (ii).*

Proof. We apply formula (2) to the map $\Psi = \Psi_{J,K}$ and the volume form $\omega = \mathrm{vol}_{S^2}$ (with integral 4π) to get

$$(12) \qquad \mathrm{lk}(J,K) = \deg(\Psi) = \frac{1}{4\pi} \int_{J \times K} \Psi^*(\mathrm{vol}_{S^2}).$$

We write $\Psi = r \circ f$ with

$$f: J \times K \to \mathbb{R}^3 - \{0\}; \quad f(q_1, q_2) = q_2 - q_1$$
$$r: \mathbb{R}^3 - \{0\} \to S^2; \quad r(x) = x/\|x\|.$$

For $x \in \mathbb{R}^3 - \{0\}$, $r^*(\mathrm{vol}_{S^2})_x \in \mathrm{Alt}^2(\mathbb{R}^3)$ is given by

$$r^*(\mathrm{vol}_{S^2})_x(v, w) = \det(x, v, w)/\|x\|^3$$

(cf. Example 9.18). The tangent space $T_{(q_1,q_2)}(J \times K)$ has a basis $\{v(q_1), \; w(q_2)\}$ and

$$Df_{(q_1,q_2)}(v(q_1)) = -v(q_1), \quad Df_{(q_1,q_2)}(w(q_2)) = w(q_2).$$

Therefore

$$(13) \quad \Psi^*(\mathrm{vol}_{S^2})_{(q_1,q_2)}(v(q_1)w(q_2)) = r^*(\mathrm{vol}_{S^2})_{q_2-q_1}(-v(q_1), w(q_2))$$
$$= \|q_1 - q_2\|^{-3} \det(q_1 - q_2, v(q_1), w(q_2)).$$

The integral of (12) can be calculated by integrating $(\alpha \times \beta)^* \Psi^*(\mathrm{vol}_{S^2})$ over the period rectangle $[0, a] \times [0, b]$. This yields Gauss's integral.

For $p \in S^2$, $I(p)$ is exactly the pre-image under Ψ. Thus p is a regular value of Ψ if and only if the determinant in (13) is non-zero for all $(q_1, q_2) \in I(p)$, and the sign $\delta(q_1, q_2)$ is determined by whether $D_{(q_1,q_2)}\Psi$ preserves or reverses orientation. Assertions (ii) and (iii) now follow from Theorems 11.5 and 11.9. $\square$

Remark 11.15 In Theorem 11.14.(ii), after a rotation of $\mathbb{R}^3$, the regular value p can be assumed to be the north pole $(0, 0, 1)$. The projections of J and K on the x_1, x_2-plane may be drawn indicating over- and undercrossings and orientations, e.g.

Figure 2

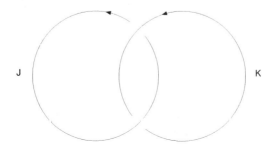

J $\qquad\qquad\qquad\qquad\qquad\qquad$ K

There is one element in $I(p)$ for every place where K crosses *over* (and not under) J. The corresponding sign δ is determined by the orientation of the curves and of the standard orientation of the plane as shown in the picture

Figure 3

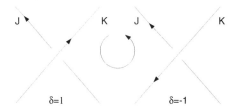

In Fig. 2, $\mathrm{lk}(J, K) = -1$. In Fig. 1,

$$\mathrm{lk}(J, K_2) = \mathrm{lk}(J, K_4) = 0 \,, \quad \mathrm{lk}(J, K_3) = 1 \,, \quad \mathrm{lk}(J, K_1) = -1.$$

The sum of these linking numbers is 0. This is in accordance with (iv) of Lemma 11.13.

We now apply the concept of degree to study singularities of vector fields.

Consider a vector field $F \in C^\infty(U, \mathbb{R}^n)$ on the open set $U \subseteq \mathbb{R}^n$, $n \geq 2$, and let us assume that $0 \in U$ is an isolated zero for F. A zero for F is also called a *singularity* for the vector field. We can choose a $\rho > 0$ with

$$\rho D^n = \{x \in \mathbb{R}^n \mid \|x\| \leq \rho\} \subseteq U$$

and such that 0 is the only zero for F in ρD^n. Define a smooth map $F_\rho \colon S^{n-1} \to S^{n-1}$ by

$$F_\rho(x) = \frac{F(\rho x)}{\|F(\rho x)\|}$$

The homotopy class of F_ρ is independent of the choice of ρ, and by Corollary 11.2 and Theorem 11.9, $\deg F_\rho \in \mathbb{Z}$ is independent of ρ.

Definition 11.16 The degree of F_ρ is called *local index* of F at 0, and is denoted $\iota(F; 0)$.

Lemma 11.17 *Suppose $F \in C^\infty(\mathbb{R}^n, \mathbb{R}^n)$ has the origin as its only zero. Then*

$$F \colon \mathbb{R}^n - \{0\} \to \mathbb{R}^n - \{0\}$$

induces multiplication by $\iota(F; 0)$ on $H^{n-1}(\mathbb{R}^n - \{0\}) \cong \mathbb{R}$.

Proof. Let $i\colon S^{n-1} \to \mathbb{R}^n - \{0\}$ be the inclusion map and $r\colon \mathbb{R}^{n-1} - \{0\} \to S^{n-1}$ the retraction $r(x) = x/\|x\|$. We have $\iota(F; 0) = \deg F_1$, where $F_1 = r \circ F \circ i$. The lemma follows from the commutative diagram below, where $H^{n-1}(i)$ and $H^{n-1}(r)$ are inverse isomorphisms:

$$
\begin{array}{ccc}
H^{n-1}(\mathbb{R}^n - \{0\}) & \xrightarrow{H^{n-1}(F)} & H^{n-1}(\mathbb{R} - \{0\}) \\
\Big\uparrow{\scriptstyle H^{n-1}(r)} & & {\scriptstyle H^{n-1}(r)}\Big\updownarrow{\scriptstyle H^{n-1}(i)} \\
H^{n-1}(S^{n-1}) & \xrightarrow{H^{n-1}(F_1)} & H^{n-1}(S^{n-1})
\end{array}
$$

$\square$

Given a diffeomorphism $\phi\colon U \to V$ to an open set $V \subseteq \mathbb{R}^n$ and a vector field on U, we can define the direct image $\phi_* F \in C^\infty(V, \mathbb{R}^n)$ by

$$
\phi_* F(q) = D_p \phi(F(p)), \quad p = \phi^{-1}(q).
$$

Lemma 11.18 *If $F \in C^\infty(U, \mathbb{R}^n)$ has 0 as an isolated singularity and $\phi\colon U \to V$ is a diffeomorphism to an open set $V \subseteq \mathbb{R}^n$ with $\phi(0) = 0$, then*

$$
\iota(\phi_* F; 0) = \iota(F, 0).
$$

Proof. By shrinking U and V we can restrict ourselves to considering the case where 0 is the only zero for F in U, and where there exists a diffeomorphism $\psi\colon V \to \mathbb{R}^n$. The assertion about ϕ will follow from the corresponding assertions about ψ and $\psi \circ \phi$, since

$$
\psi_*(\phi_* F) = (\psi \circ \phi)_* F.
$$

Thus it suffices to treat the case where $\phi\colon U \to \mathbb{R}^n$ is a diffeomorphism and where $Y = \phi_* F \in C^\infty(\mathbb{R}^n, \mathbb{R}^n)$ has the origin as its only singularity.

Let $U_0 \subseteq U$ be open and star-shaped around 0. We define a homotopy

$$
\Phi : U_0 \times [0, 1] \to \mathbb{R}^n; \quad \Phi_t(x) = \Phi(x, t) = \begin{cases} (D_0 \phi) x & \text{if } t = 0 \\ \phi(tx)/t & \text{if } t \neq 0. \end{cases}
$$

For $x \in U_0$,

$$
\phi(x) = \int_0^1 \frac{d}{dt} \phi(tx) dt = \int_0^1 \left(\sum_{i=1}^n x_i \frac{\partial \phi}{\partial x_i}(tx) \right) dt = \sum_{i=1}^n x_i \phi_i(x),
$$

where $\phi_i \in C^\infty(U_0, \mathbb{R}^n)$ is given by

$$
\phi_i(x) = \int_0^1 \frac{\partial \phi}{\partial x_i}(tx) \, dt.
$$

It follows that

$$\Phi(x,t) = \sum_{i=1}^{n} x_i \phi_i(tx),$$

and in particular that Φ has a smooth extension to an open set W with $U_0 \times [0,1] \subseteq W \subseteq U_0 \times \mathbb{R}$.

For each $t \in [0,1]$, Φ_t is a diffeomorphism from U_0 to an open subset of $\mathbb{R}^n$. Consider the direct image under Φ_t^{-1} of Y restricted to $\Phi_t(U_0)$:

$$X_t = \left(\Phi_t^{-1}\right)_* Y \in C^\infty(U_0, \mathbb{R}^n); \qquad X_t(x) = (D_x \Phi_t)^{-1} Y(\Phi_t(x)).$$

The function $X_t(x)$ is smooth on W. Now $X_1 = F_{|U_0}$ and $X_0 = (A^{-1})_* Y$, where $A = D_0\phi$.

Choose $\rho > 0$ such that $\rho D^n \subseteq U_0$. The homotopy $S^{n-1} \times [0,1] \to S^{n-1}$ given by

$$X_t(\rho x) / \|X_t(\rho x)\|, \quad 0 \le t \le 1,$$

and Corollary 11.2 shows that

$$\iota(F;0) = \iota(X_1;0) = \iota(X_0;0) = \iota\left((A^{-1})_* Y;0\right).$$

Since $A\colon \mathbb{R}^n \to \mathbb{R}^n$ is linear we have $(A^{-1})_* Y = A^{-1} \circ Y \circ A\colon \mathbb{R}^n \to \mathbb{R}^n$. This yields the commutative diagram

$$
\begin{array}{ccc}
\mathbb{R}^n - \{0\} & \xrightarrow{(A^{-1})_* Y} & \mathbb{R}^n - \{0\} \\
\downarrow{\scriptstyle A} & & \downarrow{\scriptstyle A} \\
\mathbb{R}^n - \{0\} & \xrightarrow{\ Y\ } & \mathbb{R}^n - \{0\}
\end{array}
$$

Now use the functor H^{n-1} and apply Lemma 11.17 to both Y and $(A^{-1})_* Y$ to get $\iota\left((A^{-1})_* Y;0\right) = \iota(Y;0)$. Hence $\iota(F;0) = \iota(Y;0)$. $\square$

Definition 11.19 Let X be a smooth tangent vector field on the manifold M^n, $n \ge 2$ wit $p_0 \in M$ as an isolated zero. The *local index* $\iota(X;p_0) \in \mathbb{Z}$ of X is defined by

$$\iota(X;p_0) = \iota(h_* X_{|U};0),$$

where (U,h) is an arbitrary smooth chart around p_0 with $h(p_0) = 0$.

We note that Lemma 11.18 shows that the local index does not depend on the choice of (U,h). One can picture vector fields in the plane by drawing their integral curves, e.g.

Figure 4

$\iota = -1$ $\qquad\qquad\qquad\qquad \iota = +2$

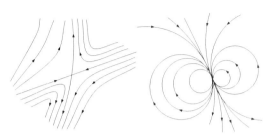

Let X be a smooth tangent vector field on M^n and let $p_0 \in M^n$ be a zero. Let

$$F = h_*(X_{|U}) \in C^\infty(h(U), \mathbb{R}^n)$$

for a chart (U, h) with $h(p_0) = 0$. If $D_0 F \colon \mathbb{R}^n \to \mathbb{R}^n$ is an isomorphism, then p_0 is said to be a *non-degenerate* singularity or zero. Note that by the inverse function theorem F is a local diffeomorphism around 0, such that 0 is an isolated zero for F. Hence $p_0 \in M^n$ is also an isolated zero for X.

Lemma 11.20 *If p_0 is a non-degenerate singularity, then*

$$\iota(X, p_0) = \operatorname{sign}(\det D_0 F) \in \{\pm 1\}.$$

Proof. By shrinking U we may assume that h maps U diffeomorphically onto an open set $U_0 \subseteq \mathbb{R}^n$, which is star-shaped around 0, and that F is a diffeomorphism from U_0 to an open set. As in the proof of Lemma 11.18 we can define a homotopy

$$G \colon U_0 \times [0, 1] \to \mathbb{R}^n; \quad G(x, t) = \begin{cases} D_0 F & \text{if } t = 0 \\ F(tx)/t & \text{if } t \neq 0, \end{cases}$$

where G can be extended smoothly to an open set W in $U_0 \times \mathbb{R}$ that contains $U_0 \times [0, 1]$. Choose $\rho > 0$ so that $\rho D^n \subseteq U_0$. We get a homotopy $\hat{G} \colon S^{n-1} \times [0, 1] \to S^{n-1}$,

$$\hat{G}(x, t) = G(\rho x, t) \, / \, \|G(\rho x, t)\|,$$

between the map F_ρ in Definition 11.16 and the analogous map A_ρ with $A = D_0 F$. It follows from Corollary 11.2 that

$$\iota(X; p_0) = \iota(F; 0) = \deg(F_\rho) = \deg A_\rho = \iota(A; 0).$$

The map $f_A \colon \mathbb{R}^n - \{0\} \to \mathbb{R}^n - \{0\}$ induced by A operates on $H^{n-1}(\mathbb{R}^n - \{0\})$ by multiplication by $\iota(X; p_0)$; cf. Lemma 11.17. The result now follows from Lemma 6.14. $\qquad\qquad\square$

Definition 11.21 Let X be a smooth vector field on M^n, with only isolated singularities. For a compact set $R \subseteq M$ we define the total index of X over R to be

$$\text{Index}\,(X; R) = \sum \iota(X; p),$$

where the summation runs over the finite number of zeros $p \in R$ for X. If M is compact we write $\text{Index}\,(X)$ instead of $\text{Index}\,(X; M)$.

Theorem 11.22 *Let $F \in C^\infty(U, \mathbb{R}^n)$ be a vector field on an open set $U \subseteq \mathbb{R}^n$, with only isolated zeros. Let $R \subseteq U$ be a compact domain with smooth boundary ∂R, and assume that $F(p) \neq 0$ for $p \in \partial R$. Then*

$$\text{Index}(F; R) = \deg f,$$

where $f \colon \partial R \to S^{n-1}$ is the map $f(x) = F(x) \,/\, \|F(x)\|$.

Proof. Let $p_1, \ldots, p_k$ be the zeros in R for F, and choose disjoint closed balls $D_j \subseteq R - \partial R$, with centers p_j. Define

$$f_j : \partial D_j \to S^{n-1}; \quad f_j(x) = F(x) \,/\, \|F(x)\|.$$

We apply Proposition 11.11 with $X = R - \bigcup_j \mathring{D}_j$. The boundary ∂X is the disjoint union of ∂R and the $(n-1)$-spheres $\partial D_1, \ldots, \partial D_k$. Here ∂D_j, considered as boundary component of X, has the opposite orientation to the one induced from D_j. Thus

$$\deg(f) + \sum_{j=1}^{k} -\deg(f_j) = 0.$$

Finally $\deg(f_j) = \iota(F; p_j)$ by the definition of local index and Corollary 11.3. $\square$

Corollary 11.23 *In the situation of Theorem 11.22, $\text{Index}(F; R)$ depends only on the restriction of F to ∂R.* $\square$

Corollary 11.24 *In the situation of Theorem 11.22, suppose for every $p \in \partial R$ that the vector $F(p)$ points outward. Let $g \colon \partial R \to S^{n-1}$ be the Gauss map which to $p \in \partial R$ associates the outward pointing unit normal vector to ∂R. Then*

$$\text{Index}(F; R) = \deg g.$$

Proof. By Corollary 11.2 it suffices to show that f and g are homotopic. Since $f(p)$ and $g(p)$ belong to the same open half-space of $\mathbb{R}^n$, the desired homotopy can be defined by

$$\frac{(1-t)f(p) + tg(p)}{\|(1-t)f(p) + tg(p)\|} \quad (0 \leq t \leq 1). \qquad \square$$

Lemma 11.25 *Suppose $F \in C^\infty(\mathbb{R}^n, \mathbb{R}^n)$ has the origin as its only zero. Then there exists an $\tilde{F} \in C^\infty(\mathbb{R}^n, \mathbb{R}^n)$, with only non-degenerate zeros, that coincides with F outside a compact set.*

Proof. We choose a function $\phi \in C^\infty(\mathbb{R}^n, [0, 1])$ with

$$\phi(x) = \begin{cases} 1 & \text{if } \|x\| \leq 1 \\ 0 & \text{if } \|x\| \geq 2. \end{cases}$$

We want to define $\tilde{F}(x) = F(x) - \phi(x)w$ for a suitable $w \in \mathbb{R}^n$. For $\|x\| > 2$ we have $\tilde{F}(x) = F(x)$. Set

$$c = \inf_{1 \leq \|x\| \leq 2} \|F(x)\| > 0$$

and choose w with $\|w\| < c$. For $1 \leq \|x\| \leq 2$, $\|\tilde{F}(x)\| \geq c - \|w\| > 0$. Thus all zeros of $\tilde{F}$ belong to the open unit ball $\mathring{D}^n$. Since $\tilde{F}$ coincides with $F - w$ on $\mathring{D}^n$,

$$\tilde{F}^{-1}(0) = \mathring{D}^n \cap F^{-1}(w).$$

We can pick w as a regular value of F with $\|w\| < c$ by Sard's theorem. Then $D_p\tilde{F} = D_pF$ will be invertible for all $p \in \tilde{F}^{-1}(0)$, and $\tilde{F}$ has the desired properties. $\qquad\square$

Note, by Corollary 11.23, that

$$(14) \qquad \iota(F; 0) = \sum_{p \in \tilde{F}^{-1}(p)} \iota\left(\tilde{F}, p\right)$$

Here is a picture of F and $\tilde{F}$ in a simple case:

Figure 5

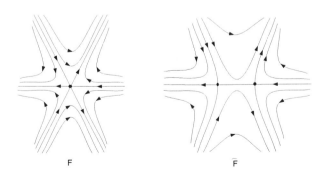

The zero for F of index -2 has been replaced by two non-degenerate zeros for $\tilde{F}$, both of index -1.

Corollary 11.26 *Let X be a smooth vector field on the compact manifold M^n with isolated singularities. Then there exists a smooth vector field $\tilde{X}$ on M having only non-degenerate zeros and with*

$$\mathrm{Index}(X) = \mathrm{Index}(\tilde{X}).$$

Proof. We choose disjoint coordinate patches which are diffeomorphic to $\mathbb{R}^n$ around the finitely many zeros of X, and apply Lemma 11.25 on the interior of each of them to obtain $\tilde{X}$. The formula then follows from (14).

Theorem 11.27 *Let $M^n \subseteq \mathbb{R}^{n+k}$ be a compact smooth submanifold and let N_ϵ be a tubular neighborhood of radius $\epsilon > 0$ around M. Denote by $g\colon \partial N_\epsilon \to S^{n+k-1}$ the outward pointing Gauss map. If X is an arbitrary smooth vector field on M^n with isolated singularities, then*

$$\mathrm{Index}(\mathrm{X}) = \deg g.$$

Proof. By Corollary 11.26 one may assume that X only has non-degenerate zeros. From the construction of the tubular neighborhood we have a smooth projection $\pi\colon N \to M$ from an open tubular neighborhood N with $N_\epsilon \subseteq N \subseteq \mathbb{R}^{n+k}$, and can define a smooth vector field F on N by

$$(15) \qquad F(q) = X(\pi(q)) + (q - \pi(q)).$$

Since the two summands are orthogonal, $F(q) = 0$ if and only if $q \in M$ and $X(q) = 0$. For $q \in \partial N_\epsilon$, $q - \pi(q)$ is a vector normal to $T_q \partial N_\epsilon$ pointing outwards. Hence $X(\pi(q)) \in T_q \partial N_\epsilon$, and $F(q)$ points outwards. By Corollary 11.24

$$\mathrm{Index}(F; N_\epsilon) = \deg g.$$

and it suffices to show that $\iota(X; p) = \iota(F; p)$ for an arbitrary zero of X. In local coordinates around p in M, with p corresponding to $0 \in \mathbb{R}^n$, X can be written in the form

$$(16) \qquad X = \sum_{i=1}^{n} f_i(x) \frac{\partial}{\partial x_i},$$

where $f_i(0) = 0$, and by Lemma 11.20 $\iota(X; p)$ is the sign of

$$(17) \qquad \det\left(\frac{\partial f_i}{\partial x_j}(0) \right).$$

By differentiating (16) and substituting 0 one gets

$$(18) \qquad \frac{\partial X}{\partial x_j}(0) = \sum_{i=1}^{n} \frac{\partial f_i}{\partial x_j}(0) \frac{\partial}{\partial x_i}.$$

It follows from (15) that $D_p F\colon \mathbb{R}^{n+k} \to \mathbb{R}^{n+k}$ is the identity on $T_p M^\perp$, and by (18) $D_p F$ maps $T_p M$ into itself by the linear map with matrix $(\partial f_i / \partial x_j(0))$ (with respect to the basis $(\partial/\partial x_i)_0$). It follows that p is a non-degenerate zero for F and that $\det D_p F$ has the same sign as the Jacobian in (17). $\qquad \square$

12. THE POINCARÉ–HOPF THEOREM

In the following, $M^n \subseteq \mathbb{R}^{n+k}$ will denote a fixed smooth submanifold. If the cohomology of M^n is finite-dimensional (e.g. when M^n is compact), the i-th *Betti number* is given by

$$(1) \qquad\qquad b_i(M) = \dim_{\mathbb{R}} H^i(M^n).$$

The *Euler characteristic* of M^n is defined to be

$$(2) \qquad\qquad \chi(M^n) = \sum_{i=0}^{n} (-1)^i \, b_i(M).$$

This chapter's main result is:

Theorem 12.1 (Poincaré–Hopf) *Let X be a smooth vector field on a compact manifold M. If X has only isolated zeros then*

$$\mathrm{Index}(X) = \chi(M).$$

By the final result of Chapter 11 it is sufficient to show the formula for just one such vector field X on M. We shall do so by making use of a Morse function on M^n.

Given $f \in C^\infty(M, \mathbb{R})$, a point $p \in M$ is a *critical point* for f if $d_p f = 0$.

Proposition 12.2 *Suppose that $p \in M$ is a critical point for $f \in C^\infty(M, \mathbb{R})$.*

   (i) *There exists a quadratic form $d_p^2 f$ on $T_p M$ characterized by the equation*

$$d_p^2 f\big(\alpha'(0)\big) = (f \circ \alpha)''(0)$$

     where $\alpha\colon (-\delta, \delta) \to M$ is any smooth curve with $\alpha(0) = p$.

  (ii) *Let $h\colon U \to \mathbb{R}^n$ be a chart around p and let $q = h(p)$. Then the composition*

$$\mathbb{R}^n \xrightarrow{D_q h^{-1}} T_p M \xrightarrow{d_p^2 f} \mathbb{R}$$

     is the quadratic form associated to the symmetric matrix

$$\left(\frac{\partial^2 f \circ h^{-1}}{\partial x_i \, \partial x_j}(q) \right).$$

Proof. Set $h \circ \alpha(t) = \gamma(t) = (\gamma_1(t), \ldots, \gamma_n(t))$ and $\phi = f \circ h^{-1}$. A direct calculation yields

$$(f \circ \alpha)'(t) = (\phi \circ \gamma)'(t) = \sum_{i=1}^{n} \frac{\partial \phi}{\partial x_i}(\gamma(t))\gamma_i'(t).$$

Since p is critical, $(\partial \phi / \partial x_i)(\gamma(0)) = 0$. By differentiating once again and substituting $t = 0$, we get

$$(f \circ \alpha)''(0) = \sum_{i=1}^{n} \sum_{j=1}^{n} \frac{\partial^2 \phi}{\partial x_i \partial x_j}(q)\gamma_i'(0)\gamma_j'(0).$$

This is the value at $\gamma'(0) = D_p h(\alpha'(0)) \in \mathbb{R}^n$ of the quadratic form from (ii). Both (i) and (ii) follow. $\square$

Consider another chart $\tilde{h} \colon \tilde{U} \to \mathbb{R}^n$ around p with $\tilde{q} = \tilde{h}(p)$ and let $F = \tilde{h} \circ h^{-1}$ defined in a neighborhood of q. The last formula in the proof above can be compared with

$$(f \circ \alpha)''(0) = \sum_{i=1}^{n} \sum_{j=1}^{n} \frac{\partial^2 \tilde{\phi}}{\partial x_i \partial x_j}(\tilde{q}) \, \tilde{\gamma}_i'(0)\tilde{\gamma}_j'(0),$$

where $\tilde{\phi} = f \circ \tilde{h}^{-1}$ and $\tilde{\gamma} = \tilde{h} \circ \alpha$. Let J denote the Jacobi matrix associated with F in q. Then $\tilde{\gamma}'(0) = J\gamma'(0)$ for the column vectors $\tilde{\gamma}'(0)$ and $\gamma'(0)$. By substituting this and comparing, one obtains the matrix identity

(3)
$$\left(\frac{\partial^2 \phi}{\partial x_i \partial x_j}(q) \right) = J^t \left(\frac{\partial^2 \tilde{\phi}}{\partial x_i \partial y_j}(\tilde{q}) \right) J.$$

Definition 12.3 A critical point $p \in M$ of $f \in C^\infty(M, \mathbb{R})$ is said to be *non-degenerate*, if the matrix in Proposition 12.2.(ii) is invertible. We call f a *Morse function*, if all critical points of f are non-degenerate. The *index* of a non-degenerate critical point p is the maximal dimension of a subspace $V \subseteq T_p M$ for which the restriction of $d_p^2 f$ to V is negative definite.

For smooth submanifolds $M^n \subseteq \mathbb{R}^{n+k}$ one can get Morse functions by:

Theorem 12.4 *For almost all $p_0 \in \mathbb{R}^{n+k}$ the function $f \colon M \to \mathbb{R}$ defined by*

$$f(p) = \frac{1}{2}\|p - p_0\|^2$$

is a Morse function.

Proof. Let $g: \mathbb{R}^n \to M$ be a local parametrization and $Y_j: \mathbb{R}^n \to \mathbb{R}^{n+k}$, $j = 1, \ldots, k$ smooth maps, such that $Y_1(\mathbf{x}), \ldots, Y_k(\mathbf{x})$ is a basis of $T_{g(\mathbf{x})}M^\perp$ for all $\mathbf{x} \in \mathbb{R}^n$. By Lemma 9.21 we know that M can be covered by at most countably many coordinate patches $g(U)$ of this type. Therefore it suffices to prove the assertion for $g(U)$ instead of M.

We define $\Phi: \mathbb{R}^{n+k} \to \mathbb{R}^{n+k}$ by

$$
(4) \qquad \Phi(\mathbf{x}, \mathbf{t}) = g(\mathbf{x}) + \sum_{j=1}^{k} t_j Y_j(\mathbf{x}) \quad (\mathbf{x} \in \mathbb{R}^n, \mathbf{t} \in \mathbb{R}^k).
$$

By Sard's theorem it suffices to prove that if p_0 is a regular value of Φ, then f becomes a Morse function on $g(U)$. Set $k = f \circ g: \mathbb{R}^n \to \mathbb{R}$; we can show instead that k becomes a Morse function on $\mathbb{R}^n$. We have

$$
(5) \qquad k(\mathbf{x}) = \frac{1}{2} \langle g(\mathbf{x}) - p_0, \, g(\mathbf{x}) - p_0 \rangle,
$$

where $\langle \, , \, \rangle$ denotes the usual inner product on $\mathbb{R}^{n+k}$.

Since $\langle \partial g / \partial x_i(\mathbf{x}), \, Y_\nu(\mathbf{x}) \rangle = 0$, it follows by differentiation with respect to x_j that

$$
(6) \qquad \left\langle \frac{\partial^2 g}{\partial x_i \partial x_j}, \, Y_\nu \right\rangle = - \left\langle \frac{\partial g}{\partial x_i}, \, \frac{\partial Y_\nu}{\partial x_j} \right\rangle.
$$

From (5) we have

$$
(7) \qquad \frac{\partial k}{\partial x_i} = \left\langle g(\mathbf{x}) - p_0, \, \frac{\partial g}{\partial x_i} \right\rangle,
$$

and therefore

$$
(8) \qquad \frac{\partial^2 k}{\partial x_i \partial x_j} = \left\langle \frac{\partial g}{\partial x_j}, \, \frac{\partial g}{\partial x_i} \right\rangle + \left\langle g(\mathbf{x}) - p_0, \, \frac{\partial^2 g}{\partial x_i \partial x_j} \right\rangle.
$$

Furthermore, by (4),

$$
(9) \qquad \frac{\partial \Phi}{\partial x_j} = \frac{\partial g}{\partial x_j} + \sum_{\nu=1}^{k} t_\nu \frac{\partial Y_\nu}{\partial x_j}, \qquad \frac{\partial \Phi}{\partial t_\nu} = Y_\nu.
$$

Assume that p_0 is a regular value of Φ and let $\mathbf{x}$ be a critical point of k. It follows from (7) that $g(\mathbf{x}) - p_0 \in T_{g(\mathbf{x})}M^\perp$. Hence there exists a unique $\mathbf{t} \in \mathbb{R}^k$ with

$$
(10) \qquad g(\mathbf{x}) - p_0 = - \sum_{\nu=1}^{k} t_\nu Y_\nu(\mathbf{x}),
$$

and $(\mathbf{x}, \mathbf{t}) \in \Phi^{-1}(p_0)$. The $n + k$ vectors in (9) are linearly independent at the point $(\mathbf{x},\mathbf{t})$. At this point, the equations (8), (10), (6) and (9) give

$$
\frac{\partial^2 k}{\partial x_i \partial x_j} = \left\langle \frac{\partial g}{\partial x_j}, \frac{\partial g}{\partial x_i} \right\rangle - \left\langle \sum_{\nu=1}^{k} t_\nu Y_\nu, \frac{\partial^2 g}{\partial x_i \partial x_j} \right\rangle
$$

$$
= \left\langle \frac{\partial g}{\partial x_j}, \frac{\partial g}{\partial x_i} \right\rangle + \left\langle \sum_{\nu=1}^{k} t_\nu \frac{\partial Y_\nu}{\partial x_j}, \frac{\partial g}{\partial x_i} \right\rangle = \left\langle \frac{\partial \Phi}{\partial x_j}, \frac{\partial g}{\partial x_i} \right\rangle.
$$

Let A denote the invertible $(n + k) \times (n + k)$ matrix with the vectors from (9) as rows. Then $AD_{\mathbf{x}}g$ takes the following form:

$$
\begin{pmatrix}
\frac{\partial^2 k}{\partial x_1 \partial x_1} & \cdots & \frac{\partial^2 k}{\partial x_1 \partial x_n} \\
\vdots & & \vdots \\
\frac{\partial^2 k}{\partial x_n \partial x_1} & \cdots & \frac{\partial^2 k}{\partial x_n \partial x_n} \\
0 & \cdots & 0 \\
\vdots & & \vdots \\
0 & \cdots & 0
\end{pmatrix}.
$$

Since $D_x g$ has rank n, so does $AD_{\mathbf{x}}g$. Hence the $n \times n$ matrix

$$
\left(\frac{\partial^2 k}{\partial x_i \partial x_j} (\mathbf{x}) \right)
$$

is invertible. This shows that $\mathbf{x}$ is a non-degenerate critical point.   □

Example 12.5 Let $f: \mathbb{R}^n \to \mathbb{R}$ be the function

$$
f(x) = c - x_1^2 - x_2^2 - \ldots - x_\lambda^2 + x_{\lambda+1}^2 + \ldots x_n^2,
$$

where $c \in \mathbb{R}$, $\lambda \in \mathbb{Z}$ and $0 \leq \lambda \leq n$. Since

$$
\mathrm{grad}_{\mathbf{x}}(f) = 2(-x_1, \ldots, -x_\lambda, x_{\lambda+1}, \ldots, x_n),
$$

0 is the only critical point of f. We find that

$$
\left(\frac{\partial^2 f}{\partial x_i \partial x_j} (0) \right) = \mathrm{diag}(-2, \ldots, -2, 2, \ldots, 2)
$$

with exactly λ diagonal entries equal to -2. Thus the origin is non-degenerate of index λ. We note that the vector field $\mathrm{grad}(f)$ has the origin as its only zero and that it is non-degenerate of index $(-1)^\lambda$.

Theorem 12.6 *Let $p \in M^n$ be a non-degenerate critical point for $f \in C^\infty(M, \mathbb{R})$. There exists a C^∞-chart $h: U \to h(U) \subseteq \mathbb{R}^n$ with $p \in U$ and $h(p) = 0$ such that*

$$f \circ h^{-1}(\mathbf{x}) = f(p) + \sum_{i=1}^{n} \delta_i x_i^2, \quad \mathbf{x} \in h(U),$$

where $\delta_i = \pm 1$ $(1 \leq i \leq n)$ (By an additional permutation of coordinates we can put f into the standard form given in Example 12.5.)

Proof. After replacing f with $f - f(p)$ we may assume that $f(p) = 0$. Since the problem is local and diffeomorphism invariant, we may also assume that $f \in C^\infty(W, \mathbb{R})$, where W is an open convex neighborhood of 0 in $\mathbb{R}^n$ and that 0 is the considered non-degenerate critical point with $f(0) = 0$.

We write f in the form

$$f(\mathbf{x}) = \sum_{i=1}^{n} x_i \, g_i(\mathbf{x}); \quad g_i(\mathbf{x}) = \int_0^1 \frac{\partial f(t\mathbf{x})}{\partial x_i} \, dt.$$

Since $g_i(0) = \frac{\partial f}{\partial x_i}(0) = 0$, we may repeat to get

$$g_i(\mathbf{x}) = \sum_{j=1}^{n} x_j \, g_{ij}(\mathbf{x}); \quad g_{ij}(\mathbf{x}) = \int_0^1 \frac{\partial g_i(s\mathbf{x})}{\partial x_j} \, ds.$$

On W we now have that

$$f(\mathbf{x}) = \sum_{i=1}^{n} \sum_{j=1}^{n} x_i \, x_j \, g_{ij}(\mathbf{x}),$$

where $g_{ij} \in C^\infty(W, \mathbb{R})$. If we introduce $h_{ij} = \frac{1}{2}(g_{ij} + g_{ji})$ then (h_{ij}) becomes a symmetric $n \times n$ matrix of smooth functions on W, and

(11) $$f(\mathbf{x}) = \sum_{i=1}^{n} \sum_{j=1}^{n} x_i \, x_j \, h_{ij}(\mathbf{x}).$$

By differentiating (11) twice and substituting 0, we get

$$\frac{\partial^2 f}{\partial x_i \partial x_j}(0) = 2h_{ij}(0).$$

In particular the matrix $(h_{ij}(0))$ is invertible.

Let us return to the original $f \in C^\infty(M, \mathbb{R})$. By induction on k, we attempt to show that the C^∞-chart h from the theorem can be choosen such that $f \circ h^{-1}$ is given by (11) with

$$(h_{ij}) = \begin{pmatrix} D & 0 \\ 0 & E \end{pmatrix},$$

with D a $(k-1) \times (k-1)$ matrix of the form $\mathrm{diag}(\pm 1, \ldots, \pm 1)$, and E some symmetric $(n-k+1) \times (n-k+1)$ matrix of smooth functions. So suppose inductively that

$$(12) \qquad f(\mathbf{x}) = \sum_{i=1}^{k-1} \delta_i x_i^2 + \sum_{i=k}^{n} \sum_{j=k}^{n} x_i\, x_j\, h_{ij}(\mathbf{x}), \qquad \delta_i = \pm 1$$

for $\mathbf{x}$ in a neighborhood W of the origin. We know that the minor E is invertible at 0. To start off we can perform a linear change of variables in $x_k, \ldots, x_n$, so that our new variables satisfy (12) with $h_{kk}(0) \neq 0$. By continuity we may assume that $h_{kk}(\mathbf{x})$ has constant sign $\delta_k = \pm 1$ on the entire W. Set

$$q = \sqrt{|h_{kk}|} \in C^\infty(W, \mathbb{R}),$$

and introduce new variables:

$$y_k = q(\mathbf{x})\left(x_k + \sum_{i=k+1}^{n} x_i \frac{h_{ik}(\mathbf{x})}{h_{kk}(\mathbf{x})} \right)$$

$$y_j = x_j \quad \text{for } j \neq k, 1 \leq j \leq n.$$

The Jacobi determinant for $\mathbf{y}$ as function of $\mathbf{x}$ is easily seen to be $\partial y_k/\partial x_k(0) = q(0) \neq 0$. The change of variables thus defines a local diffeomorphism Ψ around 0. In a neighborhood around 0 we have for $\mathbf{y} = \Psi(\mathbf{x})$:

$$f \circ \Psi^{-1}(\mathbf{y}) = f(\mathbf{x})$$

$$= \sum_{i=1}^{k-1} \delta_i x_i^2 + x_k^2 h_{kk}(\mathbf{x}) + 2x_k \sum_{j=k+1}^{n} x_j h_{jk}(\mathbf{x}) + \sum_{i=k+1}^{n} \sum_{j=k+1}^{n} x_i x_j h_{ij}(\mathbf{x})$$

$$= \sum_{i=1}^{k-1} \delta_i x_i^2 + h_{kk}(\mathbf{x})\left(x_k + \sum_{j=k+1}^{n} x_j \frac{h_{jk}(\mathbf{x})}{h_{kk}(\mathbf{x})} \right)^2$$

$$- h_{kk}(\mathbf{x})\left(\sum_{j=k+1}^{n} x_j \frac{h_{jk}(\mathbf{x})}{h_{kk}(\mathbf{x})} \right)^2 + \sum_{i=k+1}^{n} \sum_{j=k+1}^{n} x_i x_j h_{ij}(\mathbf{x})$$

$$= \sum_{i=1}^{k} \delta_i y_i^2 + \sum_{i=k+1}^{n} \sum_{j=k+1}^{n} x_i x_j \tilde{h}_{ij}(\mathbf{x})$$

$$= \sum_{i=1}^{k} \delta_i y_i^2 + \sum_{i=k+1}^{n} \sum_{j=k+1}^{n} y_i y_j \tilde{h}_{ij} \circ \Psi^{-1}(\mathbf{y}),$$

where $\tilde{h}_{ij} \in C^\infty(W, \mathbb{R})$. This completes the induction step. $\square$

We point out that with the assumptions of Theorem 12.6 p is the only critical point in U. If M is compact and $f \in C^\infty(M, \mathbb{R})$ is a Morse function then f has only finitely many critical points. Among them there will always be at least one local minimum (index $\lambda = 0$) and at least one local maximum (index $\lambda = n$).

Definition 12.7 Let $f \in C^\infty(M, \mathbb{R})$ be a Morse function. A smooth tangent vector field X on M is said to be *gradient-like* for f, if the following conditions are satisfied:

(i) For every non-critical point $p \in M$, $d_p f\left(X(p)\right) > 0$.

(ii) If $p \in M^n$ is a critical point of f then there exists a C^∞-chart $h : U \to h(U) \subseteq \mathbb{R}^n$ with $p \in U$ and $h(p) = 0$ such that

$$f \circ h^{-1}(\mathbf{x}) = f(p) - x_1^2 - \cdots - x_\lambda^2 + x_{\lambda+1}^2 + \cdots + x_n^2, \quad \mathbf{x} \in h(U),$$

and $\quad h_* X_{|U} = \mathrm{grad}\left(f \circ h^{-1}\right)$.

A smooth parametrized curve $\alpha : I \to M$ is an *integral curve* for X, if

$$\alpha'(t) = X(\alpha(t)) \quad \text{for } t \in I.$$

Hence one gets $(f \circ \alpha)'(t) = d_{\alpha(t)} f(X(\alpha(t)))$. If $\alpha(I)$ does not contain any critical points, then $f \circ \alpha : I \to \mathbb{R}$ is a monotone increasing function by condition (i).

Lemma 12.8 *Every Morse function on M admits a gradient-like vector field.*

Proof. We can find a C^∞-atlas $(U_\alpha, h_\alpha)_{\alpha \in A}$ for M which satisfies the following two conditions:

(i) Every critical point of f belongs to just one of the coordinate patches U_α.

(ii) For any $\alpha \in A$ either f has no critical point in U_α or f has precisely one critical point p in U_α, $h_\alpha(p) = 0$, and $f \circ h_\alpha^{-1}$ has the form listed in Example 12.5.

Let X_α be a tangent vector field on U_α determined by $X_\alpha = (h_\alpha^{-1})_*\left(\mathrm{grad}\,(f \circ h_\alpha^{-1})\right)$. Choose a smooth partition of unity $(\rho_\alpha)_{\alpha \in A}$ subordinate to $(U_\alpha)_{\alpha \in A}$, and define a smooth tangent vector field on M by

$$X = \sum_{\alpha \in A} \rho_\alpha X_\alpha,$$

where $\rho_\alpha X_\alpha$ is taken to be 0 outside U_α. If $p \in M$ is not a critical point for f then, for every $\alpha \in A$ with $p \in U_\alpha$ and $q = h_\alpha(p)$, we have

$$d_p f(X_\alpha(p)) = d_q(f \circ h_\alpha^{-1})\left(\mathrm{grad}_q(f \circ h_\alpha^{-1})\right) > 0.$$

Indeed, there is at least one α with $\rho_\alpha(p) > 0$ and

$$d_p f(X(p)) = \sum_\alpha \rho_\alpha(p)\, d_p f(X_\alpha(p)).$$

We see that X satisfies condition (i) in Definition 12.7.

If p is a critical point of f then there exists a unique $\alpha \in A$ with $p \in U_\alpha$. It follows from (a) that X coincides with X_α on a neighborhood of p, and condition (b) above shows that assertion (ii) in Definition 12.7 is satisfied. $\square$

The next lemma relates the index of a Morse function to the local index of vector fields as defined in Chapter 11.

Lemma 12.9 *Let f be a Morse function on M and X a smooth tangent vector field such that $d_p f(X(p)) > 0$ for every $p \in M$ that is not a critical point for f. Let $p_0 \in M$ be a critical point for f of index λ. If $X(p_0) = 0$, then*

$$\iota(X; p_0) = (-1)^\lambda.$$

Proof. We choose a gradient-like vector field $\widetilde{X}$. By Definition 12.7.(ii) and Example 12.5,

$$\iota(\widetilde{X}; p_0) = (-1)^\lambda.$$

Let U be an open neighborhood of p_0 that is diffeomorphic to $\mathbb{R}^n$ and chosen so small that p_0 is the only critical point in U. The inequalities

$$d_p f(X(p)) > 0, \qquad d_p f(\widetilde{X}(p)) > 0,$$

valid for $p \in U - \{p_0\}$, show that $X(p)$ and $\widetilde{X}(p)$ belong to the same open half-space in $T_p M$. Thus

$$(1 - t)X(p) + t\widetilde{X}(p) \quad (0 \le t \le 1)$$

defines a homotopy between X and $\widetilde{X}$ considered as maps from $U - \{p_0\}$ to $\mathbb{R}^n - \{0\}$, and $\iota(X; p_0) = \iota(\widetilde{X}; p_0)$. $\square$

Remark 12.10 Given a Riemannian metric on M and $f \in C^\infty(M, \mathbb{R})$, one can define the gradient vector field $\operatorname{grad} f$ by the equation

$$\langle \operatorname{grad}_p(f), X_p \rangle = d_p f(X_p)$$

for all $X_p \in T_p M$. Then Lemma 12.9 holds for $\operatorname{grad}(f)$.

Theorem 12.11 *Let M^n be a compact differentiable manifold and X a smooth tangent vector field on M^n with isolated singularities. Let $f \in C^\infty(M, \mathbb{R})$ be a Morse function and c_λ the number of critical points of index λ for f. Then we have that*

$$\operatorname{Index}(X) = \sum_{\lambda=0}^{n} (-1)^\lambda c_\lambda.$$

Proof. It is a consequence of Theorem 11.27 that any two tangent vector fields with isolated singularities have the same index. Thus we may assume that X is gradient-like for f. The zeros for X are exactly the critical points of f, and the claimed formula follows from Lemma 12.9.   □

It is a consequence of the above theorem that the sum

(13) $$\sum_{\lambda=0}^{n} (-1)^{\lambda} c_{\lambda}$$

is independent of the choice of Morse function $f \in C^{\infty}(M, \mathbb{R})$. Given Theorem 12.11, the Poincaré–Hopf theorem is the statement that the sum (13) is equal to the Euler characteristic; cf. (2).

We will give a proof of this based on the two lemmas below, whose proofs in turn involve methods from dynamical systems and ordinary differential equations, and will be postponed to Appendix C.

Let us fix a compact manifold M^n and a Morse function f on M. For $a \in \mathbb{R}$ we set

$$M(a) = \{p \in M \mid f(p) < a\}.$$

Recall that a number $a \in \mathbb{R}$ is a *critical value* if $f^{-1}(a)$ contains at least one critical point.

Lemma 12.12 *If there are no critical values in the interval $[a_1, a_2]$, then $M(a_1)$ and $M(a_2)$ are diffeomorphic.*

Lemma 12.13 *Suppose that a is a critical value and that $p_1, \ldots, p_r$ are the critical points in $f^{-1}(a)$. Let p_i have index λ_i. There exists an $\epsilon > 0$, and disjoint open neighbourhoods U_i of p_i, such that*

   (i) *$p_1, \ldots, p_r$ are the only critical points in $f^{-1}([a - \epsilon, a + \epsilon])$.*
   (ii) *U_i is diffeomorphic to an open contractible subset of $\mathbb{R}^n$.*
   (iii) *$U_i \cap M(a - \epsilon)$ is diffeomorphic to $S^{\lambda_i - 1} \times V_i$, where V_i is an open contractible subset of $\mathbb{R}^{n-\lambda_i+1}$ (in particular $U_i \cap M(a - \epsilon) = \emptyset$ if $\lambda_i = 0$).*
   (iv) *$M(a + \epsilon)$ is diffeomorphic to $U_1 \cup \ldots \cup U_r \cup M(a - \epsilon)$.*   □

Proposition 12.14 *In the situation of Lemma 12.13 suppose that $M(a - \epsilon)$ has finite-dimensional cohomology. Then the same will be true for $M(a + \epsilon)$, and*

$$\chi(M(a + \epsilon)) = \chi(M(a - \epsilon)) + \sum_{i=1}^{r} (-1)^{\lambda_i}.$$

Proof. For $U = U_1 \cup \ldots \cup U_r$, Lemma 12.13.(ii) and Corollary 6.10 imply that

$$H^p(U) \simeq \begin{cases} 0 & \text{if } p \neq 0 \\ \mathbb{R}^r & \text{if } p = 0. \end{cases}$$

This gives $\chi(U) = r$. Condition (iii) of Lemma 12.13 shows that $U_i \cap M(a - \epsilon)$ is homotopy equivalent to $S^{\lambda_i - 1}$, and Example 9.29 gives

$$\chi(U_i \cap M(a - \epsilon)) = 1 + (-1)^{\lambda_i - 1}.$$

Since $U \cap M(a - \epsilon)$ is a disjoint union of the sets $U_i \cap M(a - \epsilon)$, it has a finite-dimensional de Rham cohomology, and

$$\chi(U \cap M(a - \epsilon)) = \sum_{i=1}^{r} \left(1 + (-1)^{\lambda_i - 1}\right) = r - \sum_{i=1}^{r} (-1)^{\lambda_i} = \chi(U) - \sum_{i=1}^{r} (-1)^{\lambda_i}.$$

The claimed formula now follows from Lemma 12.13.(iv) and the lemma below, applied to U and $V = M(a - \epsilon)$. $\square$

Lemma 12.15 *Let U and V be open subsets of a smooth manifold. If U, V and $U \cap V$ have finite dimensional de Rham cohomology, the same is true for $U \cup V$, and*

$$\chi(U \cup V) = \chi(U) + \chi(V) - \chi(U \cap V).$$

Proof. We use the long exact Mayer–Vietoris sequence

$$\cdots \to H^{p-1}(U \cap V) \to H^p(U \cup V) \to H^p(U) \oplus H^p(V) \to H^p(U \cap V) \to \cdots$$

First we conclude that $\dim H^p(U \cup V) < \infty$. Second, the alternating sum of the dimensions of the vector spaces in an exact sequence is equal to zero; cf. Exercise 4.4. $\square$

Theorem 12.16 *If f is a Morse function on the compact manifold M^n, then*

$$\chi(M^n) = \sum_{\lambda=0}^{n} (-1)^\lambda c_\lambda,$$

where c_λ denotes the number of critical points for f of index λ.

Proof. Let $a_1 < a_2 < \ldots < a_{k-1} < a_k$ be the critical values. Choose real numbers $b_0 < a_1$, $b_j \in (a_j, a_{j+1})$ for $1 \leq j \leq k - 1$ and $b_k > a_k$. Lemma 12.12 shows that the dimensions of $H^d(M(b_j))$ are independent of the choice of b_j from the relevant interval. If $M(b_{j-1})$ has finite-dimensional de Rham cohomology, the same will be true for $M(b_j)$ according to Proposition 12.14, and

$$(14) \qquad \chi(M(b_j)) - \chi(M(b_{j-1})) = \sum_{p \in f^{-1}(a_j)} (-1)^{\lambda(p)}$$

Here the sum runs over the critical points $p \in f^{-1}(a_j)$, and $\lambda(p)$ denotes the index of p. We can start from $M(b_0) = \emptyset$. An induction argument shows that $\dim H^d(M(b_j)) < \infty$ for all j and d. The sum of the formulas of (14) for $1 \leq j \leq k$ gives

$$\chi(M) = \chi(M(b_k)) = \sum_p (-1)^{\lambda(p)}$$

where p runs over the critical points. □

The Poincaré–Hopf theorem 12.1 follows by combining Theorems 12.11 and 12.16.

Corollary 12.17 *If M^n is compact and of odd dimension n then $\chi(M^n) = 0$.*

Proof. Let f be a Morse function on M. Then $-f$ is also a Morse function, and $-f$ has the same critical points as f. If a critical point p has index λ with respect to f, then p has index $n - \lambda$ with respect to $-f$. Theorem 12.16 applied to both f and $-f$ gives

$$\chi(M) = \sum_{\lambda=0}^{n} (-1)^{\lambda} c_\lambda = \sum_{\lambda=0}^{n} (-1)^{n-\lambda} c_\lambda.$$

The two sums differ by the factor $(-1)^n$, and the assertion follows. □

Example 12.18 (Gauss–Bonnet in $\mathbb{R}^3$). We consider a compact regular surface $S \subseteq \mathbb{R}^3$, oriented by means of the Gauss map $N \colon S \to S^2$. The Gauss curvature of S at the point p is

$$K(p) = \det(d_pN); \qquad T_pS = \{p\}^{\perp} = T_pS^2.$$

Sard's theorem implies that we can find a pair of antipodal points in S^2 that are both regular values of N. After a suitable rotation of the entire situation we can assume that $p_{\pm} = (0, 0, \pm 1)$ are regular values of N.

Let $f \in C^\infty(S, \mathbb{R})$ be the projection on the third coordinate axis of $\mathbb{R}^3$. The critical points $p \in S$ of f are exactly the points for which T_pS is parallel with the x_1, x_2-plane, i.e. $N(p) = p_{\pm}$. At such a point p, the differential of the Gauss map d_pN is an isomorphism. Hence $K(p) \neq 0$. A neighborhood of p in S can be parametrized by $(u, v, f(u, v))$, and in these local coordinates the Gauss curvature has the following expression:

$$K = \left(1 + \left(\frac{\partial f}{\partial u}\right)^2 + \left(\frac{\partial f}{\partial v}\right)^2\right)^{-1} \det \begin{pmatrix} \dfrac{\partial^2 f}{\partial u^2} & \dfrac{\partial^2 f}{\partial u \partial v} \\[2ex] \dfrac{\partial^2 f}{\partial u \partial v} & \dfrac{\partial^2 f}{\partial v^2} \end{pmatrix}$$

(see e.g. [do Carmo] page 163). Since $K(p) \neq 0$, the determinant in the expression does not vanish at p, so p is a non-degenerate critical point for f. If $K(p) > 0$ the determinant is positive and p has index 0 or 2. If $K(p) < 0$ the determinant is negative and p has index 1. We apply Theorem 12.16 to get

$$\chi(S) = \#\{p \in S \mid N(p) = p_{\pm}, K(p) > 0\} - \#\{p \in S \mid N(p) = p_{\pm}, K(p) < 0\}.$$

Since p_+ is a regular value for N, we have by Theorem 11.9

$$\deg(N) = \#\{p \in N^{-1}(p_+) \mid K(p) > 0\} - \#\{p \in N^{-1}(p_+) \mid K(p) < 0\}$$

and analogously with p_- instead of p_+. It follows that

(15) $$\chi(S) = 2 \deg(N).$$

The map

$$\mathrm{Alt}^2(dN_p)\colon \mathrm{Alt}^2\left(T_{N(p)}S^2\right) = \mathrm{Alt}^2(T_pS) \to \mathrm{Alt}^2(T_pS)$$

is multiplication by $\det(d_pN) = K(p)$, so $N^*(\mathrm{vol}_{S^2}) = K(p)\mathrm{vol}_S$. Hence

$$\int_S K\mathrm{vol}_S = \int_{S^2} N^*(\mathrm{vol}_{S^2}) = (\deg N)\int_{S^2} \mathrm{vol}_{S^2} = 4\pi \deg N.$$

Combined with (15) this yields the *Gauss–Bonnet formula*

(16) $$\frac{1}{2\pi}\int_S K\mathrm{vol}_S = \chi(S).$$

Example 12.19 Consider the torus T in $\mathbb{R}^3$. The height function $f\colon T \to \mathbb{R}$ is a Morse function with the four indicated critical points. The Gauss curvature of T is

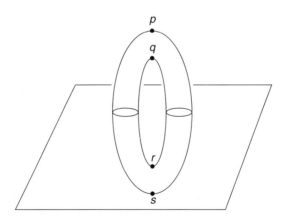

positive at p and s, but negative at q and r (cf. Example 12.18). Hence f is a Morse function on T. The index at p, q, r and s is 0, 1, 1 and 2, respectively. Theorem 12.16 gives $\chi(T) = 0$. Since we know that $\dim H^0(T) = \dim H^2(T) = 1$, we can calculate $\dim H^1(T) = 2$.

Example 12.20 (Morse function on $\mathbb{RP}^n$) Real functions on $\mathbb{RP}^n$ are equivalent to even functions $f\colon S^n \to \mathbb{R}$, i.e. $f(x) = f(-x)$ for all $x \in S^n$. Let us try

$$f(\mathbf{x}) = \sum_{i=0}^{n} a_i x_i^2$$

for $\mathbf{x} = (x_0, x_1, \ldots, x_n) \in S^n \subseteq \mathbb{R}^{n+1}$, and real numbers a_i. The differential of f at $\mathbf{x}$ is given by

$$d_x f(v_0, \ldots, v_n) = 2 \sum_{i=0}^{n} a_i x_i v_i,$$

where $v = (v_0, v_1, \ldots, v_n) \in T_x S^n$ so that

$$\sum_{i=0}^{n} x_i v_i = 0.$$

Thus $\mathbf{x}$ is a critical point for f if and only if the vectors $\mathbf{x}$ and $(a_0 x_0, a_1 x_1, \ldots, a_n x_n)$ are linearly independent. If the coefficients a_i are distinct, this occurs precisely for $\mathbf{x} = \pm e_i = (0, \ldots, \pm 1, \ldots, 0)$, and f has exactly $2n + 2$ critical points. The induced smooth map $\tilde{f}\colon \mathbb{RP}^n \to \mathbb{R}$ then has $n + 1$ critical points $[e_i]$. In a neighborhood of $\pm e_0 \in S^n$ we have the charts h with

$$h_{\pm}^{-1}(u_1, \ldots, u_n) = \left(\pm\sqrt{1 - \sum_{i=1}^{n} u_i^2}, \; u_1, \ldots, u_n \right),$$

and in a neighborhood of $0 \in \mathbb{R}^n$,

$$f \circ h_{\pm}^{-1}(u_1, \ldots, u_n) = a_0 \left(1 - \sum_{i=1}^{n} u_i^2 \right) + \sum_{i=1}^{n} a_i u_i^2 = a_0 + \sum_{i=1}^{n} (a_i - a_0) u_i^2.$$

The matrix of the second-order partial derivatives for $f \circ h_{\pm}^{-1}$ (at 0) is the diagonal matrix

$$\mathrm{diag}(2(a_1 - a_0), 2(a_2 - a_0), \ldots, 2(a_n - a_0)).$$

Hence $\pm e_0$ are non-degenerate critical points for f; the index for each is equal to the number of indices i with $1 \le i \le n$ and $a_i < a_0$. An analogous result holds for the other critical points $\pm e_j$. For simplicity, suppose that $a_0 < a_1 < a_2 < \ldots < a_n$. Then the two critical points $\pm e_j$ for $f\colon S^n \to \mathbb{R}$ have index j. The induced function $\tilde{f}\colon \mathbb{RP}^n \to \mathbb{R}$ is a Morse function with critical points $[e_j]$ of index j. We apply Theorem 12.16 to $\tilde{f}$. Since $c_\lambda = 1$ for $0 \le \lambda \le n$, we get

$$\chi(\mathbb{RP}^n) = \begin{cases} 1 & \text{if } n \text{ is even} \\ 0 & \text{if } n \text{ is odd.} \end{cases}$$

This agrees with Example 9.31.

13. POINCARÉ DUALITY

Given a compact oriented smooth manifold M^n of dimension n, Poincaré duality is the statement that

(1) $$H^p(M) \cong H^{n-p}(M)^*, \quad p \in \mathbb{Z}$$

where $H^{n-p}(M)^*$ denotes the dual vector space of linear forms on $H^{n-p}(M)$. The proof we give below is based upon induction over an open cover of M. Thus we need a generalization of (1) to oriented manifolds that are not necessarily compact. The general statement we will prove is that

(2) $$H^p(M) \cong H_c^{n-p}(M)^*$$

where the subscript c refers to de Rham cohomology with compact support.

For a smooth manifold M we let $\Omega_c^*(M)$ be the subcomplex of the de Rham complex that in degree p consists of the vector space $\Omega_c^p(M)$ of p-forms with compact support. The cohomology groups of $(\Omega_c^*(M), d)$ are denoted $H_c^*(M)$, i.e.

$$H_c^p(M) = \frac{\mathrm{Ker}\big(d \colon \Omega_c^p(M) \to \Omega_c^{p+1}(M)\big)}{\mathrm{Im}\big(d \colon \Omega_c^{p-1}(M) \to \Omega_c^p(M)\big)}.$$

Note that when M is compact $\Omega_c^*(M) = \Omega^*(M)$, so that $H_c^*(M) = H^*(M)$ in this case.

The vector spaces $H_c^*(M)$ are not in general a (contravariant) functor on the category of all smooth maps. However, if $\varphi \colon M \to N$ is *proper*, i.e. if $\varphi^{-1}(K)$ is compact whenever K is, then the induced form $\varphi^*(\omega)$ will have compact support when ω has. Indeed

$$\mathrm{supp}_M \varphi^*(\omega) \subset \varphi^{-1}(\mathrm{supp}_N \omega)$$

and φ^* becomes a chain map from $\Omega_c^*(N)$ to $\Omega_c^*(M)$. Hence by Lemma 4.3 there is an induced map

$$H_c^p(\varphi) \colon H_c^p(N) \to H_c^p(M)$$

and $H_c^p(-)$ becomes a contravariant functor on the category of smooth manifolds and smooth proper maps. A diffeomorphism is proper, so $H_c^p(M) \cong H_c^p(N)$ when M and N are diffeomorphic.

Remarks 13.1

(i) The vector space $H_c^0(M)$ consists of the locally constant functions $f: M \to \mathbb{R}$ with compact support. Such an f must be identically zero on every non-compact connected component of M. In particular $H_c^0(M) = 0$ for non-compact connected M. In contrast $H^0(M) = \mathbb{R}$ for such a manifold.

(ii) If M^n is connected, oriented and n-dimensional, then we have the isomorphism from Theorem 10.13,

$$\int_M : H_c^n(M) \overset{\cong}{\to} \mathbb{R}$$

whereas by (i) and (2), $H^n(M) = 0$ if M is non-compact.

Lemma 13.2

$$H_c^q(\mathbb{R}^n) = \begin{cases} \mathbb{R} & \text{if } q = n \\ 0 & \text{otherwise.} \end{cases}$$

Proof. The above remarks give the result for $q = 0$ and $q = n$, so we may assume that $0 < q < n$. We identify $\mathbb{R}^n$ with $S^n - \{p_0\}$, e.g. by stereographic projection, and can thus instead prove that

$$H_c^q(S^n - \{p_0\}) = 0.$$

Now the chain complex $\Omega_c^*(S^n - \{p_0\})$ is the subcomplex of $\Omega^*(S^n)$ consisting of differential forms which vanish in a neighborhood of p_0.

Let $\omega \in \Omega_c^q(S^n - \{p_0\})$ be a closed form. Since $H^q(S^n) = 0$, by Example 9.29, ω is exact in $\Omega^*(S^n)$, so there is a $\tau \in \Omega^{q-1}(S^n)$ with $d\tau = \omega$. We must show that τ can be chosen to vanish in a neighborhood of p_0. Suppose W is an open neighborhood of $\{p_0\}$, diffeomorphic to $\mathbb{R}^n$, where $\omega|_W = 0$.

If $q = 1$, then τ is a function on S^n that is constant on W, say $\tau|_W = a$. But then $\kappa = \tau - a \in \Omega_c^0(S^n - \{p_0\})$ and $d\kappa = \omega$.

If $2 \leq q < n$ then we use that $H^{q-1}(W) \cong H^{q-1}(\mathbb{R}^n) = 0$, and that $\tau|_W$ is a closed form, to find a $\sigma \in \Omega^{q-2}(W)$ with $d\sigma = \tau|_W$. Now choose a smooth function $\varphi: S^n \to [0, 1]$ with $\operatorname{supp}(\varphi) \subset W$ taking the value 1 in a smaller open neighborhood $U \subset W$ of $\{p_0\}$. The form $\varphi\sigma$ on W can be extended to all of S^n, assigning the value zero on $S^n - W$. Let $\tilde{\sigma}$ be the extended form and let $\kappa = \tau - d\tilde{\sigma}$. Then $\kappa|_U = 0$ and $d\kappa = d\tau + dd\tilde{\sigma} = \omega$. $\square$

Let $V \subset U$ be open subsets of a smooth manifold M, and let $i: V \to U$ be the inclusion. There is an induced chain map

$$i_* : \Omega_c^*(V) \to \Omega_c^*(U),$$

defined by setting

$$i_*(\omega)|_V = \omega, \quad i_*(\omega)|_{U - \mathrm{supp}\,(\omega)} = 0.$$

for $\omega \in \Omega_c^p(V)$. We get a linear map

(3) $$\qquad\qquad\qquad i_*: H_c^p(V) \to H_c^p(U)$$

which is called the *direct image homomorphism*. Given a second inclusion $j: W \to V$, $(i \circ j)_*(\omega) = i_* \circ j_*(\omega)$, so that $\left(H_c^p(-), i_*\right)$ becomes a *covariant* functor on the category of open subsets and inclusions of a given manifold.

There is also a Mayer–Vietoris theorem for this functor. Indeed, if U_1 and U_2 are open subsets of M with union U, and $i_\nu: U_\nu \to U$, $j_\nu: U_1 \cap U_2 \to U_\nu$ are the inclusions, then the sequence

(4) $$\qquad 0 \to \Omega_c^q(U_1 \cap U_2) \xrightarrow{J_q} \Omega_c^q(U_1) \oplus \Omega_c^q(U_2) \xrightarrow{I_q} \Omega_c^q(U) \to 0$$

is exact, where

$$I_q(\omega_1, \omega_2) = i_{1*}(\omega_1) + i_{2*}(\omega_2) \quad \text{and} \quad J_q(\omega) = (j_{1*}(\omega), -j_{2*}(\omega)).$$

We leave the verification of the exactness to the reader, with the remark that surjectivity of I_q uses a smooth partition of unity on U subordinate to the covering $\{U_\nu\}$; cf. Theorem 5.1.

Theorem 13.3 (Mayer–Vietoris) *With the above notation there is an exact sequence*

$$\cdots \to H_c^q(U_1 \cap U_2) \xrightarrow{J_*} H_c^q(U_1) \oplus H_c^q(U_2) \xrightarrow{I_*} H_c^q(U) \xrightarrow{\partial_*} H_c^{q+1}(U_1 \cap U_2) \to \cdots.$$

Proof. This follows from Theorem 4.9 applied to (4). $\qquad\qquad\square$

In comparing Theorem 13.3 with Theorem 5.2 the reader will notice that the directions of all arrows have been reversed.

For later use, let us explicate Definition 4.5. $\partial_*[\omega] \in H_c^{q+1}(U_1 \cap U_2)$ is defined as follows: write $\omega = \omega_1 + \omega_2$ with $\omega_\nu \in \Omega_c^q(U)$ and $\mathrm{supp}_U(\omega_\nu) \subset U_\nu$. Then $d\omega_1$ and $-d\omega_2$ agree on $U_1 \cap U_2$ and the common value $\tau = d\omega_1|_{U_1 \cap U_2} = -d\omega_2|_{U_1 \cap U_2}$ is a closed form in $\Omega_c^{q+1}(U_1 \cap U_2)$ that represents $\partial_*[\omega]$.

Proposition 13.4 *Suppose $\{U_\alpha \mid \alpha \in A\}$ is a family of pairwise disjoint open subsets of the smooth manifold M with union U. Then there are isomorphisms*

(i) $H^q(U) \to \prod_{\alpha \in A} H^q(U_\alpha); \quad [\omega] \mapsto [i_\alpha^*(\omega)]$
(ii) $\bigoplus_{\alpha \in A} H_c^q(U_\alpha) \to H_c^q(U); \quad ([\omega_\alpha])_{\alpha \in A} \mapsto \sum_{\alpha \in A} i_{\alpha*}([\omega_\alpha])$

where $i_\alpha: U_\alpha \to U$ denotes the inclusion.

Proof. There are isomorphisms

$$\Phi^q: \Omega^q(U) \to \prod_{\alpha \in A} \Omega^q(U_\alpha); \quad \Phi^q(\omega) = (i_\alpha^*(\omega))_{\alpha \in A}$$

$$\Psi_q: \bigoplus_{\alpha \in A} \Omega_c^q(U_\alpha) \to \Omega_c(U); \quad \Psi_q\big((\omega_\alpha)_{\alpha \in A}\big) = \sum i_{\alpha*}(\omega_\alpha)$$

which define isomorphisms of chain complexes when we give $\prod \Omega^*(U_\alpha)$ the differential

$$d\big((\tau_\alpha)_{\alpha \in A}\big) = (d\tau_\alpha)_{\alpha \in A}$$

and view $\bigoplus \Omega_c^*(U_\alpha) \subset \prod \Omega_c^*(U_\alpha) \subset \prod \Omega^*(U_\alpha)$ as a subcomplex. $\qquad\square$

The contravariant functor that sends a vector space A to its dual vector space $A^* = \mathrm{Hom}_{\mathbb{R}}(A, \mathbb{R})$ is exact, i.e. if $A \xrightarrow{\varphi} B \xrightarrow{\psi} C$ is an exact sequence of vector spaces, then $C^* \xrightarrow{\psi^*} B^* \xrightarrow{\varphi^*} A^*$ is exact. It is clear that $\varphi^* \circ \psi^* = 0$. The other inclusion $\mathrm{Ker}\,\varphi^* \subseteq \mathrm{Im}\,\psi^*$ follows because a linear map defined on $\mathrm{Im}\,\psi$ can be extended to a linear map on all of C. (This uses Zorn's lemma when C is infinite-dimensional.)

We can therefore dualize the exact sequence of Theorem 13.3 to get the exact sequence

(5)
$$\cdots \to H_c^{q+1}(U_1 \cap U_2)^* \xrightarrow{\partial^!} H_c^q(U)^* \xrightarrow{I^!} H_c^q(U_1)^* \oplus H_c^q(U_2)^* \xrightarrow{J^!} H_c^q(U_1 \cap U_2) \to \cdots$$

with

$$I^!(\alpha) = \Big(i_1^!(\alpha), i_2^!(\alpha)\Big) \quad \text{and} \quad J^!(\alpha_1, \alpha_2) = j_1^!(\alpha_1) - j_2^!(\alpha_2)$$

where we have written $i_1^!: H_c^q(U)^* \to H_c^q(U_1)^*$ for the vector space dual of i_{1*}, etc. The dual of a direct sum is a direct product, so Proposition 13.4.(ii) implies the isomorphism

(6)
$$H_c^q(U)^* \xrightarrow{\cong} \prod_{\alpha \in A} H_c^q(U_\alpha)^*; \quad \beta \mapsto (i_\alpha^!(\beta))_{\alpha \in A}.$$

For an oriented n-dimensional manifold the exterior product defines a bilinear map

$$\Omega^p(M) \times \Omega_c^{n-p}(M) \to \Omega_c^n(M)$$

since $\mathrm{supp}\,(\omega_1 \wedge \omega_2) \subseteq \mathrm{supp}(\omega_1) \cap \mathrm{supp}(\omega_2)$. It induces a bilinear map

$$H^p(M) \times H_c^{n-p}(M) \to H_c^n(M)$$

and we may compose with integration (cf. Remark 13.1.(ii)) to obtain a bilinear pairing

$$H^p(M) \times H_c^{n-p}(M) \to \mathbb{R}; \quad ([\omega_1], [\omega_2]) \mapsto \int_M \omega_1 \wedge \omega_2$$

which in turn defines a linear map $D_M^p: H^p(M) \to H_c^{n-p}(M)^*$.

Theorem 13.5 (Poincaré duality) *For an oriented smooth n-dimensional manifold,* D_M^p *is an isomorphism for all* p.

The proof is based upon a series of lemmas.

Lemma 13.6 *Suppose* $V \subseteq U \subseteq M^n$ *are open subsets. Then the diagram*

$$
\begin{array}{ccc}
H^p(U) & \xrightarrow{H^p(i)} & H^p(V) \\
\big\downarrow{\scriptstyle D_U^p} & & \big\downarrow{\scriptstyle D_V^p} \\
H_c^{n-p}(U)^* & \xrightarrow{i^!} & H_c^{n-p}(V)^*
\end{array}
$$

commutes.

Proof. Let $\omega \in \Omega^p(U)$, $\tau \in \Omega_c^{n-p}(V)$ be closed forms representing cohomology classes $[\omega]$ and $[\tau]$. Then

$$
D_V^p \circ H^p(i)([\omega])([\tau]) = D_V^p([i^*(\omega)])[\tau] = \int_V i^*(\omega) \wedge \tau
$$

$$
i^! \circ D_U^p([\omega])([\tau]) = D_U^p([\omega])[i_*(\tau)] = \int_U \omega \wedge i_*(\tau).
$$

Since $\mathrm{supp}_U(\omega \wedge i_*(\tau)) \subset \mathrm{supp}_U(i_*(\tau)) = \mathrm{supp}_V(\tau)$ we may as well in the second integral just integrate over V. But the n-forms $i^*(\omega) \wedge \tau$ and $\omega \wedge i_*(\tau)$ agree on V. $\qquad\square$

Lemma 13.7 *For open subsets* U_1 *and* U_2 *of* M^n *with union* U *the diagram*

$$
\begin{array}{ccc}
H^p(U_1 \cap U_2) & \xrightarrow{\partial^*} & H^{p+1}(U) \\
\big\downarrow{\scriptstyle D_{U_1 \cap U_2}^p} & & \big\downarrow{\scriptstyle D_U^{p+1}} \\
H_c^{n-p}(U_1 \cap U_2)^* & \xrightarrow{(-1)^{p+1}\partial^!} & H_c^{n-p-1}(U)^*
\end{array}
$$

is commutative. Here ∂^* *is the boundary in the Mayer–Vietoris sequence of Theorem 5.2, and* $\partial^!$ *is from* (5).

Proof. Let $\omega \in \Omega^p(U_1 \cap U_2)$ and $\tau \in \Omega_c^{n-p-1}(U)$ be closed forms. We write $\omega = j_1^*(\omega_1) - j_2^*(\omega_2)$ with $\omega_\nu \in \Omega^p(U_\nu)$ and $j_\nu : U_1 \cap U_2 \to U_\nu$ the inclusions. Let $\kappa \in \Omega^{p+1}(U)$ be the $(p+1)$-form with $i_\nu^*(\kappa) = d\omega_\nu$, where $i_\nu : U_\nu \to U$ are the inclusions. Then κ represents $\partial^*([\omega])$ so that

$$
D_U^{p+1} \partial^*([\omega])([\tau]) = D_U^{p+1}([\kappa])[\tau] = \int_U \kappa \wedge \tau.
$$

It was pointed out after the proof of Theorem 13.3 that a representative for $\partial_*[\tau] \in H_c^{n-p}(U_1 \cap U_2)$ can be obtained by the following procedure: write $\tau = \tau_1 + \tau_2$ with

$$
\tau_\nu \in \Omega_c^{n-p-1}(U) \quad \text{and} \quad \mathrm{supp}_U(\tau_\nu) \subset U_\nu
$$

and let $\sigma = j_1^*(d\tau_1) = -j_2^*(d\tau_2)$. Then σ is a closed $(n-p)$-form that represents $\partial_*([\tau])$. Hence

$$\partial^! D^p_{U_1 \cap U_2}([\omega])([\tau]) = D^p_{U_1 \cap U_2}([\omega])[\sigma] = \int_{U_1 \cap U_2} \omega \wedge \sigma.$$

We must show that the two integrals are equal up to the sign $(-1)^{p+1}$. We have

$$\int_U \kappa \wedge \tau = \int_U \kappa \wedge \tau_1 + \int_U \kappa \wedge \tau_2 = \int_{U_1} d\omega_1 \wedge \tau_1 + \int_{U_2} d\omega_2 \wedge \tau_2$$

since $\mathrm{supp}\,(\tau_\nu) \subseteq U_\nu$. Now $d(\omega_\nu \wedge \tau_\nu) = d\omega_\nu \wedge \tau_\nu + (-1)^p \omega_\nu \wedge d\tau_\nu$, and by Corollary 10.9,

$$\int_{U_\nu} d(\omega_\nu \wedge \tau_\nu) = 0,$$

so that

$$(-1)^{p+1} \int_U \kappa \wedge \tau = \int_{U_1} \omega_1 \wedge d\tau_1 + \int_{U_2} \omega_2 \wedge d\tau_2.$$

On the other hand, $d\tau_1|_{U_1} = j_{1*}(\sigma)$ and $d\tau_2|_{U_2} = -j_{2*}(\sigma)$, and we have

$$\int_{U_1} \omega_1 \wedge d\tau_1 + \int_{U_2} \omega_2 \wedge d\tau_2 = \int_{U_1} \omega_1 \wedge j_{1*}(\sigma) - \int_{U_2} \omega_2 \wedge j_{2*}(\sigma)$$

$$= \int_{U_1 \cap U_2} j_1^*(\omega_1) \wedge \sigma - \int_{U_1 \cap U_2} j_2^*(\omega_2) \wedge \sigma$$

$$= \int_{U_1 \cap U_2} \omega \wedge \sigma. \qquad \square$$

Corollary 13.8

  (i) Let U_1 and U_2 be open subsets of M^n, and suppose that U_1, U_2 and $U_1 \cap U_2$ satisfy Poincaré duality. Then so does $U = U_1 \cup U_2$.
  (ii) Let $(U_\alpha)_{\alpha \in A}$ be a family of pairwise disjoint open subsets of M^n. If each U_α satisfies Poincaré duality then so does the union $U = \bigcup_\alpha U_\alpha$.
  (iii) Every open subset $U \subseteq M^n$ that is diffeomorphic to $\mathbb{R}^n$ satisfies Poincaré duality.

Proof. Consider the diagram

$$
\begin{array}{ccccccccc}
\longrightarrow & H^p(U) & \xrightarrow{I^*} & H^p(U_1) \oplus H^p(U_2) & \xrightarrow{J^*} & H^p(U_1 \cap U_2) & \xrightarrow{\partial^*} & H^{p+1}(U) & \longrightarrow \\
& \downarrow{\scriptstyle D^p_U} & & \downarrow{\scriptstyle D^p_{U_1} \oplus D^p_{U_2}} & & \downarrow{\scriptstyle D^p_{U_1 \cap U_2}} & & \downarrow{\scriptstyle D^{p+1}_U} & \\
\longrightarrow & H_c^{n-p}(U)^* & \xrightarrow{I^!} & H_c^{n-p}(U_1)^* \oplus H_c^{n-p}(U_2)^* & \xrightarrow{J^!} & H_c^{n-p}(U_1 \cap U_2) & \xrightarrow{(-1)^{p+1}\partial^!} & H_c^{n-p-1}(U) & \longrightarrow
\end{array}
$$

This is a commutative diagram according to the two lemmas above. Our assumptions are that $D_{U_1}^p \oplus D_{U_2}^p$ and $D_{U_1 \cap U_2}^p$ are isomorphisms for all p, and it follows by the 5-lemma (cf. Exercise 4.1) that so is D_U^p.

The proof of (ii) uses the commutative diagram

$$
\begin{array}{ccc}
H^p(U) & \longrightarrow & \prod_{\alpha \in A} H^p(U_\alpha) \\
\Big\downarrow{\scriptstyle D_U^p} & & \Big\downarrow{\scriptstyle \Pi D_{U_\alpha}^p} \\
H_c^{n-p}(U)^* & \longrightarrow & \prod_{\alpha \in A} H_c^{n-p}(U_\alpha)^*
\end{array}
$$

The horizontal maps are isomorphisms by Proposition 13.4 and the right-hand vertical map is an isomorphism by assumption.

To prove (iii), we use that $H^p(U) \cong H^p(\mathbb{R}^n)$ and $H_c^{n-p}(U) \cong H_c^{n-p}(\mathbb{R}^n)$ together with Theorem 3.15 and Lemma 13.2. We only need to check that $D_U^0 : H^0(U) \to H_c^n(U)^*$ is an isomorphism. The constant function 1 on U is mapped to the basis element

$$
\int_U : H_c^n(U) \to \mathbb{R}
$$

of $H_c^n(U)^* \cong \mathbb{R}$. $\qquad\square$

Theorem 13.9 (Induction on open sets). *Let M^n be a smooth n-dimensional manifold equipped with an open cover $\mathcal{V} = (V_\beta)_{\beta \in B}$. Suppose $\mathcal{U}$ is a collection of open subsets of M that satisfies the following four conditions*

(i) *$\emptyset \in \mathcal{U}$.*
(ii) *Any open subset $U \subseteq V_\beta$ diffeomorphic with $\mathbb{R}^n$ belongs to $\mathcal{U}$.*
(iii) *If $U_1, U_2, U_1 \cap U_2$ belong to $\mathcal{U}$ then $U_1 \cup U_2 \in \mathcal{U}$.*
(iv) *If $U_1, U_2, \ldots$ is a sequence of pairwise disjoint open subsets with $U_i \in \mathcal{U}$ then their union $\bigcup_i U_i \in \mathcal{U}$.*

Then M^n belongs to $\mathcal{U}$.

The proof is based upon the following lemma, where the term *relatively compact* means that the closure is compact.

Lemma 13.10 *In the situation of Theorem 13.9, suppose $U_1, U_2, \ldots$ is a sequence of open, relatively compact subsets of M with*

(i) *$\bigcap_{j \in J} U_j \in \mathcal{U}$ for any finite subset J,*
(ii) *$(U_j)_{j \in \mathbb{N}}$ is locally finite.*

Then the union $U_1 \cup U_2 \cup \ldots$ belongs to $\mathcal{U}$.

Proof. First we show by induction on m that $U_{j_1} \cup \ldots \cup U_{j_m} \in \mathcal{U}$ for every set of indices $j_1, \ldots, j_m$. The cases $m = 1, 2$ follow from (i) and condition (iii) of Theorem 13.9, so suppose $m \geq 3$ and that the claim is true for sets of $m - 1$ indices. Then setting $V = U_{j_2} \cup \ldots \cup U_{j_m}$,

$$U_{j_1} \cap V = \bigcup_{\nu=2}^{m} U_{j_1} \cap U_{j_\nu} \in \mathcal{U}$$

by the induction hypothesis applied to the new sequence $(U_{j_1} \cap U_j)_{j \in \mathbb{N}}$, and condition (iii) of Theorem 13.9 implies that $U_{j_1} \cup \ldots \cup U_{j_m} \in \mathcal{U}$. Since $U_i \cap U_j \in \mathcal{U}$ by (i), we also have

(7)
$$\bigcup_{\nu=1}^{m} (U_{i_\nu} \cap U_{j_\nu}) \in \mathcal{U}$$

for any set of $2m$ indices $i_1, j_1, \ldots, i_m, j_m$.

Inductively we define index sets I_m and open sets $W_m \subseteq M$ as follows: $I_1 = \{1\}$, $W_1 = U_1$ and for $m \geq 2$

(8)
$$I_m = \{m\} \cup \{i \mid i > m, \ U_i \cap W_{m-1} \neq \emptyset\} - \bigcup_{j=1}^{m-1} I_j$$
$$W_m = \bigcup_{i \in I_m} U_i$$

If I_{m-1} is finite, then W_{m-1} is relatively compact and (ii) implies that W_{m-1} only intersects finitely many of the sets U_i. This shows inductively that I_m is indeed finite for all m. Moreover, if $m \geq 2$ does not belong to any I_j with $j < m$ then it certainly, by definition of I_m, belongs to I_m. Thus $\mathbb{N}$ is the disjoint union of the finite sets I_m.

Since we already know that finite unions are in $\mathcal{U}$ we have $W_m \in \mathcal{U}$ (if $I_m = \emptyset$, $W_m = \emptyset \in \mathcal{U}$). Similarly, (7) shows that

$$W_m \cap W_{m+1} = \bigcup_{(i,j) \in I_m \times I_{m+1}} U_i \cap U_j \in \mathcal{U}.$$

Note also from (8) that $W_m \cap W_k = \emptyset$ if $k \geq m + 2$. Indeed, if the intersection were non-empty, then there would exist $i \in I_k$ with $W_m \cap U_i \neq \emptyset$, and by (8) $i \in I_j$ for some $j \leq m + 1$.

One now uses condition (iv) of Theorem 13.9 to see that the following three sets are in $\mathcal{U}$:

$$W^{(0)} = \bigcup_{j=1}^{\infty} W_{2j}, \quad W^{(1)} = \bigcup_{j=1}^{\infty} W_{2j-1}, \quad W^{(0)} \cap W^{(1)} = \bigcup_{j=1}^{\infty} (W_m \cap W_{m+1}).$$

But then Theorem 13.9.(iii) implies that the union $\bigcup_{i=1}^{\infty} U_i = W^{(0)} \cup W^{(1)} \in \mathcal{U}$. $\square$

Proof of Theorem 13.9. Consider first the special case where $M = W \subseteq \mathbb{R}^n$ is an open subset. We consider the maximum norm on $\mathbb{R}^n$

$$\|x\|_{\infty} = \max_{1 \leq i \leq n} |x_i|$$

whose open balls are the cubes $\Pi_{i=1}^{n}(a_i, b_i)$. The argument of Proposition A.6 gives us a sequence of open $\| \ \|_{\infty}$-balls U_j such that

(i) $W = \bigcup_{j=1}^{\infty} U_j = \bigcup_{j=1}^{\infty} \overline{U}_j$.
(ii) $(U_j)_{j \in \mathbb{N}}$ is locally finite.
(iii) Each $\overline{U}_j$ is contained in at least one V_β.

A finite intersection $U_{j_1} \cap \ldots \cap U_{j_m}$ is either empty or of the form $\Pi_{i=1}^{n}(a_i, b_i)$, so is diffeomorphic to $\mathbb{R}^n$. Thus we can conclude from Lemma 13.10 that $W \in \mathcal{U}$.

In the general case, consider first an open coordinate patch (U, h) of M, $h : U \to W$ a diffeomorphism onto an open subset $W \subseteq \mathbb{R}^n$. We apply the above special case to W with cover $\left(h^{-1}(V_\beta) \right)_{\beta \in B}$, and with $\mathcal{U}^h$ the open sets of W whose images by h^{-1} belong to the given $\mathcal{U}$. The conclusion is that $U \in \mathcal{U}$ for each coordinate patch.

If M is compact then we apply the argument of Lemma 13.10 to a finite cover of M by coordinate patches to conclude that $M \in \mathcal{U}$. In the non-compact case we make use of a locally finite cover of M by a sequence of relatively compact coordinate patches; cf. Theorem 9.11. (One may alternatively imitate the proof of Proposition A.6 using relatively compact coordinate patches instead of discs to construct the desired cover). $\square$

Proof of Theorem 13.5. Let

$$\mathcal{U} = \left\{ U \subseteq M^n \mid U \text{ open}, D_U^p \text{ an isomorphism for all } p \right\}$$

and let $\mathcal{V} = (V_\beta)_{\beta \in B}$ be the trivial cover consisting only of M. Corollary 13.8 tells us that the assumptions in Theorem 13.9 are satisfied, so $M \in \mathcal{U}$. $\square$

We close the chapter with an exact sequence associated to a smooth compact manifold pair (N, M), i.e. a smooth compact submanifold M of a smooth compact manifold N. Let U be the complement $U = N - M$, and let

$$i : U \to N, \quad j : M \to N$$

be the inclusions. Then we have

Proposition 13.11 *There is a long exact sequence*

$$\cdots \to H^{q-1}(M) \xrightarrow{\delta} H_c^q(U) \xrightarrow{i_*} H^q(N) \xrightarrow{j^*} H^q(M) \to \cdots$$

The proof of this result is based upon the following:

Lemma 13.12

(i) $j^* = \Omega^q(j) \colon \Omega^q(N) \to \Omega^q(M)$ *is an epimorphism.*

(ii) *If* $\omega \in \Omega^q(M)$ *is closed, there exists a* q*-form* $\tau \in \Omega^q(N)$ *such that* $j^*(\tau) = \omega$ *and such that* $d\tau$ *is identically zero on some open set in* N *containing* M.

(iii) *If* $\tau \in \Omega^q(N)$ *has* $\operatorname{supp}_N(d\tau) \cap M = \emptyset$ *and* $j^*(\tau)$ *is exact, there exists a form* $\sigma \in \Omega^{q-1}(N)$ *such that* $\tau - d\sigma$ *is identically zero on some open set in* N *containing* M.

Proof. We can assume by Theorem 8.11 that N is a smooth submanifold of $\mathbb{R}^k$. Theorem 9.23 gives us tubular neighborhoods in $\mathbb{R}^k$ with corresponding smooth inclusions and retractions (V_N, i_N, r_N), (V_M, i_M, r_M) for N and M respectively; we may arrange that $V_M \subseteq V_N$. Let $\varphi \colon N \to [0,1]$ be a smooth function such that $\operatorname{supp}_N(\varphi) \subseteq N \cap V_M$ and such that φ is constantly equal to 1 on some open set $W \subseteq N \cap V_M$ with $M \subseteq W$.

Let $\omega \in \Omega^q(M)$ be closed, and let $\widetilde{\omega} = r_M^*(\omega) \in \Omega^q(N \cap V_M)$. Then (ii) follows upon defining $\tau \in \Omega^q(N)$ to be equal $\varphi\widetilde{\omega}$ on $N \cap V_M$ and extended trivially over the remaining part of N. The same argument proves (i).

To prove (iii) we set

$$\widetilde{\tau} = r_N^*(\tau)_{|V_M} \in \Omega^q(V_M).$$

The assumption that $\operatorname{supp}_N(d\tau) \cap M = \emptyset$ implies that $dr_N^*(\tau) = r_N^*(d\tau)$ vanishes on a neighborhood of M. Hence V_M may be chosen so small that $d\widetilde{\tau} = 0$. Observe that

$$i_M^*(\widetilde{\tau}) = (i_N \circ j)^*(r_N^*(\tau)) = j^* \circ i_N^* \circ r_N^*(\tau) = j^*(\tau)$$

so $[i_M^*(\widetilde{\tau})] = 0$ in $H^q(M)$. It follows from Corollary 9.28 that $[\widetilde{\tau}] = 0$ in $H^q(V_M)$. Pulling this back by the inclusion $N \cap V_M \to V_M$ we find that $\tau_{|N \cap V_M}$ is exact. Now choose $\sigma_0 \in \Omega^{q-1}(N \cap V_M)$ with $d\sigma_0 = \tau_{|N \cap V_M}$ and define $\sigma \in \Omega^{q-1}(N)$ to be $\varphi\sigma_0$ on $N \cap V_M$ and extended trivially over the remaining part of N. Then $\tau - d\sigma$ vanishes on W. $\qquad\square$

Proof of Proposition 13.11. By Lemma 13.12.(i) we have a short exact sequence of chain complexes

$$0 \to \Omega^*(N, M) \to \Omega^*(N) \xrightarrow{j^*} \Omega^*(M) \to 0$$

where $\Omega^q(N, M)$ is defined to be the kernel of $\Omega^q(j)$. Let us denote the cohomology of $(\Omega^*(N, M), d)$ by $H^*(N, M)$. Then Theorem 4.9 gives us a long exact sequence

$$\cdots \to H^{q-1}(M) \to H^q(N, M) \to H^q(N) \xrightarrow{j^*} H^q(M) \to \cdots.$$

The chain map $i_*: \Omega_c^*(U) \to \Omega^*(N)$ has image in $\Omega^*(N, M)$, so it suffices to show that

$$i_*: \Omega_c^*(U) \to \Omega^*(N, M)$$

induces an isomorphism in cohomology. We can then substitute $H_c^q(U)$ for $H^q(N, M)$ in the above exact sequence.

Consider an element $[\omega]$ in the kernel of

$$H^q(i_*): H_c^q(U) \to H^q(N, M)$$

represented by a closed q-form $\omega \in \Omega_c^q(U)$. Then $i_*(\omega) = d\tau$ for some $\tau \in \Omega^{q-1}(N, M)$. Since $j^*(\tau) = 0$ and $\operatorname{supp}_N(d\tau) \subseteq U$ we may apply Lemma 13.12.(iii) to find $\sigma \in \Omega^{q-2}(N)$ such that $\tau - d\sigma$ vanishes on an open set around M. This gives us $\kappa = (\tau - d\sigma)_{|U} \in \Omega_c^{q-1}(U)$ with $d\kappa = \omega$, so $[\omega] = 0$.

Let $[\omega] \in H^q(N, M)$ be represented by the closed q-form $\omega \in \Omega^q(N, M)$. We use Lemma 13.12.(iii) to find $\sigma \in \Omega^{q-1}(N)$, with the property that $\omega - d\sigma$ vanishes on an open set containing M. Observe that

$$d(j^*(\sigma)) = j^*(d\sigma) = j^*(\omega) = 0.$$

By Lemma 13.12.(ii) we may choose $\tau \in \Omega^{q-1}(N)$ with $j^*(\sigma) = j^*(\tau)$ and such that $d\tau$ vanishes on a neighborhood of M. Thus $\sigma - \tau \in \Omega^{q-1}(N, M)$, and defining

$$\kappa = (\omega - d(\sigma - \tau))_{|U} = (\omega - d\sigma)_{|U} + d\tau_{|U} \in \Omega_c^q(U),$$

we obtain $[\omega] = [\omega - d(\sigma - \tau)] = H^q(i_*)[\kappa]$. $\qquad\square$

Let us finally introduce the important *signature invariant* of oriented compact manifolds of dimension congruent to zero modulo 4. Given a $2k$-dimensional compact smooth manifold M^{2k}, its *intersection form* is the bilinear form

$$\mu: H^k(M) \times H^k(M) \to \mathbb{R}$$

defined by

(9)
$$\mu([\alpha], [\beta]) = \int_M \alpha \wedge \beta.$$

This is $(-1)^k$-symmetric, i.e.

$$\mu([\alpha], [\beta]) = (-1)^k \mu([\alpha], [\beta]).$$

We focus on $k \equiv 0 \pmod 2$, where μ becomes symmetric. All such bilinear forms can be diagonalized; i.e. there exists a basis $e_1, ..., e_m$ such that

(10)
$$\mu(e_i, e_j) = \begin{cases} 0 & \text{if } i \neq j \\ \alpha_i & \text{if } i = j. \end{cases}$$

Given a diagonalization of μ, we define the *signature* by

$$\sigma(\mu) = \#\{i \mid \alpha_i > 0\} - \#\{i \mid \alpha_i < 0\}.$$

This number is independent of the chosen diagonalization. In the case of the intersection form we know that $\alpha_i \neq 0$ for all i, since by Poincaré duality the adjoint D_M^k of μ is an isomorphism.

Definition 13.13 The signature of an oriented closed $4k$-dimensional manifold is the signature of its intersection form.

14. THE COMPLEX PROJECTIVE SPACE $\mathbb{CP}^n$

The set of 1-dimensional complex subspaces of $\mathbb{C}^{n+1}$ is denoted by $\mathbb{CP}^n$, and is called the complex projective n-dimensional space.

For $\mathbf{z} = (z_0, z_1, \ldots, z_n) \in \mathbb{C}^{n+1} - \{0\}$ let $\pi(\mathbf{z}) = [z_0, z_1, \ldots, z_n]$ denote the "point" $\mathbb{C}\mathbf{z} \in \mathbb{CP}^n$ spanned by $\mathbf{z}$. We give $\mathbb{CP}^n$ the quotient topology with respect to π: a set $U \subseteq \mathbb{CP}^n$ is open if and only if $\pi^{-1}(U) \subseteq \mathbb{C}^{n+1} - \{0\}$ is open. In particular there are open subsets of $\mathbb{CP}^n$,

$$U_j = \{[z_0, \ldots, z_n] \in \mathbb{CP}^n \mid z_j \neq 0\}$$

and the homeomorphisms $h_j: U_j \to \mathbb{C}^n$,

(1) $$h_j([z_0, \ldots, z_n]) = \left(z_0/z_j, \ldots, \widehat{z_j/z_j}, \ldots, z_n/z_j\right)$$

with inverses

(2) $$h_j^{-1}(w_1, \ldots, w_n) = [w_1, \ldots, 1, \ldots, w_n].$$

The transition functions $h_k \circ h_j^{-1}$ have coordinate functions of the form w_l/w_m or $1/w_m$. The atlas $\mathcal{H} = \{(U_j, h_j)\}$ gives $\mathbb{CP}^n$ the structure of a *complex* (analytic or holomorphic) manifold, because the transition functions are holomorphic. In the following, however, we shall mainly use the underlying structure as a smooth manifold, by interpreting $\mathcal{H}$ as a C^∞-atlas upon identifying $\mathbb{C}^n$ with $\mathbb{R}^{2n}$.

Example 14.1 (The Riemann sphere and Hopf fibration) Since $\mathbb{C} \times \mathbb{R}$ can be identified with $\mathbb{R}^3$, the unit sphere S^2 can be written as

$$S^2 = \{(z, t) \in \mathbb{C} \times \mathbb{R} \mid |z|^2 + t^2 = 1\}$$

with north pole $p_+ = (0, 1)$ and south pole $p_- = (0, -1)$. The equatorial plane $\mathbb{C} \times \{0\}$ is identified with $\mathbb{C}$. The stereographic projections $\psi_\pm: S^2 - \{p_\pm\} \to \mathbb{C}$ map p to the points of intersection between the equatorial plane and the line through $p_\pm$ and p. A straightforward calculation gives, for $(z, t) \in S^2 - \{p_\pm\}$,

$$\psi_+(z, t) = \frac{z}{1 - t}, \quad \psi_-(z, t) = \frac{z}{1 + t}.$$

The $\psi_\pm$ are diffeomorphisms with inverses

$$\psi_\pm^{-1}(w) = \left(\frac{2w}{|w|^2 + 1}, \pm \frac{|w|^2 - 1}{|w|^2 + 1}\right).$$

The transition function $\psi_- \circ \psi_+^{-1}: \mathbb{C} - \{0\} \to \mathbb{C} - \{0\}$ is easily calculated to be

$$\psi_- \circ \psi_+^{-1}(w) = (\overline{w})^{-1}.$$

If we orient $\mathbb{C}$ in the usual manner and consider S^2 as the boundary of D^3 with standard orientation from $\mathbb{R}^3$, then ψ_- is orientation-preserving and ψ_+ orientation-reversing (check at the poles!). This inspires us to replace ψ_+ by its conjugate $\overline{\psi}_+ \colon S^2 - \{p_+\} \to \mathbb{C}$. Then the transisition functions between $\overline{\psi}_+$ and ψ_- are inversion in the multiplicative group of $\mathbb{C}$, so the atlas on S^2 consisting of $\overline{\psi}_+$ and ψ_- gives S^2 the structure of a 1-dimensional complex manifold (Riemann surface).

The classical Hopf map $\eta \colon S^3 \to S^2$ is given by

$$(3) \qquad\qquad\qquad \eta(z_0, z_1) = (2\bar{z}_0 z_1, |z_0|^2 - |z_1|^2).$$

Hopf discovered (in 1931) that η is not homotopic to a constant map.

The Riemann surface of Example 14.1 is holomorphically equivalent to $\mathbb{CP}^1$. Indeed, by (2), the compositions

$$S^2 - \{p_-\} \xrightarrow{\psi_-} \mathbb{C} \to \mathbb{CP}^1 \quad \text{and} \quad S^2 - \{p_+\} \xrightarrow{\overline{\psi}_+} \mathbb{C} \xrightarrow{h_1^{-1}} \mathbb{CP}^1$$

are

$$h_0^{-1} \circ \psi_-(z, t) = [1, z/(1+t)], \quad h_1^{-1} \circ \overline{\psi}_+(z, t) = [\bar{z}/(1-t), 1].$$

These expressions agree when $t \in (-1, 1)$, so we can define a homeomorphism

$$(4) \qquad \Psi : S^2 \to \mathbb{CP}^1; \quad \Psi(z, t) = \begin{cases} [1+t, z], & (z, t) \neq (0, -1) \\ [\bar{z}, 1-t], & (z, t) \neq (0, 1). \end{cases}$$

Actually Ψ is holomorphic with holomorphic inverse. The complex projective plane $\mathbb{CP}^1$ is often identified with $\mathbb{C} \cup \{\infty\}$ by letting $[z_0, z_1] \in \mathbb{CP}^1$ correspond to $z_0^{-1} z_1$, and assigning ∞ to $z_0 = 0$. In this identification $\Psi \colon S^2 \to \mathbb{C} \cup \{\infty\}$ becomes the stereographic projection ψ_- from the south pole p_-, extended by mapping p_- to ∞.

One may generalize the Hopf fibration to the map

$$(5) \qquad\qquad \pi \colon S^{2n+1} \to \mathbb{CP}^n; \quad \pi(z_0, \ldots, z_n) = [z_0, \ldots, z_n].$$

Its fiber $\pi^{-1}(p)$ is the unit circle in the complex line $p \in \mathbb{CP}^n$. For $n = 1$ this is nothing but the Hopf fibration of Example 14.1. Indeed with the notation of (3), (4)

$$\Psi^{-1}([z_0, z_1]) = \Psi^{-1}\big([1, z_0^{-1} z_1]\big) = \psi_-^{-1}\big(z_0^{-1} z_1\big) = \big(2\bar{z}_0 z_1, |z_0|^2 - |z_1|^2\big).$$

The unit circle $S^1 \subseteq \mathbb{C}$ acts on the sphere $S^{2n+1} \subseteq \mathbb{C}^{n+1}$ by coordinatewise multiplication,

$$(6) \qquad\qquad S^1 \times S^{2n+1} \to S^{2n+1}; \quad (\lambda, \mathbf{z}) \mapsto \lambda \mathbf{z}.$$

The action is free, and the set of orbits of this action is precisely $\mathbb{CP}^n$ in the sense that the orbit space S^{2n+1}/S^1 is homeomorphic to $\mathbb{CP}^n$.

Theorem 14.2 *The cohomology of $\mathbb{CP}^n$ is*

$$H^{2j}(\mathbb{CP}^n) = \mathbb{R} \quad \text{for} \quad 0 \leq j \leq n$$
$$H^k(\mathbb{CP}^n) = 0 \quad \text{otherwise.}$$

Proof. The embedding $\mathbb{C}^n \subset \mathbb{C}^{n+1}$ induces an embedding of $\mathbb{CP}^{n-1}$ into $\mathbb{CP}^n$, and we can use Proposition 13.11 on the pair $(\mathbb{CP}^n, \mathbb{CP}^{n-1})$. We can assume the result for $\mathbb{CP}^{n-1}$ and that $n \geq 2$, since the cohomology of $\mathbb{CP}^1 = S^2$ was given in Example 9.29. The complement

$$\mathbb{CP}^n - \mathbb{CP}^{n-1} = U_n$$

is by (1) and (2) homeomorphic with $\mathbb{R}^{2n}$. The exact cohomology sequence takes the form

$$\cdots \to H^q_c(\mathbb{R}^{2n}) \xrightarrow{i_*} H^q(\mathbb{CP}^n) \xrightarrow{j^*} H^q(\mathbb{CP}^{n-1}) \xrightarrow{\delta} H^{q+1}_c(\mathbb{R}^{2n}) \to \cdots.$$

We know from Lemma 13.2 that $H^q_c(\mathbb{R}^{2n})$ is non-zero only for $q = 2n$, when it is a copy of $\mathbb{R}$, and the result follows easily. $\qquad \square$

It follows from Theorem 14.2 that the general Hopf map $\pi \colon S^{2n+1} \to \mathbb{CP}^n$ cannot have a section $s \colon \mathbb{CP}^n \to S^{2n+1}$: if s existed with $\pi \circ s = \mathrm{id}$, then

$$H^2(\mathbb{CP}^n) \xrightarrow{\pi^*} H^2(S^{2n+1}) \xrightarrow{s^*} H^2(\mathbb{CP}^n)$$

would be the identity, but this is impossible as $H^2(\mathbb{CP}^n) \neq 0$ and $H^2(S^{2n+1}) = 0$. Since $H^{2n}(\mathbb{CP}^n) = \mathbb{R}$ we know from Exercise 10.4 that $\mathbb{CP}^n$ is an oriented manifold, and hence from Theorem 13.5 that the bilinear map

$$H^q(\mathbb{CP}^n) \times H^{2n-q}(\mathbb{CP}^n) \to \mathbb{R}$$

given by the wedge product (followed by the integration isomorphism) is a dual (non-singular) pairing. In particular, the generator of $H^{2p}(\mathbb{CP}^n)$ and the generator of $H^{2n-2p}(\mathbb{CP}^n)$ has non-trivial product.

Theorem 14.3 *The cohomology algebra $H^*(\mathbb{CP}^n)$ is a truncated polynomial algebra*
$$H^*(\mathbb{CP}^n) = \mathbb{R}[c]/(c^{n+1})$$
where c is a non-zero class in degree 2, and (c^{n+1}) the ideal generated by c^{n+1}.

Proof. We use induction over n, so suppose the theorem proved for $\mathbb{CP}^{n-1}$. The inclusion

$$j \colon \mathbb{CP}^{n-1} \to \mathbb{CP}^n$$

induces the map

$$j^*: H^i(\mathbb{CP}^n) \to H^i(\mathbb{CP}^{n-1}).$$

The proof of Theorem 14.2 shows that j^* is an isomorphism for $i \leq 2n - 2$. Hence $c^{n-1} \neq 0$ in $H^{2n-2}(\mathbb{CP}^n)$, and the pairing

$$H^2(\mathbb{CP}^n) \times H^{2n-2}(\mathbb{CP}^n) \to H^{2n}(\mathbb{CP}^n)$$

implies that $c^n \neq 0$. $\qquad\qquad\square$

It is a consequence of the above theorem that $\mathbb{CP}^n$ supports an orientation-reversing diffeomorphism only if n is odd. Indeed, a map $f: \mathbb{CP}^n \to \mathbb{CP}^n$ induces a homomorphism

$$f^*: H^*(\mathbb{CP}^n) \to H^*(\mathbb{CP}^n).$$

If $f^*(c) = ac$, then

$$f^*(c^j) = a^j c^j \quad (0 \leq j \leq n).$$

For $j = n$ we get $\deg(f) = a^n$. If n is even then $\deg(f) \geq 0$. In this case there are no orientation-reversing diffeomorphisms of $\mathbb{CP}^n$. If n is odd, on the other hand, then complex conjugation is an orientation-reversing diffeomorphism of $\mathbb{CP}^n$:

$$f([z_0, z_1, \dots, z_n]) = [\bar{z}_0, \bar{z}_1, \dots, \bar{z}_n]$$

has $a = -1$ as one sees by restricting to $S^2 = \mathbb{CP}^1$.

We close this chapter by constructing closed differential forms which represent a basis for $H^{2j}(\mathbb{CP}^n)$. This requires some preparations.

Consider a unit vector $v \in S^{2n+1} \subset \mathbb{C}^{n+1}$ with image $p = \pi(v)$. Now iv is a unit vector tangent to the S^1-orbit of v (which was precisely the unit circle in the 1-dimensional $\mathbb{C}$-subspace $\mathbb{C}v$). The orthogonal complement $(\mathbb{C}v)^\perp$ with respect to the usual hermitian inner product on $\mathbb{C}^{n+1}$ is a real $2n$-dimensional subspace of $T_v S^{2n+1}$, and is orthogonal to iv with respect to the real inner product on $T_v S^{2n+1}$ induced from $\mathbb{C}^{n+1} = \mathbb{R}^{2n+2}$.

Lemma 14.4

(i) *Let $p \in \mathbb{CP}^n$ and $v \in \pi^{-1}(p) \subseteq S^{2n+1}$. There is an open neighborhood U around p in $\mathbb{CP}^n$ and a smooth map $s: U \to S^{2n+1}$ such that $s(p) = v$ and $\pi \circ s = \mathrm{id}_U$.*

(ii) *Let $v \in S^{2n+1}$ and $p = \pi(v)$. The differential $D_v \pi$ induces an $\mathbb{R}$-linear isomorphism from $(\mathbb{C}v)^\perp$ to $T_p \mathbb{CP}^n$.*

(iii) *There exists a well-defined structure on $T_p \mathbb{CP}^n$ as an n-dimensional $\mathbb{C}$-vector space with hermitian inner product, which makes the isomorphisms of (ii) into $\mathbb{C}$-linear isometries.*

Proof. Choose $U = U_j$ such that $p \in U$, cf. (1). Consider the map $s_j : U_j \to S^{2n+1}$ given by

(7) $$s_j([z_0, \ldots, z_j, \ldots, z_n]) = \left(\sum_{k=0}^{n} |z_k|^2 \right)^{-1/2} (z_0, \ldots, z_j, \ldots, z_n)$$

where $z_j = 1$. We let s be this map composed with multiplication by the unique $\lambda \in S^1$ such that $\lambda s_j(p) = v$. The chain rule implies that

$$D_v \pi : T_v S^{2n+1} \to T_p \mathbb{C}\mathrm{P}^n$$

is surjective. The kernel has real dimension 1, and since it contains iv, assertion (ii) follows.

Let $\phi : S^{2n+1} \to S^{2n+1}$ be the diffeomorphism given by multiplication by $\lambda \in S^1$. The chain rule gives a commutative diagram

$$T_v S^{2n+1} \xrightarrow{D_v \phi} T_{\lambda v} S^{2n+1}$$

$$D_v \pi \searrow \qquad \swarrow D_{\lambda v} \pi$$

$$T_p \mathbb{C}\mathrm{P}^n$$

Multiplication by λ can be considered as an $\mathbb{R}$-linear map $\mathbb{C}^{n+1} \to \mathbb{C}^{n+1}$, and $D_v \phi$ is its restriction to $T_v S^{2n+1}$. We can conclude that $D_v \phi$ acts on $(\mathbb{C}v)^{\perp}$ by multiplying by λ. Since this is a $\mathbb{C}$-linear isometry, assertion (iii) follows. $\square$

Let V be a $\mathbb{C}$-vector space and let rV denote the underlying $\mathbb{R}$-vector space. A $\mathbb{C}$-linear map $F : V \to W$ induces an $\mathbb{R}$-linear map $rF : rV \to rW$.

Lemma 14.5 *If V is a finite-dimensional $\mathbb{C}$-vector space and $F : V \to V$ is a $\mathbb{C}$-linear map, then*

$$\det(rF) = |\det F|^2.$$

Proof. We use induction on $m = \dim_{\mathbb{C}} V$. If $m = 1$ then F is multiplication by some $z \in \mathbb{C}$. The matrix for rF, with respect to a basis of the form b, ib for rV, is

$$\begin{pmatrix} x & -y \\ y & x \end{pmatrix}$$

where $x = \mathrm{Re}\, z$ and $y = \mathrm{Im}\, z$. Since $\det(rF) = x^2 + y^2 = |z|^2$ the formula holds in this case.

If $m \geq 2$ we can choose a complex line $V_0 \subset V$ with $F(V_0) \subseteq V_0$ (generated by an eigenvector of F). F induces $\mathbb{C}$-linear maps

$$F_0 : V_0 \to V_0, \qquad F_1 : V/V_0 \to V/V_0,$$

and we may assume the formula for both F_0 and F_1. Since

$$\det F = (\det F_0)(\det F_1), \qquad \det rF = (\det rF_0)(\det rF_1),$$

we are done. $\square$

Corollary 14.6 *If V is an m-dimensional $\mathbb{C}$-vector space then rV has a natural orientation with the property that any basis $b_1, \ldots, b_m$ over $\mathbb{C}$ gives rise to a positive basis $\{b_1, ib_1, b_2, ib_2, \ldots, b_m, ib_m\}$ for rV.*

Proof. Let $b'_1, \ldots, b'_m$ be another basis of V. We can apply Lemma 14.5 to the $\mathbb{C}$-linear map F determined by $F(b_j) = b'_j$ $(1 \leq j \leq m)$. Since $\det(rF) > 0$, the assertion follows. $\qquad\square$

Proposition 14.7 *Let V be an m-dimensional $\mathbb{C}$-vector space with hermitian inner product $\langle\,,\,\rangle$. Then*

(i) $g(v_1, v_2) = \mathrm{Re}\langle v_1, v_2\rangle$ *defines an inner product on rV, and*

$$\omega(v_1, v_2) = g(iv_1, v_2) = -\mathrm{Im}\langle v_1, v_2\rangle$$

defines an element of $\mathrm{Alt}^2(rV)$.

(ii) *If $\mathrm{vol} \in \mathrm{Alt}^{2m}(rV)$ denotes the volume element determined by g and the orientation from Corollary 14.6, then $\omega^m = m!\,\mathrm{vol}$, where $\omega^m = \omega \wedge \omega \wedge \ldots \wedge \omega$ (m factors).*

Proof. We leave (i) to the reader. An orthonormal basis $b_1, \ldots, b_m$ of V with respect to $\langle\,,\,\rangle$ gives rise to the positively oriented orthonormal basis of rV with respect to g,

$$b_1, \ ib_1, \ b_2, \ ib_2, \ldots, b_m, \ ib_m.$$

Let $\epsilon_1, \ \tau_1, \ \epsilon_2, \ \tau_2, \ldots, \epsilon_m, \ \tau_m$ denote the dual basis for $\mathrm{Alt}^1(rV)$. Since $\omega(b_j, \ ib_j) = -\omega(ib_j, \ b_j) = 1$, and ω vanishes on all other pairs of vectors, Lemma 2.13 shows that

$$(8) \qquad\qquad \omega = \sum_{j=1}^{m} \epsilon_j \wedge \tau_j.$$

Furthermore, $\mathrm{vol} = \epsilon_1 \wedge \tau_1 \wedge \epsilon_2 \wedge \tau_2 \wedge \ldots \wedge \epsilon_m \wedge \tau_m$, because both sides are 1 on the basis above. Direct computation gives $\omega^m = m!\,\mathrm{vol}$. (See Appendix B.) $\qquad\square$

Note that if $V = \mathbb{C}^{n+1}$, with the usual hermitian scalar product and standard basis $e_0, \ldots, e_n$, then (8) takes the form

$$(9) \qquad\qquad \omega_{\mathbb{C}^{n+1}} = \sum_{j=0}^{n} dx_j \wedge dy_j \in \Omega^2(r\mathbb{C}^{n+1})$$

where x_j and y_j are the real and the imaginary components of the coordinate z_j.

We can apply Proposition 14.7 to $T_p\mathbb{C}P^n$ with the complex structure from Lemma 14.4.(iii). This gives us a real scalar product g_p on $T_p\mathbb{C}P^n$ and $\omega_p \in \mathrm{Alt}^2 T_p\mathbb{C}P^n$ for each $p \in \mathbb{C}P^n$.

Theorem 14.8 *The $\omega = \{\omega_p\}_{p\in\mathbb{C}\mathbf{P}^n}$ define a closed 2-form on $\mathbb{C}\mathbf{P}^n$ and $g = \{g_p\}_{p\in\mathbb{C}\mathbf{P}^n}$ is a Riemannian metric on $\mathbb{C}\mathbf{P}^n$ (the Fubini–Study metric). Moreover,*

$$\omega^n = n!\,\mathrm{vol}_{\mathbb{C}\mathbf{P}^n},$$

where $\mathrm{vol}_{\mathbb{C}\mathbf{P}^n}$ is the volume form determined by g and the natural orientation from Corollary 14.6.

Proof. Let $p \in \mathbb{C}\mathbf{P}^n$ and $v \in S^{2n+1}$ with $\pi(v) = p$. Choose $s: U \to S^{2n+1}$ with $\pi \circ s = \mathrm{id}_U$ and $s(p) = v$ as in Lemma 14.4. We will show that

(10) $$\omega_{|U} = s^*(\omega_{\mathbb{C}^{n+1}}).$$

By (9) we have $d\omega_{\mathbb{C}^{n+1}} = 0$. Hence (10) will show that ω is a closed 2-form on $\mathbb{C}\mathbf{P}^n$. If $w_\nu \in T_p\mathbb{C}\mathbf{P}^n$, $\nu = 1,2$, and $D_p s(w_\nu) = t_\nu + u_\nu$, where t_ν is a tangent vector to the fiber in S^{2n+1} over p and $u_\nu \in (\mathbb{C}v)^\perp$, then

$$w_\nu = D_v\pi \circ D_p s(w_\nu) = D_v\pi(t_\nu + u_\nu) = D_v\pi(u_\nu).$$

Since $\mathrm{Alt}^2(D_v\pi)(\omega_p)$ is the restriction to $r(\mathbb{C}v)^\perp$ of $\omega_{\mathbb{C}^{n+1}}$, we have

$$\omega_p(w_1, w_2) = \omega_{\mathbb{C}^{n+1}}(u_1, u_2),$$

and (10) follows from

$$s^*(\omega_{\mathbb{C}^{n+1}})(w_1, w_2) = \omega_{\mathbb{C}^{n+1}}(D_p s(w_1), D_p s(w_2))$$
$$= \omega_{\mathbb{C}^{n+1}}(t_1 + u_1, t_2 + u_2) = \omega_{\mathbb{C}^{n+1}}(u_1, u_2).$$

In the final equality we used that t_1 and t_2 are orthogonal to respectively u_2 and u_1 in $\mathbb{C}^{n+1}$, and the fact that t_1 and t_2 are linearly dependent over $\mathbb{R}$.

When showing the smoothness of g, it suffices, since $g_p(w_1, w_2) = -\omega_p(iw_1, w_2)$, to show for a smooth tangent vector field X on an open set $U \subseteq \mathbb{C}\mathbf{P}^n$ that iX is smooth too. This is left to the reader. The last part of the theorem follows directly from Proposition 14.7. $\square$

Corollary 14.9 *Let ω be the closed 2-form on $\mathbb{C}\mathbf{P}^n$ constructed in Theorem 14.8. The j-th exterior power ω^j represents a basis element of $H^{2j}(\mathbb{C}\mathbf{P}^n)$ when $1 \leq j \leq n$.*

Proof. The class in $H^{2n}(\mathbb{C}\mathbf{P}^n) \cong \mathbb{R}$ determined by $\mathrm{vol}_{\mathbb{C}\mathbf{P}^n}$ is non-trivial. Since $[\omega] \in H^2(\mathbb{C}\mathbf{P}^n)$ we have

$$[\omega]^n = n![\mathrm{vol}_{\mathbb{C}\mathbf{P}^n}] \in H^{2n}(\mathbb{C}\mathbf{P}^n).$$

Therefore $[\omega]^n \neq 0$ and thus $[\omega]^j \neq 0$, for $j \leq n$. The assertion now follows from Theorem 14.2. $\square$

Example 14.10 (The Hopf fibration again) Let $z_\nu = x_\nu + iy_\nu$, $\nu = 0, 1$. The Hopf fibration η from (3) is the restriction to $S^3 \subseteq \mathbb{R}^4$ of the map $h : \mathbb{R}^4 \to \mathbb{R}^3$ given by

$$h(x_0, y_0, x_1, y_1) = \begin{pmatrix} 2(x_0 x_1 + y_0 y_1) \\ 2(x_0 y_1 - x_1 y_0) \\ x_0^2 + y_0^2 - x_1^2 - y_1^2 \end{pmatrix}$$

with Jacobian matrix

$$2 \begin{pmatrix} x_1 & y_1 & x_0 & y_0 \\ y_1 & -x_1 & -y_0 & x_0 \\ x_0 & y_0 & -x_1 & -y_1 \end{pmatrix}.$$

If $v \in S^3$ has real coordinates (x_0, y_0, x_1, y_1), then iv will have coordinates $(-y_0, x_0, -y_1, x_1)$. In $(\mathbb{C}v)^\perp$ we have the positively oriented real orthonormal basis given by

$$\mathbf{b} = (-x_1, y_1, x_0, -y_0)$$
$$i\mathbf{b} = (-y_1, -x_1, y_0, x_0).$$

Their images under $D_v\eta \colon T_v S^3 \to T_{\eta(v)} S^2$ can be found by taking the matrix product with the Jacobian matrix above:

$$\frac{1}{2} D_v\eta(\mathbf{b}) = \begin{pmatrix} x_0^2 - y_0^2 - x_1^2 + y_1^2 \\ -2x_0 y_0 - 2x_1 y_1 \\ -2x_0 x_1 + 2y_0 y_1 \end{pmatrix}, \quad \frac{1}{2} D_v\eta(i\mathbf{b}) = \begin{pmatrix} 2x_0 y_0 - 2x_1 y_1 \\ x_0^2 - y_0^2 + x_1^2 - y_1^2 \\ -2x_0 y_1 - 2y_0 x_1 \end{pmatrix}.$$

A straightforward calculation (use that $\left(x_0^2 + y_0^2 + x_1^2 + y_1^2\right)^2 = 1$) shows that $\frac{1}{2} D_v\eta(\mathbf{b})$ and $\frac{1}{2} D_v\eta(i\mathbf{b})$ define an orthonormal basis of $T_{\eta(v)} S^2$ with respect to the Riemannian metric inherited from $\mathbb{R}^3$. Since $\Psi \circ \eta = \pi \colon S^3 \to \mathbb{CP}^1$ with $\Psi \colon S^2 \to \mathbb{CP}^1$ the holomorphic equivalence from (4), the chain rule gives that

$$D_{\eta(v)} \Psi \colon T_{\eta(v)} S^2 \to T_p \mathbb{CP}^1 \quad (p = \pi(v))$$

with respect to the listed orthonormal bases has matrix $\mathrm{diag}\,(1/2, 1/2)$. Hence

$$\Psi^*(\omega) = \frac{1}{4} \,\mathrm{vol}_{S^2},$$

where $\omega = \mathrm{vol}_{\mathbb{CP}^1}$ by Theorem 14.8. In particular we have

$$\mathrm{Vol}\left(\mathbb{CP}^1\right) = \frac{1}{4} \,\mathrm{Vol}\left(S^2\right) = \pi.$$

It follows furthermore that $\mathbb{CP}^1$ with the Riemannian metric g is isometric with the sphere of radius $\frac{1}{2}$ in $\mathbb{R}^3$.

15. FIBER BUNDLES AND VECTOR BUNDLES

Definition 15.1 A *fiber bundle* consists of three topological spaces E, B, F and a continuous map $\pi \colon E \to B$, such that the following condition is satisfied: Each $b \in B$ has an open neighborhood U_b and a homeomorphism

$$h \colon U_b \times F \to \pi^{-1}(U_b)$$

such that $\pi \circ h = \mathrm{proj}_1$.

The space E is called the *total space*, B the *base space* and F the *(typical) fiber*. The pre-image $\pi^{-1}(x)$, frequently denoted by F_x, is called *the fiber* over x. A fiber bundle is said to be *smooth*, if E, B and F are smooth manifolds, π is a smooth map and the h above can be chosen to be diffeomorphisms. One may think of a fiber bundle as a continuous (smooth) family of topological spaces F_x (all of them homeomorphic to F), indexed by $x \in B$.

The most obvious example is the *product* fiber bundle $\varepsilon_B^F = (B \times F, B, F, \mathrm{proj}_1)$. In general, the condition of Definition 15.1 expresses that the "family" is *locally trivial*.

Example 15.2 (The canonical line bundle). At the beginning of Chapter 14 we considered the action of S^1 on S^{2n+1} with orbit space $\mathbb{CP}^n$. We view this as an action from the right, $\mathbf{z}.\lambda = (z_0\lambda, \dots, z_n\lambda)$. The circle acts also on $S^{2n+1} \times \mathbb{C}^k$, $(\mathbf{z}, \mathbf{u})\lambda = (\mathbf{z}\lambda, \lambda^{-1}\mathbf{u})$. The associated orbit space is denoted $S^{2n+1} \times_{S^1} \mathbb{C}^k$. The projection on the first factor gives a continuous map

$$\pi \colon S^{2n+1} \times_{S^1} \mathbb{C}^k \to \mathbb{CP}^n$$

with fiber $\mathbb{C}^k$. Similarly, if S^1 acts continuously (from the left) on any topological space F we get

$$\pi \colon S^{2n+1} \times_{S^1} F \to \mathbb{CP}^n.$$

This is a fiber bundle with fiber F. Indeed, we have the open sets U_j displayed at the beginning of Chapter 14 which cover $\mathbb{CP}^n$, and the smooth sections

$$s_j \colon U_j \to S^{2n+1}$$

from (14.7). We can define local trivializations by

$$\widehat{s}_j([\mathbf{z}], \mathbf{u}) = (s_j([\mathbf{z}]), \mathbf{u}) \in \pi^{-1}(U_j)$$

for $[\mathbf{z}] \in U_j$ and $\mathbf{u} \in F$. If F is a smooth manifold with smooth S^1-action then we obtain a smooth fiber bundle.

If we take $F = \mathbb{C}$ with its usual action of S^1 then we obtain the *dual* Hopf bundle, or canonical line bundle, over $\mathbb{CP}^n$. It will be denoted

$$H_n = S^{2n+1} \times_{S^1} \mathbb{C}$$

It is a vector bundle in the sense of Definition 15.4 below.

Example 15.3 Over the real projective space $\mathbb{RP}^n$ we have similar bundles $S^n \times_{S^0} F$, where $S^0 = \{\pm 1\}$. In particular $D(H) = S^1 \times_{S^0} D^1$ is the Möbius band.

Definition 15.4 A *vector bundle* $\xi = (E, B, V, \pi)$ is a fiber bundle where the typical fiber V and each $\pi^{-1}(x)$ are vector spaces, and where the local homeomorphism $h: U_b \times V \to \pi^{-1}(U_b)$ can be chosen so that $h(x, -): V \to \pi^{-1}(x)$ is a linear isomorphism for each $x \in U_b$.

Vector bundles can be real, complex or quaternion depending on which category V and $h(x, -)$ in Definition 15.4 belong to. For the time being we concentrate on real vector bundles. A smooth vector bundle is a vector bundle that is also a smooth fiber bundle. The dimension of a vector bundle is the dimension of the fiber. Vector bundles of dimension 1 are called line bundles.

We mostly denote vector bundles by small Greek letters. If ξ is a vector bundle then $E(\xi)$ will denote its total space and $F_b(\xi)$, or just ξ_b, its fiber over b. If $W \subset B$ then we write $\xi_{|W}$ for the restriction of ξ to W, i.e. $E(\xi_{|W}) = \pi_\xi^{-1}(W)$.

Example 15.5 (The tangent bundle) Let $M^n \subset \mathbb{R}^{n+k}$ be a smooth manifold. Consider

$$TM = \{(p, v) \in M \times \mathbb{R}^{n+k} \mid v \in T_pM\}, \qquad \pi(p, v) = p.$$

The fiber over $p \in M$ is the tangent space T_pM. We show that the triple $\tau_M = (TM, M, \pi)$ is a vector bundle. Let $b \in M$. Choose a parametrization (U, g) around $b, g: W \to U, \quad W \subseteq \mathbb{R}^n$ and let

$$h: U \times \mathbb{R}^n \to \pi^{-1}(U); \quad h(x, v) = Dg_{g^{-1}(x)}(v).$$

This gives the required local triviality.

Example 15.6 (The normal bundle) Let $M^n \subset \mathbb{R}^{n+k}$ be a smooth manifold. Let

$$N_p(M) = (T_pM)^\perp$$

be the orthogonal complement to $T_pM \subset \mathbb{R}^{n+k}$. Set

$$E(\nu_M) = \bigcup_{p \in M} N_p(M) \subset M \times \mathbb{R}^{n+k}; \qquad \pi(v) = p \text{ when } v \in N_p(M).$$

We must show that ν_M is locally trivial. For each $p_0 \in M$, the proof of Lemma 9.21 produced vector fields

$$X_1, \ldots, X_n, Y_1, \ldots, Y_k: W \to \mathbb{R}^{n+k}$$

defined in a neighborhood W of p_0, with the property that they are orthonormal for each $p \in W$ and such that $X_1(p), \ldots, X_n(p) \in T_pM$. Hence $Y_1(p), \ldots, Y_k(p) \in N_p(M)$ is a basis and the map

$$h: W \times \mathbb{R}^k \to \pi^{-1}(W); \quad h(p, \mathbf{t}) = \sum_{i=1}^{k} t_i Y_i(p)$$

is a local trivialization.

Definition 15.7

    (i) A map $(f, \hat{f})$ between (smooth) fiber bundles (E, B, π) and (E', B', π') is a pair of (smooth) continuous maps

$$f: B \to B', \quad \hat{f}: E \to E'$$

such that $\pi' \circ \hat{f} = f \circ \pi$.

    (ii) A homomorphism between (smooth) vector bundles ξ and ξ' is a (smooth) fiber bundle map such that $\hat{f}: \pi^{-1}(x) \to (\pi')^{-1}(f(x))$ is linear for all $x \in B$.

Example 15.8 A smooth map $f: M \to M'$ between smooth manifolds induces a map of tangent bundles (f, Tf), where

$$T_pf = D_pf: T_pM \to T_pM'$$

is the derivative of f.

Definition 15.9 Vector bundles ξ and η over the same base space B are called *isomorphic*, if there exist homomorphisms $(\mathrm{id}_B, \hat{f})$ and $(\mathrm{id}_B, \hat{g})$ between them such that $\hat{f} \circ \hat{g} = \mathrm{id} = \hat{g} \circ \hat{f}$. A vector bundle which is isomorphic to a product bundle is called *trivial*, and a specific isomorphism is called a *trivialization*.

In the above definition the homomorphisms $\hat{f}$ and $\hat{g}$ are assumed to be smooth when the vector bundles are smooth. The next lemma is a convenient tool for deciding if two bundles are isomorphic.

Lemma 15.10 *A (smooth) continuous map $\hat{f}: E(\xi) \to E(\eta)$ of (smooth) vector bundles over B, which map the fiber $F_b(\xi)$ isomorphically onto the fiber $F_1(\eta)$ is a (smooth) isomorphism.*

Proof. Since $\hat{f}$ is a bijection, it is sufficient to show that $\hat{f}^{-1}$ is a (smooth) homomorphism of vector bundles (over id_B). We need to check that $\hat{f}^{-1}$ is continuous (smooth). Since $\hat{f}$ is a fiberwise isomorphism, it is enough to examine

$$\hat{f}^{-1}: \pi_\eta^{-1}(U) \to \pi_\xi^{-1}(U)$$

where ξ and η are trivial over U. Let
$$h\colon U \times \mathbb{R}^n \to \pi_\xi^{-1}(U) \quad \text{and} \quad k\colon U \times \mathbb{R}^n \to \pi_\eta^{-1}(U)$$
be isomorphisms. Then
$$F = k \circ \hat{f} \circ h^{-1}\colon U \times \mathbb{R}^n \to U \times \mathbb{R}^n$$
is an isomorphism of trivial bundles and it has the form
$$F(x, v) = (x, F_2(x, v)), \quad x \in U.$$
The map $x \to F_2(x, -)$ defines a map
$$\mathrm{ad}(F_2)\colon U \to GL_n(\mathbb{R}).$$
Conversely such a map induces a homomorphism $\hat{f}\colon \pi_\xi^{-1}(U) \to \pi_\eta^{-1}(U)$. Note that
$$\mathrm{ad}(F_2)^{-1}\colon U \to GL_n(\mathbb{R})$$
determines F^{-1}. Finally, it is easy to see that f is continuous (smooth) if and only if $\mathrm{ad}(F_2)$ is. The lemma now follows because matrix inversion $(-)^{-1}\colon GL_n(\mathbb{R}) \to GL_n(\mathbb{R})$ is a smooth map. $\square$

Definition 15.11 The direct sum $\xi \oplus \eta$ of two vector bundles over the same base space B is the vector bundle over B with total space
$$E(\xi \oplus \eta) = \big\{(v, w) \in E(\xi) \times E(\eta) \;\big|\; \pi_\xi(v) = \pi_\eta(w)\big\}$$
and projection $\pi_{\xi \oplus \eta}(v, w) = \pi_\xi(v) = \pi_\eta(w)$. The fiber $(\xi \oplus \eta)_b$ is equal to $\xi_b \oplus \eta_b$.

Definition 15.12 An inner product on a (smooth) vector bundle ξ is a (smooth) map $\phi\colon E(\xi \oplus \xi) \to \mathbb{R}$ such that $\phi\colon F_b(\xi) \oplus F_b(\xi) \to \mathbb{R}$ is an inner product on each fiber $F_b(\xi)$.

Proposition 15.13 *Every vector bundle over a compact B has an inner product.*

Proof. Choose local trivializations
$$h_i\colon U_i \times \mathbb{R}^n \to \pi_\xi^{-1}(U_i)$$
where $U_1, \ldots, U_r$ cover B, and choose a partition of unity $\{\alpha_i\}_{i=1}^r$ with $\mathrm{supp}(\alpha_i) \subset U_i$.
The usual inner product in $\mathbb{R}^n$ induces an inner product in $U_i \times \mathbb{R}^n$ and hence, via h_i, an inner product $\phi_i \; \pi_\xi^{-1}(U_i)$. Now
$$\phi(v, w) = \sum \phi_i(v, w) \alpha_i(\pi(v))$$
is an inner product in ξ. $\square$

An inner product in the tangent bundle τ_M of a smooth manifold is the same as a *Riemannian metric* on M (cf. Definition 9.15).

Remark 15.14 In the above proposition we used the existence of a continuous partition of unity, i.e. continuous functions $\alpha_i: B \to [0,1]$ with $\operatorname{supp}(\alpha_i) \subset U_i$ and $\sum \alpha_i(b) = 1$ for all $b \in B$. When B is a smooth manifold the existence of a smooth partition of unity is proved in Appendix A. More generally, a Hausdorff space B is called *paracompact* if every open covering $\{U_\alpha\}$ has an open refinement $\{V_\beta\}$ which is locally finite. For a given open cover $\{U_\alpha\}$ of a paracompact space there exists a partition of unity subordinated to $\{U_\alpha\}$, i.e. continuous functions $s_\alpha: B \to [0,1]$ with $\operatorname{supp}(s_\alpha) \subset U_\alpha$ and such that each $b \in B$ has an open neighborhood V_b for which $\{\alpha \mid s_{\alpha|V_b} \neq 0\}$ is a finite set.

Definition 15.15 A (smooth) *section* in a (smooth) fiber bundle $(E, B, F; \pi)$ is a (smooth) map $s: B \to E$ such that $\pi \circ s = E$.

The set of sections of a vector bundle ξ is a vector space $\Gamma(\xi)$. One adds sections by using the vector space structure of each fiber. The origin in $\Gamma(\xi)$ is the *zero section* which to $b \in B$ assigns the origin in the fiber ξ_b. If ξ is a smooth vector bundle then we let $\Omega^0(\xi) \subset \Gamma(\xi)$ denote the subspace of smooth sections.

It follows from local triviality that in a neighbourhood U of each point of the base, we can find sections $s_1, \ldots, s_n \in \Gamma(\xi_{|_U})$ (or in $\Omega^0(\xi_{|U})$) such that $\{s_1(x), \ldots, s_n(x)\}$ is a basis of ξ_x. We call this a *frame*. If ξ has an inner product we may even choose sections locally so that $\{s_1(x), \ldots, s_n(x)\}$ is an orthogonal basis (Gram–Schmidt). We say that $\{s_1, \ldots, s_n\}$ is an orthonormal frame for ξ over U.

Let $(\hat{f}, \operatorname{id}_B)$ be a homomorphism from ξ to η, and let $\{s_i\}$, $\{t_i\}$ be frames over U. Then $\hat{f}_x: \xi_x \to \eta_x$ is represented by a matrix, and we obtain a map

(1) $$\operatorname{ad}(\hat{f}): U \to M_n(\mathbb{R})$$

depending on the given frames. In the smooth situation $\operatorname{ad}(\hat{f})$ is smooth. Note that $\hat{f}_x: \xi_x \to \eta_x$ is an isomorphism if and only if $\operatorname{ad}(\hat{f}_x) \in GL_n(\mathbb{R})$. If ξ and η have an inner product, and $\{s_i\}, \{t_i\}$ are orthonormal frames, then $\hat{f}_x$ is isometric precisely if $\operatorname{ad}(\hat{f}_x) \in O_n$, the orthogonal subgroup of $GL_n(\mathbb{R})$.

Lemma 15.16 *Let ξ and η be (smooth) vector bundles with inner product over the compact space B, and let $\hat{f}: \xi \to \eta$ be an isomorphism. Then there exists an $\epsilon > 0$ such that every homomorphism $\hat{g}: \xi \to \eta$ that satisfies $\|\hat{f}_b - \hat{g}_b\| < \epsilon$ for $b \in B$ is also an isomorphism.*

Proof. If ξ and η are trivial, then after choice of frames, $\hat{f}$ and $\hat{g}$ are represented by maps $\operatorname{ad}(\hat{f}): B \to GL_n(\mathbb{R})$ and $\operatorname{ad}(\hat{g}): B \to M_n(\mathbb{R})$. Since B is compact and $GL_n(\mathbb{R})$ is open, some ϵ-neighborhood of $\operatorname{ad}(\hat{f})(B)$ in $M_n(\mathbb{R})$ is still contained in $GL_n(\mathbb{R})$. But then $\operatorname{ad}(\hat{g})(B) \subset GL_n(\mathbb{R})$ when g satisfies the condition of the lemma. In general, we can cover B with a finite number of compact neighborhoods

over which the bundles are trivial, and take the minimum of the resulting epsilons.
$\hspace{14cm}\square$

Given two smooth vector bundles ξ and η, one might wonder if there is any essential distinction between the notions of continuous and smooth isomorphism. The next result shows that this is not the case. Along the same lines one may ask if each isomorphism class of continuous vector bundles over a compact manifold contains a smooth representative. This is indeed the case (cf. Exercise 15.8).

Lemma 15.17 *If two smooth vector bundles ξ and η over the compact manifold B are isomorphic as continuous bundles, then they are smoothly isomorphic.*

Proof. We choose a cover $U^1, \ldots, U^r$ of B and smooth local orthonormal frames $\mathbf{s}^i = (s_1^i, \ldots, s_n^i)$ and $\mathbf{t}^i = (t_1^i, \ldots, t_n^i)$ for ξ and η, over U^i.
A continuous isomorphism $\hat{f} \colon \xi \to \eta$ gives continuous maps

$$\mathrm{ad}(\hat{f}^i) \colon U^i \to GL_n(\mathbb{R}).$$

Let $G^i \colon U^i \to GL_n(\mathbb{R})$ be a smooth ϵ-approximation with $\|G^i(x) - \mathrm{ad}(\hat{f}_x^i)\| < \epsilon$ for $x \in U^i$. Construct a smooth homomorphism $\hat{g}^i \colon \pi_\xi^{-1}(U^i) \to \pi_\eta^{-1}(U^i)$ with $\mathrm{ad}(\hat{g}^i) = G^i$ by the formula

$$\hat{g}_b^i \left(\sum \lambda_k s^k(b) \right) = \sum_{k,\nu} t^k(b) \cdot G_{k\nu}^i(b) \cdot \lambda_\nu.$$

Then $\mathrm{ad}(\hat{g}^i) = G^i$, and $\|\hat{f}_b - \hat{g}_b^i\| < \epsilon$ for $b \in U^i$. We can then use a smooth partition of unity $\{\alpha_i\}_{i=1}^r$ with $\mathrm{supp}(\alpha_i) \subset U^i$ to define

$$\hat{g} \colon \xi \to \eta, \quad \hat{g}_b = \sum_{i=1}^r \alpha_i(b)\, \hat{g}_b^i.$$

Then $\|\hat{f}_b - \hat{g}_b\| = \|f_b - \Sigma\alpha_i(b)\hat{g}_b^i\| \le \Sigma\alpha_i(b)\|f_b - \hat{g}_b^i\| \le \Sigma\alpha_i(b)\epsilon = \epsilon$. With ϵ as in Lemma 15.16, $\hat{g}$ becomes an isomorphism on every fiber and hence a smooth isomorphism by Lemma 15.10.
$\hspace{14cm}\square$

In Examples 15.5 and 15.6 we constructed two vector bundles associated to a submanifold $M^n \subset \mathbb{R}^{n+k}$, namely the tangent bundle τ and the normal bundle ν. It is obvious that $\tau \oplus \nu$ is a trivial vector bundle. Indeed, by construction $\tau_p \oplus \nu_p = \mathbb{R}^{n+k}$ for every $p \in M$, and there is a globally defined frame for $\tau \oplus \nu$. Hence $\tau \oplus \nu \cong \varepsilon^{n+k}$, where ε^{n+k} is the trivial bundle over M of dimension $n + k$. We now give the general construction of *complements* of vector bundles.

Theorem 15.18 *Every vector bundle ξ over a compact base space B has a complement η, i.e. $\xi \oplus \eta \cong \varepsilon^N$ (for a suitably large N).*

Proof. Choose an open cover $U^1, \ldots, U^r$ of B admitting trivializations h_i of $\xi_{|U_i}$, and let $\{\alpha_i\}$ be a partition of unity with $\mathrm{supp}(\alpha_i) \subset U^i$. Denote by f^i the composite

$$\pi_\xi^{-1}(U^i) \xrightarrow{h_i^{-1}} U^i \times \mathbb{R}^n \xrightarrow{\mathrm{proj}_2} \mathbb{R}^n,$$

and define

(2)
$$S \colon E(\xi) \to B \times \mathbb{R}^{nr};$$
$$S(v) = \big(\pi_\xi(v), \alpha_1\big(\pi_\xi(v)\big)f^1(v), \ldots, \alpha_r\big(\pi_\xi(v)\big)f^r(v)\big).$$

This is a fiberwise map and gives a homomorphism $S \colon \xi \to \varepsilon^{nr}$ which is an inclusion on each fiber. We give $\mathbb{R}^{nr}$ the usual inner product and let

$$E(\eta) = \{(b, v) \mid v \in S(F_b(\xi))^\perp\}.$$

It is easy to see that

$$\eta = \big(E(\eta), B, \mathbb{R}^{nr-n}, \mathrm{proj}_1\big)$$

is a vector bundle (cf. Example 15.6) and by definition $\xi \oplus \eta = \varepsilon^{nr}$. $\square$

If ξ in Theorem 15.18 is smooth then so is the constructed complement η, provided h_i and α_i are choosen smooth. The above proof uses that B is compact to ensure $r < \infty$. The theorem is not true without the compactness condition; see however Exercise 15.10 when ξ is a smooth vector bundle – it is the finite-dimensionality which counts.

Let $\mathrm{Vect}_n(B)$ denote the isomorphism classes of vector bundles over B of dimension n. Direct sum induces a map

$$\mathrm{Vect}_n(B) \times \mathrm{Vect}_m(B) \xrightarrow{\oplus} \mathrm{Vect}_{n+m}(B)$$

such that

$$\mathrm{Vect}(B) = \coprod_{n=0}^{\infty} \mathrm{Vect}_n(B)$$

becomes an abelian semigroup. The zero dimensional bundle $\varepsilon^0 = B \times \{0\}$ is the unit element.

To any abelian semigroup $(V, +)$ one can associate an abelian group $(K(V), +)$ defined as the formal differences $a - b$, or pairs (a, b), subject to the relation

$$(a + x) - (b + x) = a - b$$

where $x \in V$ is arbitrary. The construction has the *universal* property that any homomorphism from V to an abelian group A factors over $K(V)$, i.e. is induced from a homomorphism from $K(V)$ to A. The construction $V \to K(V)$, often called the Grothendieck construction, corresponds to the way the integers are constructed from the natural numbers, except that we do not demand that cancellation "$x + a = y + a \Rightarrow x = y$" holds in V. When B is *compact*, we define

$$(3) \qquad\qquad KO(B) = K(\mathrm{Vect}(B)).$$

By Theorem 15.18 every element of $KO(B)$ has the form $[\xi] - [\varepsilon^k]$, where $[\xi]$ denotes the isomorphism class of the vector bundle $[\xi]$. Indeed

$$[\xi_1] - [\xi_2] = ([\xi_1] + [\eta_2]) - ([\xi_2] + [\eta_2]) = [\xi_1 \oplus \eta_2] - [\xi_2 \oplus \eta_2] = [\xi] - [\varepsilon^k]$$

if we choose η_2 to be a complement to ξ_2.

Example 15.19 The normal bundle to the unit sphere $S^2 \subseteq \mathbb{R}^3$ is trivial, since the outward directed unit normal vector defines a global frame. We also know that $\tau_{S^2} \oplus \nu_{S^2} = \varepsilon^3$, such that

$$[\tau_{S^2}] + [\varepsilon^1] = [\varepsilon^3]$$

in $\mathrm{Vect}(S^2)$. However, $[\tau_{S^2}] \neq [\varepsilon^2]$ in $\mathrm{Vect}(S^2)$. Indeed, if $[\tau_{S^2}]$ were equal to $[\varepsilon^2]$, then there would exist a section $s \in \Gamma(\tau_{S^2})$ with $s(x) \neq 0$ for all $x \in S^2$. However, Theorem 7.3 implies that τ_{S^2} does not have a non-zero section. We see that cancellation does *not* hold in $\mathrm{Vect}(S^2)$.

Definition 15.20 Let $f \colon X \to B$ be a continuous (smooth) map and ξ a (smooth) vector bundle over B. The pre-image or pull-back $f^*(\xi)$ is the vector bundle over X given by

$$E(f^*(\xi)) = \{(x, v) \in X \times E(\xi) \mid f(x) = \pi_\xi(v)\}, \quad \pi_{f^*(\xi)} = \mathrm{proj}_1.$$

We note the homomorphism $(f, \hat{f}) \colon f^*(\xi) \to \xi$ given by $\hat{f}(x, v) = v$. It is obvious that the pull-backs of isomorphic bundles are isomorphic, so f^* induces homomorphisms

$$f^* \colon \mathrm{Vect}(B) \to \mathrm{Vect}(X) \quad \text{and} \quad f^* \colon KO(B) \to KO(X),$$

and $(g \circ f)^* = f^* \circ g^*$, $\mathrm{id}^* = \mathrm{id}$. Thus $\mathrm{Vect}(B)$ and $KO(B)$ become contravariant functors.

Theorem 15.21 *If f_0 and f_1 are homotopic maps, then $f_0^*(\xi)$ and $f_1^*(\xi)$ are isomorphic.*

Proof. Let $F: X \times I \to B$, $I = [0,1]$, be a homotopy between f_0 and f_1, $f_0(x) = F(x,0)$ and $f_1(x) = F(x,1)$. When $t \in I$ we get $[f_t^*(\xi)] \in \mathrm{Vect}(X)$. It is sufficient to see that the function $t \to [f_t^*(\xi)]$ is locally constant, and thus constant.

Fix t and consider the bundles

$$\zeta = \mathrm{proj}_1^* f_t^*(\xi) \quad \text{and} \quad \eta = F^*(\xi)$$

over $X \times I$. Since $F = f_t \circ \mathrm{proj}_1$ on $X \times \{t\}$, $\zeta = \eta$ on $X \times \{t\}$. We can choose a fiberwise isomorphism

$$E(\zeta) \xrightarrow{\hat{h}} E(\eta)$$
$$\searrow \qquad \swarrow$$
$$X \times \{t\}$$

The first step is to extend $\hat{h}$ to a homomorphism of vector bundles on $X \times [t - \epsilon, t + \epsilon]$ for some $\epsilon > 0$. This can be done as follows. Since X is compact, there exists a finite cover $U_1, \ldots, U_r$ of X with $\epsilon_i > 0$ such that both ζ and η are trivial on $U_i \times [t - \epsilon_i, t + \epsilon_i]$. We can extend $\hat{h}$ to

$$E(\zeta) \xrightarrow{\hat{h}_i} E(\eta)$$
$$\searrow \qquad \swarrow$$
$$U_i \times [t - \epsilon_i, t + \epsilon_i]$$

Let $\alpha_1, \ldots, \alpha_r$ be a partition of unity on X with $\mathrm{supp}(\alpha_i) \subset U_i$. We define

$$E(\zeta) \xrightarrow{\hat{k}} E(\eta)$$
$$\searrow \qquad \swarrow$$
$$X \times [t - \epsilon, t + \epsilon]$$

where $\epsilon = \min(\epsilon_i)$ by setting

$$\hat{k}(v) = \sum \alpha_i \big(\mathrm{proj}_1 \circ \pi_\zeta(v)\big) \cdot \hat{h}_i(v).$$

Since $\hat{h}_i(v) = \hat{h}(v)$ when $\pi_\zeta(v) \in X \times \{t\}$, and since $\sum \alpha_i(x) \equiv 1$, we have $\hat{k}(v) = \hat{h}(v)$ on $X \times \{t\}$. In particular $\hat{k}$ is an isomorphism on $X \times \{t\}$.

We finally show that $\hat{k}$ is an isomorphism in a neighborhood $X \times [t - \epsilon_1, t + \epsilon_1]$ of $X \times \{t\}$. Since X is compact, it suffices to show that $\hat{k}$ is an isomorphism on a neighborhood $V(x,t)$ of any point $(x,t) \in X \times \{t\}$. Let $\mathbf{e}$ and $\mathbf{s}$ be frames of ζ and η in a neighborhood W of (x,t), and $\mathrm{ad}(\hat{k})\colon W \to M_n(\mathbb{R})$ the resulting map, cf. (1). Since $GL_n(\mathbb{R}) \subset M_n(\mathbb{R})$ is open and $\mathrm{ad}\,(\hat{k})(x,t) \in GL_n(\mathbb{R})$, there exists a neighborhood $V(x,t)$ where $\mathrm{ad}(\hat{k}) \in GL_n(\mathbb{R})$, and $\hat{k}$ is an isomorphism.   □

The above theorem expresses that $\mathrm{Vect}(X)$ (and hence also $KO(X)$) is a *homotopy functor*: homotopic maps $f \simeq g\colon X \to Y$ induce the same map

$$f^* = g^* \colon \mathrm{Vect}(Y) \to \mathrm{Vect}(X).$$

Corollary 15.22 *Every vector bundle over a contractible base space is trivial.*

Proof. With our assumption $\mathrm{id}_B \simeq f$, where f is the constant map with value $f(B) = \{b\}$. Hence $f^*(\xi) \cong \xi$. But $f^*(\xi)$ is trivial by construction when f is constant.   □

In the above we have concentrated on *real* vector bundles. There is a completely analogous notion of *complex* (or even quaternion) vector bundles. In Definition 15.4 one simply requires V and $\pi^{-1}(x)$ to be complex vector spaces and $h(x,-)$ to be a complex isomorphism. The direct sum of complex vector bundles is a complex vector bundle. A hermitian inner product on a complex vector bundle is a map ϕ as in Definition 15.15 but such that it induces a hermitian inner product in each fiber. Proposition 15.13 and Theorem 15.18 and 15.21 have obvious analogues for complex vector bundles.

The isomorphism class of complex vector bundles over B of complex dimension n is denoted $\mathrm{Vect}_n^{\mathbb{C}}(B)$. These sets give rise to a semigroup whose corresponding group (for compact B) is traditionally denoted

$$(4) \qquad\qquad K(B) = K(\mathrm{Vect}^{\mathbb{C}}(B)).$$

It is a contravariant homotopy functor of B, often somewhat easier to calculate than its real analogue $KO(B)$.

16. OPERATIONS ON VECTOR BUNDLES AND THEIR SECTIONS

The main operations to be considered are tensor products and exterior products. We begin with a description of these operations on vector spaces, then apply them fiberwise to vector bundles, and end with the relation between the constructions on bundles and their equivalent constructions on spaces of sections.

Let R be a unital commutative ring and let V and W be R-modules. In the simplest applications $R = \mathbb{R}$ or $\mathbb{C}$ and V and W are R-vector spaces, but we present the definitions in the general setting. Denote by $R[V \times W]$ the free R-module with basis the set $V \times W$, i.e. the space of maps from the set $V \times W$ to R that are zero except for a finite number of points in $V \times W$. In $R[V \times W]$, we consider the submodule $R(V, W)$ which is generated (via finite linear combinations) by elements of the form

(1)
$$(v_1 + v_2, w) - (v_1, w) - (v_2, w)$$
$$(v, w_1 + w_2) - (v, w_1) - (v, w_2)$$
$$(rv, w) - r(v, w)$$
$$(v, rw) - r(v, w)$$

where $v_i \in V, w_i \in W$ and $r \in R$.

Definition 16.1 The *tensor product* $V \otimes_R W$ of two R-modules is the quotient module $R[V \times W]/R(V, W)$.

Let $\pi: R[V \times W] \to V \otimes_R W$ be the canonical projection and write $v \otimes_R w$ for the image of $(v, w) \in R[V \times W]$. It is clear from (1) that $\pi: V \times W \to V \otimes_R W$ is R-bilinear. Moreover, it is *universal* with this property in the following sense:

Lemma 16.2 *Let V, W and U be R-modules, and let $f: V \times W \to U$ be any R-bilinear map. Then there exists a unique R-linear map $\overline{f}: V \otimes_R W \to U$, with $f = \overline{f} \circ \pi$.*

Proof. Since the set $V \times W$ is a basis for the R-module $R[V \times W]$, f extends to an R-linear map $\hat{f}: R[V \times W] \to U$. The bilinearity of f implies that $\hat{f}(R(V, W)) = 0$, so that $\hat{f}$ induces a map $\overline{f}$ from the quotient $V \otimes_R W$ to U. By construction $f = \overline{f} \circ \pi$, $\overline{f}$ is R-linear and since $\pi(V \times W)$ generates the R-module $V \otimes_R W$, $\overline{f}$ is uniquely determined by f. $\square$

It is immediate from Lemma 16.2 that tensor product is a functor. Indeed, if $\varphi: V \to V'$ and $\psi: W \to W'$ are R-linear maps then the composition

$$V \times W \overset{\varphi \times \psi}{\to} V' \times W' \overset{\pi'}{\to} V' \otimes_R W'$$

is bilinear, so induces a unique map

(2) $\varphi \otimes_R \psi \colon V \otimes_R W \to V' \otimes_R W'.$

The uniqueness guarantees that $(\varphi' \otimes_R \psi') \circ (\varphi \otimes_R \psi) = \varphi' \circ \varphi \otimes_R \psi' \circ \psi$ when $\varphi' \colon V' \to V''$, $\psi' \colon W' \to W''$.

In the general setting of modules over a commutative ring, not every R-module V has a basis, i.e. is of the form $R[B]$ for a subset $B \subset V$, but if it does we say that V is a *free* R-module.

Lemma 16.3 *Let V and V' be free R-modules with bases B and B'. Then $V \otimes_R V'$ is a free R-module with basis $\{b \otimes_R b' \mid b \in B,\ b' \in B'\}$.*

Proof. The bilinearity of $\pi \colon V \times V' \to V \otimes_R V'$ shows that the stated set generates $V \otimes_R V'$; so suppose that

(3) $$\sum r_{ij}\, b_i \otimes_R b_j' = 0$$

is a (finite) relation. Let $\varphi_0 \colon V \to R$ and $\varphi_0' \colon V' \to R$ be the linear maps with

$$\varphi_0(b_i) = 0 \text{ if } i \neq i_0,\ \varphi_0(b_{i_0}) = 1$$
$$\varphi_0'(b_i') = 0 \text{ if } i \neq j_0,\ \varphi_0'(b_{j_0}') = 1$$

where (i_0, j_0) is a pair of indices that appear in (3). The composition

$$V \otimes_R V' \xrightarrow{\varphi_0 \otimes \varphi_0'} R \otimes_R R \xrightarrow{\text{mult}} R$$

maps the left-hand side in (3) into $r_{i_0 j_0}$, which must therefore be zero. $\square$

We note the general relations

(4)
    (i) $R \otimes_R V \cong V \cong V \otimes_R R$
    (ii) $V_1 \otimes_R V_2 \cong V_2 \otimes_R V_1$
    (iii) $V_1 \otimes_R (V_2 \otimes_R V_3) \cong (V_1 \otimes_R V_2) \otimes_R V_3$
    (iv) $(V_1 \oplus V_2) \otimes_R V_3 \cong V_1 \otimes_R V_3 \oplus V_2 \otimes_R V_3.$

They all follow from the universal property of Lemma 16.2. For example, scalar multiplication defines an R-bilinear map from $R \times V$ into V, and hence a map from $R \otimes_R V$ into V whose inverse is the map that sends $v \in V$ into $1 \otimes_R v$. The other cases are equally simple.

There is an obvious generalization of Definition 16.1 to k-variable tensor products $V_1 \otimes_R \dots \otimes_R V_k$ and a corresponding generalization of Lemma 16.2, namely

Lemma 16.4 *For every R-multilinear map $f: V_1 \times \ldots \times V_k \to W$ there exists a unique R-linear map $\overline{f}: V_1 \otimes_R \ldots \otimes_R V_k \to W$ such that the following diagram commutes.*

$$V_1 \times \cdots \times V_k \xrightarrow{\quad f \quad} W$$

$$\searrow \pi \qquad \nearrow \overline{f}$$

$$V_1 \otimes_R \cdots \otimes_R V_k$$

$\square$

It is easy to see that the k-variable tensor product is isomorphic to the iteration of the two-variable ones,

$$(\ldots((V_1 \otimes_R V_2) \otimes_R V_3) \otimes_R \ldots \otimes_R V_k) \cong V_1 \otimes_R \ldots \otimes_R V_k.$$

Let us specialize to $V_1 = \ldots = V_k = V$. Consider the sub-R-module $A^k(V) \subseteq \bigotimes_R^k V$ generated by the set

$$\{v_1 \otimes_R \ldots \otimes_R v_k \mid v_i \in V \text{ and } v_{i_0} = v_{i_0+1} \text{ for some } i_0\}.$$

Definition 16.5 The quotient module

$$\Lambda_R^k(V) = \bigotimes_R^k (V)/A^k(V)$$

is called the exterior k-th power.

The image of $v_1 \otimes_R \ldots \otimes_R v_k$ in $\Lambda_R^k(V)$ under the canonical projection

$$\pi_1 : \bigotimes_R^k (V) \to \Lambda_R^k(V)$$

is denoted $v_1 \wedge_R \ldots \wedge_R v_k$. Then for a permutation σ

(5) $$v_1 \wedge_R \ldots \wedge_R v_k = \text{sign}(\sigma)\, v_{\sigma(1)} \wedge_R \ldots \wedge_R v_{\sigma(k)}$$

as in Lemma 2.2 and Lemma 2.7, so that the composition

(6) $$\rho = \pi_1 \circ \pi: V \times \ldots \times V \to \Lambda_R^k(V)$$

is an alternating map, i.e. an R-multilinear map with $\rho(v_1, \ldots, v_k) = 0$ when $v_i = v_j$ for some $i < j$. With these definitions we get from Lemma 16.4:

Lemma 16.6 For any R-alternating map $\omega: V \times \ldots \times V \to W$ there is a unique R-linear map $\overline{\omega}: \Lambda_R^k(V) \to W$ such that $\omega = \overline{\omega} \circ \rho$.   □

In the special case where $R = \mathbb{R}$ (or $\mathbb{C}$) and where we restrict attention to finite-dimensional vector spaces the exterior powers $\Lambda_R^k(V)$ are dual to the alternating power $\text{Alt}^k(V)$ introduced in Chapter 2. To simplify notation we drop the subscript $\mathbb{R}$ and write $\bigotimes^k(V)$, $\Lambda^k(V)$, $V \otimes W$, $\text{Hom}(V,W)$ instead of $\bigotimes_\mathbb{R}^k(V)$, $\Lambda_\mathbb{R}^k(V)$, $V \otimes_\mathbb{R} W$, $\text{Hom}_\mathbb{R}(V,W)$.

Let $V^* = \text{Hom}_\mathbb{R}(V, \mathbb{R})$ be the dual vector space. The exterior product introduced in Definition 2.5 defines an alternating map

$$\varphi: V^* \times \ldots \times V^* \to \text{Alt}^k V$$

and thus by Lemma 16.6 a linear map

$$\overline{\varphi}: \Lambda^k(V^*) \to \text{Alt}^k(V).$$

Theorem 16.7

(i) The map $\overline{\varphi}$ is an isomorphism.

(ii) If $\{e_i\}_{i=1}^n$ is a basis for V then $\{e_{i_1} \wedge \ldots \wedge e_{i_k} \mid i_1 < \ldots < i_k\}$ is a basis for $\Lambda^k(V)$.

(iii) There is a natural isomorphism $\Lambda^k(V^*) \cong \Lambda^k(V)^*$.

Proof. The map $\overline{\varphi}$ is surjective by Theorem 2.15, and hence $\dim \Lambda^k(V^*) > \binom{n}{k}$. On the other hand Lemma 16.3 and (5) imply that the set in (ii) generates $\Lambda^k(V)$, so that $\dim \Lambda^k(V) \leq \binom{n}{k}$. Since $\dim \Lambda^k(V) = \dim \Lambda^k(V^*)$, the common dimension is $\binom{n}{k}$. Thus $\overline{\varphi}$ must be an isomorphism. This proves (i) and (ii). For each fixed $\omega \in \text{Alt}^k(V)$, $\omega(v_1, \ldots, v_k)$ is an alternating map, and so defines by Lemma 16.6 a linear map from $\Lambda^k(V)$ into $\mathbb{R}$. This gives a map

$$\overline{\psi}: \text{Alt}^k(V) \to \Lambda^k(V)^*$$

which is linear and injective. Since the dimensions on both sides agree, $\overline{\psi}$ is an isomorphism, and (iii) follows from (i).   □

Consider the bilinear pairing

$$\wedge: \Lambda^k(V) \times \Lambda^l(V) \to \Lambda^{k+l}(V)$$

which takes $(v_1 \wedge \ldots \wedge v_k, w_1 \wedge \ldots \wedge w_k)$ into $v_1 \wedge \ldots \wedge v_k \wedge w_1 \wedge \ldots \wedge w_l$. Then

$$\overline{\varphi}: \Lambda^k(V^*) \to \text{Alt}^k(V)$$

becomes multiplicative, $\overline{\varphi}(\omega_1 \wedge \omega_2) = \overline{\varphi}(\omega_1) \wedge \overline{\varphi}(\omega_2)$, where the product on the range is the one from Definition 2.5. The composition

$$\psi : \Lambda^k(V^*) \otimes \Lambda^k(V) \overset{\overline{\varphi} \otimes 1}{\to} \mathrm{Alt}^k(V) \otimes \Lambda^k(V) \overset{\mathrm{ev}}{\to} \mathbb{R}$$

with $\mathrm{ev}\,(\omega, v_1 \wedge \ldots \wedge v_k) = \omega(v_1, \ldots, v_k)$ induces the isomorphism $\overline{\psi} \colon \Lambda^k(V^*) \cong \Lambda^k(V)^*$ in Theorem 16.7.(iii). On the other hand, Lemma 2.13 gives that

(7) $$\psi((\epsilon_1 \wedge \ldots \wedge \epsilon_k) \otimes (v_1 \wedge \ldots \wedge v_k)) = \det((\epsilon_i(v_j))_{i,j=1}^k).$$

In particular, if V has an inner product $\langle , \rangle$, then we may identify V with V^* by sending v to $\langle v, - \rangle$, and ψ becomes the map

$$\psi(w_1 \wedge \ldots \wedge w_k, v_1 \wedge \ldots \wedge v_k) = \det\left(\langle w_i, v_j \rangle_{i,j=1}^k\right).$$

Theorem 16.7.(iii) then translates into

Addendum 16.8 (Grassmann inner product) For an inner product space V, the formula
$$\langle w_1 \wedge \ldots \wedge w_k, v_1 \wedge \ldots \wedge v_k \rangle = \det\left(\langle w_i, v_j \rangle_{i,j=1}^k\right)$$
defines an inner product on $\Lambda^k(V)$. □

For vector spaces V and W there is an obvious linear map

$$V^* \otimes W \to \mathrm{Hom}(V, W)$$

which takes $f \otimes w$ into the linear map $v \mapsto f(v)w$. It follows from Lemma 16.3 that this is an isomorphism

(8) $$V^* \otimes W \cong \mathrm{Hom}(V, W); \quad \dim W < \infty.$$

The above constructions on vector spaces induce constructions on vector bundles (over the same base space) by applying them fiberwise. So if ξ and η are (smooth) vector bundles over X then we get new (smooth) vector bundles over X:

(9) $$\xi \otimes \eta, \quad \mathrm{Hom}\,(\xi, \eta), \quad \bigotimes^k(\xi), \quad \Lambda^k(\xi), \quad \xi^*, \quad \mathrm{Alt}^k(\xi)$$

and so forth. The fiber over $x \in X$ in each case is the associated construction on ξ_x, η_x. To be more specific, let us run through the definition in one of the cases, say $\mathrm{Hom}\,(\xi, \eta)$. We define

$$E = E(\mathrm{Hom}(\xi, \eta)) = \coprod_{x \in X} \mathrm{Hom}(\xi_x, \eta_x)$$

with the obvious projection map π onto X, which maps the entire vector space $\mathrm{Hom}(\xi_x, \eta_x)$ to $x \in X$. The problem is to define a topology on E that makes this the total space of a vector bundle.

Let $\{U_j\}_{j \in J}$ be an open cover of X for which both $\xi|_{U_j}$ and $\eta|_{U_j}$ are trivial. Choose isomorphisms

$$h_j: \xi|_{U_j} \cong U_j \times \mathbb{R}^n, \qquad k_j: \eta|_{U_j} \cong U_j \times \mathbb{R}^m.$$

They induce inclusions

$$U_j \times \operatorname{Hom}(\mathbb{R}^n, \mathbb{R}^m) \overset{H_j}{\hookrightarrow} E(\operatorname{Hom}(\xi, \eta))$$

(which depend on h_j and k_j). We can use $\{H_j\}$ to define a topology on E:

(10) $A \subset E$ open $\Leftrightarrow$ $H_j^{-1}(A)$ open for all j.

where the topology on $U_j \times \operatorname{Hom}(\mathbb{R}^n, \mathbb{R}^m)$ is the product topology. It is left to the reader to show that (10) defines a topology and that it is independent of the choice of cover and isomorphisms h_j, k_j. This uses that $\operatorname{Hom}(V, W)$ is a continuous functor, i.e. that the maps

$$T: \operatorname{Hom}(V, V_1) \times \operatorname{Hom}(W_1, W) \to \operatorname{Hom}(\operatorname{Hom}(V_1, W_1), \operatorname{Hom}(V, W));$$
$$T(f, g)(\phi) = g \circ \phi \circ f$$

are continuous. In particular note that $\xi \cong X \times V$ and $\eta \cong X \times W$ give $\operatorname{Hom}(\xi, \eta) \cong X \times \operatorname{Hom}(V, W)$.

All of the constructions in (9) produce smooth vector bundles when ξ and η are smooth.

Lemma 16.9 *For (smooth) vector bundles ξ and η there are isomorphisms $\xi \cong \xi^{**}$ and $\xi^* \otimes \eta \cong \operatorname{Hom}(\xi, \eta)$.*

Proof. For finite-dimensional vector spaces there are natural isomorphisms

$$V \to V^{**}, \qquad V^* \otimes W \to \operatorname{Hom}(V, W)$$

defined without any reference to basis. This gives maps of vector bundles (over the identity)

$$\xi \to \xi^{**}, \qquad \xi^* \otimes \eta \to \operatorname{Hom}(\xi, \eta)$$

which are isomorphisms on each fiber, and we can apply Lemma 15.10. $\square$

For finite-dimensional vector spaces $V \cong V^*$, since the dimensions agree. However, it is not true in general that $\xi \cong \xi^*$; cf. Properties 18.11 below. The above proof breaks down because there is no isomorphism from V to V^* defined independently of choice of basis. As a result one cannot define a homomorphism from ξ to ξ^*.

The isomorphisms of Theorem 16.7.(iii) are again defined without reference to a basis, and so give isomorphisms

$$(11) \qquad \Lambda^k(\xi^*) \cong \Lambda^k(\xi)^* \cong \mathrm{Alt}^k(\xi)$$

for any vector bundles ξ.

We have in Theorem 16.7, Addendum 16.8, Lemma 16.9, and (11) concentrated on real vector spaces and vector bundles, and leave the reader to formulate and prove the corresponding results for the complex cases.

Every complex vector bundle ξ gives rise to a real vector bundle ξ_{R} upon forgetting part of the structure. On the other hand, every real vector bundle induces a complex vector bundle η_{C} upon complexifying each fiber, $\eta_{\mathsf{C}} = \eta \otimes_{\mathsf{R}} \varepsilon_{\mathsf{C}}^1$, where $\varepsilon_{\mathsf{C}}^1$ denotes the trivial complex line bundle. There are the following relations between these operations.

Lemma 16.10

(i) For a real vector bundle η, $(\eta_{\mathsf{C}})_{\mathsf{R}} \cong \eta \oplus \eta$.
(ii) For a complex vector bundle ξ, $(\xi_{\mathsf{R}})_{\mathsf{C}} \cong \xi \oplus \xi^*$.

Proof. We leave (i) to the reader and prove (ii). Given a complex vector space V we get a new one $V \otimes_{\mathsf{R}} \mathbb{C}$ where scalar multiplication is in the second factor. The map

$$\varphi: V \otimes_{\mathsf{R}} \mathbb{C} \to V \oplus V; \quad \varphi(v \otimes z) = (zv, \bar{z}v)$$

defines an isomorphism of the underlying real vector spaces, and φ becomes complex linear if, in the second summand of the target, we replace V by its conjugate vector space $\overline{V}$, in which multiplication by $z \in \mathbb{C}$ is replaced by multiplication with the complex conugate $\bar{z}$. Suppose that V has a hermitian product. Then we may identify $\overline{V}$ with the complex dual vector space V^* via the map $v \mapsto \langle -, v \rangle$, and φ gives a complex linear isomorphism

$$\varphi: V \otimes_{\mathsf{R}} \mathbb{C} \to V \oplus V^*.$$

We choose a hermitian product on ξ, and apply φ fiberwise to prove (ii). □

We next consider spaces of sections. We shall primarily be interested in the space of smooth sections $\Omega^0(\xi)$ of smooth vector bundles; cf. Definition 15.15. For the tangent bundle τ_M,

$$(12) \qquad \Omega^0\left(\mathrm{Alt}^k(\tau_M)\right) = \Omega^k(M).$$

Indeed, an element of the left-hand side associates to each $x \in M$ an element $\omega_x \in \mathrm{Alt}^k(T_x M)$, and since $\Omega^0(-)$ denotes the space of smooth sections,

$\{\omega_x\}_{x \in M}$ is a smooth k-form in the sense of Definition 9.15. Similarly $\Omega^0(\tau_M)$ is the space of smooth tangent vector fields on M.

The section space $\Omega^0(\xi)$ of a smooth vector bundle over M is always a module over the ring $\Omega^0(M) = C^\infty(M, \mathbb{R})$; if $s \in \Omega^0(\xi)$ and $f: M \to \mathbb{R}$ is a smooth map then $(fs)(x) = f(x)s(x)$, $x \in M$ is a new smooth section of ξ. We can apply Definitions 16.1 and 16.5 with $R = \Omega^0(M)$ and $V, W = \Omega^0(\xi), \Omega^0(\eta)$, etc.

The $\Omega^0(M)$-module $\Omega^0(\xi)$ is not in general a free module. Indeed, if it is then a choice of $\Omega^0(M)$-basis $e_1, \dots, e_k \in \Omega^0(\xi)$ has the property that $e_1(p) \dots, e_k(p)$ is a basis for ξ_p for every $p \in M$ and ξ must be trivial.

Lemma 16.11 *For every smooth vector bundle ξ over a compact smooth manifold M, $\Omega^0(\xi)$ is a direct $\Omega^0(M)$-summand in a finitely generated free $\Omega^0(M)$-module.*

Proof. By Theorem 15.18 there is a complement η to ξ, $\xi \oplus \eta \cong \varepsilon^{k+l}$. Then

$$\Omega^0(\xi) \oplus \Omega^0(\eta) \cong \Omega^0(\xi \oplus \eta) \cong \Omega^0(\varepsilon^{k+l})$$

and $\Omega^0(\varepsilon^{k+l})$ is a free $\Omega^0(M)$-module (of dimension $k + l$). $\square$

Direct summands in free modules are called *projective* modules, so the above lemma tells us that $\Omega^0(\xi)$ is always a finitely generated projective R-module, with $R = \Omega^0(M)$.

Lemma 16.12 *Let P_1 and P_2 be finitely generated projective R-modules. Then there are isomorphisms*

$$P_1 \cong P_1^{**}, \qquad \operatorname{Hom}_R(P_1, P_2) \cong P_1^* \otimes_R P_2$$

where $P_1^* = \operatorname{Hom}_R(P_1, R)$.

Proof. One first proves the assertions for finitely generated free modules, where the argument is completely similar to the case of vector spaces. The general case follows easily upon choosing complements $P_1 \oplus Q_1 = R^{n_1}$, $P_2 \oplus Q_2 = R^{n_2}$. Details are left as an exercise. $\square$

Theorem 16.13 *There are the following isomorphisms*

(i) $\Omega^0(\operatorname{Hom}(\xi, \eta)) \cong \operatorname{Hom}_{\Omega^0(M)}(\Omega^0(\xi), \Omega^0(\eta))$

(ii) $\Omega^0(\xi \otimes \eta) \cong \Omega^0(\xi) \otimes_{\Omega^0(M)} \Omega^0(\eta)$

(iii) $\Omega^0(\xi^*) \cong \operatorname{Hom}_{\Omega^0(M)}(\Omega^0(\xi), \Omega^0(M))$

(iv) $\Omega^0(\Lambda^i \xi) \cong \Lambda^i_{\Omega^0(M)}(\Omega^0(\xi))$.

Proof. By definition, $\Omega^0(\mathrm{Hom}\,(\xi, \eta))$ is the space of smooth fiberwise homomorphisms $\widehat{\varphi}\colon \xi \to \eta$, and we define an $\Omega^0(M)$-linear homomorphism

$$F\colon \Omega^0(\mathrm{Hom}\,(\xi, \eta)) \to \mathrm{Hom}_{\Omega^0(M)}\left(\Omega^0(\xi), \Omega^0(\eta)\right)$$

by $F(\widehat{\varphi})(s) = \widehat{\varphi} \circ s$, $s \in \Omega^0(\xi)$. Suppose $F(\widehat{\varphi}) = 0$. To check injectivity of F we must show that $\widehat{\varphi}_x\colon \xi_x \to \eta_x$ is the zero homomorphism for every $x \in M$. Fix $x \in M$ and $v \in \xi_x$. There is a section $s_v \in \Omega^0(\xi)$ with $s_v(x) = v$. Indeed, such an s_v can always be defined in a neighborhood U of $x \in M$ where $\xi_{|U}$ is trivial, and we can obtain a global section upon choosing a smooth function f on M with $\mathrm{supp}\,(f) \subset U$ and $f(x) = 1$ (cf. Appendix A) and replace the local section s_v by fs_v, giving it the value zero outside U.
Now $F(\widehat{\varphi})(s_v) = 0$ implies that $\widehat{\varphi}_x(v) = 0$. Since x and v were arbitrary, F is injective.
Let $\Phi \in \mathrm{Hom}_{\Omega^0(M)}\left(\Omega^0(\xi), \Omega^0(\eta)\right)$. We wish to define a fiberwise smooth homomorphism $\widehat{\varphi}\colon \xi \to \eta$ by setting $\widehat{\varphi}_x(v) = \Phi(s_v)(x)$ where $s_v \in \Omega^0(\xi)$ is a section with $s_v(x) = v$. This requires that one has

$$\Phi(s_v)(x) = \Phi\left(s_v'\right)(x)$$

for two sections of ξ with $s_v(x) = s_v'(x)$, or in other words that

(13) $\qquad\qquad$ if $\quad s(x) = 0 \quad$ then $\quad \Phi(s)(x) = 0$.

Chose sections $e_1, \ldots, e_k \in \Omega^0(\xi)$ such that $e_1(p), \ldots, e_k(p)$ form a basis for ξ_p for p in some neighborhood U of x. Again this can be done locally, and the local sections can be extended to global ones as above. Now

$$s(p) = \sum f_i(p)e_i(p) \qquad \text{for } p \in U$$

for smooth functions f_i defined on U. Let $\lambda \in \Omega^0(M)$ have $\mathrm{Supp}(\lambda) \subset U$ and $\lambda(x) = 1$. Then

$$\Phi(s) = \Phi(\lambda s + (1 - \lambda)(s)) = \Phi(\lambda s) + (1 - \lambda)\Phi(s)$$

so that $\Phi(s)(x) = \Phi(\lambda s)(x)$. But $\lambda s = \sum(\lambda f_i)e_i$ and λf_i extends to a smooth function g_i defined on all of M with $g_i(x) = f_i(x) = 0$. Since Φ is $\Omega^0(M)$-linear,

$$\Phi(\lambda s) = \sum g_i \Phi(e_i) \in \Omega^0(\eta).$$

But $g_i(x) = 0$ so $\Phi(\lambda s)(x) = 0$. This proves (i).
Assertion (iii) is the special case of (i) corresponding to $\eta = \varepsilon^1$, the trivial line bundle, and (ii) follows from Lemma 16.9 and (i):

$$\Omega^0(\xi \otimes \eta) \cong \Omega^0(\mathrm{Hom}\,(\xi^*, \eta)) \cong \mathrm{Hom}_{\Omega^0(M)}\left(\Omega^0(\xi^*), \Omega^0(\eta)\right)$$
$$\cong \mathrm{Hom}_{\Omega^0(M)}\left(\mathrm{Hom}_{\Omega^0(M)}(\Omega^0(\xi), \Omega^0(M)), \Omega^0(\eta)\right)$$
$$\cong \Omega^0(\xi) \otimes_{\Omega^0(M)} \Omega^0(\eta)$$

where the last isomorphism is from Lemma 16.11 and Lemma 16.12. Finally there is a commutative diagram

$$\Omega^0(\bigotimes^i \xi) \longrightarrow \bigotimes_{\Omega^0(M)} \Omega^0(\xi)$$

$$\downarrow \qquad\qquad\qquad \downarrow$$

$$\Omega^0(\Lambda^i \xi) \longrightarrow \Lambda^i_{\Omega^0(M)}(\Omega^0(\xi))$$

By (ii), the upper horizontal map is an isomorphism. To see that the bottom homomorphism is also an isomorphism, one can use Theorem 16.7.(ii), and local sections as in the proof of (i). The details are left as an exercise. □

We close with a weaker form of the universal property of tensor products, stated in Lemma 16.2. Let R be a unital commutative ring and S an R-algebra. In our applications in the next chapter $R = \mathbb{R}$ and $S = \Omega^0(M)$, the smooth functions on M, or their complex versions $R = \mathbb{C}$ and $S = \Omega^0(M; \mathbb{C})$. Suppose that V and W are R-modules and that S operates from the right on V and from the left on W, e.g.

$$V = \Omega^i(M), \qquad W = \Omega^0(\xi).$$

Definition 16.14 The *balanced product* (or tensor product) $V \otimes_S W$ is the cokernel of the R-linear homomorphism

$$V \otimes_R S \otimes_R W \xrightarrow{\beta} V \otimes_R W$$

given by $\beta(v \otimes_R s \otimes_R w) = vs \otimes_R w - v \otimes_R sw.$

When S is commutative, and this will be the case in our applications, then there are no distinctions between left and right actions of S, and $V \otimes_S W$ becomes an S-module upon defining $s(v \otimes_S w) = vs \otimes_S w$.

This S-module is obviously isomorphic to the one defined in Definition 16.1. We record for later use the obvious

Lemma 16.15 *Let $f: V \otimes_R W \to U$ be an R-linear map which is S-balanced in the sense that $f(vs \otimes_R w) = f(v \otimes_R sw)$ for $f \in V$, $w \in W$ and $s \in S$. Then there is an induced R-linear map $\overline{f}: V \otimes_S W \to U$.* □

17. CONNECTIONS AND CURVATURE

Let ξ be a smooth vector bundle over a smooth manifold M^n of dimension n.

Definition 17.1 A *connection* on ξ is an $\mathbb{R}$-linear map

$$\nabla : \Omega^0(\xi) \to \Omega^1(M) \otimes_{\Omega^0(M)} \Omega^0(\xi)$$

which satisfies "Leibnitz' rule" $\nabla(f \cdot s) = df \otimes s + f \cdot \nabla s$, where $f \in \Omega^0(M)$, $s \in \Omega^0(\xi)$ and d is the exterior differential. If ξ is a complex vector bundle then $\Omega^0(\xi)$ is a complex vector space and we require ∇ to be $\mathbb{C}$-linear.

Let τ be the tangent bundle of M. Then $\Omega^1(M) = \Omega^0(\tau^*)$, and by Theorem 16.13 we have the following rewritings of the range for ∇,

$$(1) \qquad \Omega^1(M) \otimes_{\Omega^0(M)} \Omega^0(\xi) \cong \Omega^0(\operatorname{Hom}(\tau, \xi)) \cong \operatorname{Hom}_{\Omega^0(M)}\big(\Omega^0(\tau), \Omega^0(\xi)\big).$$

A tangent vector field X on M is a section in the tangent bundle $X \in \Omega^0(\tau)$, and induces an $\Omega^0(M)$-linear map $\operatorname{Ev}_X : \Omega^1(M) \to \Omega^0(M)$, and hence an $\Omega^0(M)$-linear map

$$\operatorname{Ev}_X : \Omega^1(M) \otimes_{\Omega^0(M)} \Omega^0(\xi) \to \Omega^0(\xi).$$

The composition $\operatorname{Ev}_X \circ \nabla$ is an $\mathbb{R}$-linear map $\nabla_X : \Omega^0(\xi) \to \Omega^0(\xi)$ which satisfies

$$(2) \qquad \nabla_X(fs) = d_X(f)s + f \, \nabla_X(s),$$

where $d_X(f)$ is the directional derivative of f in the direction X, since $\operatorname{Ev}_X \circ df = d_X(f)$. Thus a connection allows us to take directional derivatives of sections. For fixed $s \in \Omega^0(\xi)$ the map $X \to \nabla_X(s)$ is $\Omega^0(M)$-linear in X:

$$\nabla_{gX+hY}(s) = g \, \nabla_X(s) + h \, \nabla_Y(s)$$

for smooth functions $g, h \in \Omega^0(M)$ and vector fields $X, Y \in \Omega^0(\tau)$. Moreover, the value $\nabla_X(s)(p) \in \xi_p$ depends only on the value $X_p \in T_pM$. This is clear from the second term in (1) which implies that ∇ can be considered as an $\mathbb{R}$-linear map

$$\nabla : \Omega^0(\xi) \to \operatorname{HOM}(\tau, \xi) := \Omega^0 \operatorname{Hom}(\tau, \xi).$$

Here the range is the set of smooth bundle homomorphisms from τ to ξ (over the identity). If $X_p \in T_pM$ then $\nabla_{X_p}(s) = (\nabla s)(X_p)$, and

$$(3) \qquad \begin{aligned} \nabla_{X_p}(f \cdot s) &= d_{X_p}(f) \cdot s(p) + f(p) \, \nabla_{X_p}(s) \\ \nabla_{aX_p + bY_p}(s) &= a \, \nabla_{X_p}(s) + b \, \nabla_{Y_p}(s) \end{aligned}$$

where $X_p, Y_p \in T_pM$, and a and b are real numbers. Conversely (3) guarantees that $\nabla_{X_p}(s)$ defines a connection.

Example 17.2 Let $M^n \subset \mathbb{R}^{n+k}$ be a smooth manifold. One can define a connection on its tangent bundle as follows: a section $s \in \Omega^0(\tau)$ can be considered as a smooth function $s\colon M \to \mathbb{R}^{n+k}$ with $s(p) \in T_p M$, and we set

$$\nabla_{X_p}(s) = j_p\big(d_{X_p}(s)\big) \in T_p M$$

where $j_p\colon \mathbb{R}^{n+k} \to T_p M$ is the orthogonal projection and $X_p \in T_p M$. It is easy to see that (3) is satisfied.

It is a consequence of the "Leibnitz rule" that ∇ is a *local operator* in the sense that if $s \in \Omega^0(\xi)$ is a section that vanishes on an open subset $U \subseteq M$ then so does $\nabla(s)$. A local operator between section spaces always induces an operator between the section spaces of the vector bundles restricted to open subsets. In particular a connection on ξ induces a connection on $\xi_{|U}$.

Let $e_1, \ldots, e_k \in \Omega^0(\xi)$ be sections such that $e_1(p), \ldots, e_k(p)$ is a basis for ξ_p for every $p \in U$ (a frame over U). Elements of $\Omega^1(U) \otimes_{\Omega^0(U)} \Omega^0(\xi_{|U})$ can be written uniquely as $\sum \tau_i \otimes e_i$ for some $\tau_i \in \Omega^1(U)$, so for a connection ∇ on ξ,

$$(4) \qquad\qquad \nabla(e_i) = \sum_{j=1}^{k} A_{ij} \otimes e_j$$

where $A_{ij} \in \Omega^1(U)$ is a $k \times k$ matrix of 1-forms, which is called the *connection form* with respect to e, and is denoted by A.

Conversely, given an arbitrary matrix A of 1-forms on U, and a frame for $\xi_{|U}$, then (4) defines a connection on $\Omega^0(\xi_{|U})$. Since $s \in \Omega^0(\xi_{|U})$ can be written as $s(p) = \sum s_i(p) e_i(p)$, with $s_i \in \Omega^0(U)$,

$$\nabla\left(\sum s_i e_i\right) = \sum ds_i \otimes e_i + \sum s_i \nabla e_i = \sum (ds_j + s_i A_{ij}) \otimes e_j.$$

With respect to e $= (e_1, \ldots, e_k)$, ∇ has the matrix form

$$(5) \qquad\qquad \nabla(s_1, \ldots, s_k) = (ds_1, \ldots, ds_k) + (s_1, \ldots, s_k) A.$$

Example 17.3 Suppose $\xi \oplus \eta \cong \varepsilon^{n+k}$, and let $i\colon \xi \to \varepsilon^{n+k}$ and $j\colon \varepsilon^{n+k} \to \xi$ be the inclusion and the projection on the first factor, respectively. We give the trivial bundle the connection ∇_0 from (5) with $A = 0$. There are maps

$$\Omega^0(\xi) \xrightarrow{i_*} \Omega^0(\varepsilon^{n+k}) \quad \text{and} \quad \Omega^0(\varepsilon^{n+k}) \xrightarrow{j_*} \Omega^0(\xi),$$

and the composition

$$\Omega^0(\xi) \xrightarrow{i_*} \Omega^0(\varepsilon^{n+k}) \xrightarrow{\nabla_0} \Omega^1(M) \otimes_{\Omega^0(M)} \Omega^0(\varepsilon^{n+k}) \xrightarrow{\mathrm{id} \otimes j_*} \Omega^1(M) \otimes_{\Omega^0(M)} \Omega^0(\xi)$$

defines a connection ∇ on ξ. Note that

$$\Omega^0(\varepsilon^{n+k}) \cong \Omega^0(M) \oplus \ldots \oplus \Omega^0(M)$$
$$\Omega^1(M) \otimes_{\Omega^0(M)} \Omega^0(\varepsilon^{n+k}) \cong \Omega^1(M) \oplus \ldots \oplus \Omega^1(M),$$

and that $\nabla_0 = d \oplus \ldots \oplus d$. If ξ and η are complex vector bundles and ε^{n+k} is the trivial complex bundle, then the construction gives a complex connection.

Example 17.3 shows that every smooth vector bundle over a compact base manifold has at least one connection, since bundles have complements by Theorem 15.18. We observe that Example 17.2 is a special case of Example 17.3 corresponding to $\xi = \tau_M$ and $\eta = \nu_M$; see also the exercises.

Remark 17.4 After choice of a connection ∇ on ξ one can compare the fibers ξ_p at different points $p \in M$ by a "parallel translation along curves". Let $\alpha(t)$ be a smooth curve in M and $w(t) \in \Omega^0(\xi_{\alpha(t)})$ a section along α, i.e. $w(t) = w(\alpha(t))$ for some $w \in \Omega^0(\xi)$. There exists a unique operator ("covariant differentiation") $\frac{D}{dt}$ that satisfies:

(i) $\dfrac{D(w_1 + w_2)}{dt} = \dfrac{Dw_1}{dt} + \dfrac{Dw_2}{dt}$

(ii) $\dfrac{D(f \cdot w)}{dt} = \dfrac{df}{dt} w + f \dfrac{Dw}{dt}$

(iii) $\dfrac{Dw}{dt} = \nabla_{\alpha'(t)} w.$

Suppose first that $\alpha(t) \subset U$, where $(U, \mathbf{x})$ is a chart on M. Let $\partial_i = \frac{\partial}{\partial x_i} \in \Omega^0(\tau_U)$ and let $e = (e_1, \ldots, e_k)$ be a frame of $\Omega^0(\xi_{|U})$. Then $\alpha(t) = \mathbf{x}^{-1}(u_1(t), \ldots, u_n(t))$ for smooth functions $u_i(t)$, and $w(t) = \Sigma w_i(t) e_i(t)$, where $e_i(t) = e_i(\alpha(t))$. Conditions (i), (ii) and (iii) give

$$\frac{Dw}{dt} = \sum \left(\frac{dw_i}{dt} e_i(t) + w_i(t) \nabla_{\alpha'(t)} (e_i) \right),$$

and since $\alpha'(t) = \Sigma \frac{du_i}{dt} \cdot \partial_i$, (4) implies that for certain smooth functions Γ^l_{ji} on U

$$\nabla_{\alpha'(t)} e_i = \sum_{j=1}^n \frac{du_j}{dt} \nabla_{\partial_j} (e_i) = \sum_{j,\nu} \frac{du_j}{dt} \Gamma^\nu_{ji} e_\nu.$$

This gives

$$\frac{Dw}{dt} = \sum_{l=1}^k \left(\frac{dw_l}{dt} + \sum_{i,j} \frac{du_j}{dt} \Gamma^l_{ji} w_i \right) e_l.$$

Conversely this formula defines an operator $\frac{D}{dt}$ which satisfies (i), (ii) and (iii). Since we can cover $\alpha(t)$ with coordinate patches, the assumption that α be contained in just one chart is irrelevant.

A section $w(t)$ in ξ along $\alpha(t)$ is said to be *parallel*, if $\frac{Dw}{dt} = 0$. For a given $w(0) \in \xi_{\alpha(0)}$ and smooth curve there exists a unique such section, and the assignment $w(0) \to w(1)$ is an isomorphism from $\xi_{\alpha(0)}$ to $\xi_{\alpha(1)}$.

Let us introduce the notation

(6) $$\Omega^i(\xi) = \Omega^i(M) \otimes_{\Omega^0(M)} \Omega^0(\xi).$$

Then a connection is an $\mathbb{R}$-linear operator $\nabla : \Omega^0(\xi) \to \Omega^1(\xi)$ which satisfies the Leibnitz rule. We want to extend ∇ to an operator

$$d^\nabla : \Omega^i(\xi) \to \Omega^{i+1}(\xi)$$

by requiring that d^∇ satisfy a suitable Leibnitz rule, similar in spirit to Theorem 3.7.(iii).

Let ξ and η be two vector bundles over M. There is an $\Omega^0(M)$-bilinear product

(7) $$\wedge : \Omega^i(\eta) \otimes \Omega^j(\xi) \to \Omega^{i+j}(\eta \otimes \xi)$$

defined by setting

$$(w \otimes t) \wedge (\tau \otimes s) = w \wedge \tau \otimes (s \otimes t)$$

where $w \in \Omega^i(M)$, $\tau \in \Omega^j(M)$, $s \in \Omega^0(\xi)$ and $t \in \Omega^0(\eta)$ and $w \wedge \tau$ is the exterior product; cf. Theorem 16.13.(ii).

We shall first use the product when $\eta = \varepsilon^1$, the trivial line bundle. In this case $\Omega^i(\eta) = \Omega^i(M)$, and for $i = 0$ the product in (7) is just the $\Omega^0(M)$-module structure on $\Omega^j(\xi)$. Note also for $w \in \Omega^i(M)$ and $s \in \Omega^0(\xi)$ that $w \wedge s = w \otimes s$ in $\Omega^i(\xi)$.

Given three bundles η, θ and ξ one checks from associativity of the exterior product that the product in (7) is associative, and that the constant function $1 \in \Omega^0(M)$ acts as a unit. In particular we record (for $\eta = \varepsilon^1$):

Lemma 17.5 *The product of* (7) *satisfies:*

   (i) $(w \wedge \tau) \wedge \rho = w \wedge (\tau \wedge \rho)$
   (ii) $1 \wedge \rho = \rho$

where $w \in \Omega^i(M), \tau \in \Omega^j(M)$ *and* $\rho \in \Omega^k(\xi)$.   □

Lemma 17.6 *There is a unique $\mathbb{R}$-linear operator $d^\nabla : \Omega^j(\xi) \to \Omega^{j+1}(\xi)$ that satisfies*

   (i) $d^\nabla = \nabla$ *when* $j = 0$
   (ii) $d^\nabla(w \wedge t) = dw \wedge t + (-1)^i w \wedge d^\nabla t$, *where* $w \in \Omega^i(M)$ *and* $t \in \Omega^j(\xi)$.

Proof. Let $\tau \in \Omega^j(M)$ and $s \in \Omega^0(\xi)$ and set $d^\nabla(\tau \otimes s) = dt \wedge s + (-1)^j \tau \wedge \nabla s$. One checks that d^∇ is $\Omega^0(M)$-balanced in the sense of Definition 16.14, and applies Lemma 16.15. Since $s \in \Omega^0(\xi)$ we have $d\tau \wedge s = d\tau \otimes s$, and $d^\nabla = \nabla$ when $j = 0$. We show that (ii) is satisfied. For $\omega \in \Omega^i(M)$ and $t = \tau \otimes s \in \Omega^j(\xi)$,

$$
\begin{aligned}
d^\nabla(\omega \wedge (\tau \otimes s)) &= d^\nabla((\omega \wedge \tau) \otimes s) = d(\omega \wedge \tau) \otimes s + (-1)^{i+j}(\omega \wedge \tau) \wedge \nabla s \\
&= (d\omega \wedge \tau) \otimes s + (-1)^i \omega \wedge d\tau \otimes s + (-1)^{i+j}(\omega \wedge \tau) \wedge \nabla s \\
&= d\omega \wedge (\tau \otimes s) + (-1)^i \omega \wedge d^\nabla(\tau \otimes s) .
\end{aligned}
$$
$\square$

We have now a sequence

(8) $$ 0 \to \Omega^0(\xi) \xrightarrow{\nabla} \Omega^1(\xi) \xrightarrow{d^\nabla} \Omega^2(\xi) \to \cdots $$

which when ξ is the trivial line bundle $\xi = \varepsilon^1$ and $\nabla = d$ is precisely the de Rham complex of Chapter 9. One might expect that (8) is a complex, i.e. that $d^\nabla \circ \nabla = 0$ and $d^\nabla \circ d^\nabla = 0$, but this is in general *not* the case. We do have however that

$$ F^\nabla = d^\nabla \circ \nabla : \Omega^0(\xi) \to \Omega^2(\xi) $$

is $\Omega^0(M)$-linear, since

$$
\begin{aligned}
d^\nabla \circ \nabla(fs) &= d^\nabla(df \wedge s + f \wedge \nabla s) \\
&= ddf \wedge s - df \wedge \nabla s + df \wedge \nabla s + f \wedge d^\nabla(\nabla s) \\
&= f(d^\nabla \circ \nabla(s)).
\end{aligned}
$$

On the other hand Theorem 16.13 gives

(9) $$ \mathrm{Hom}_{\Omega^0(M)}(\Omega^0(\xi), \Omega^2(\xi)) \cong \Omega^2(\mathrm{Hom}\,(\xi, \xi)). $$

Indeed, there is the following string of isomorphisms

$$
\begin{aligned}
\mathrm{Hom}_{\Omega^0(M)}(\Omega^0(\xi), \Omega^2(\xi)) &\cong \mathrm{Hom}_{\Omega^0(M)}(\Omega^0(\xi), \Omega^0(\xi)) \otimes_{\Omega^0(M)} \Omega^2(M) \\
&\cong \Omega^0(\mathrm{Hom}(\xi, \xi)) \otimes_{\Omega^0(M)} \Omega^2(M) \\
&\cong \Omega^2(\mathrm{Hom}(\xi, \xi)).
\end{aligned}
$$

Definition 17.7 The 2-form $F^\nabla \in \Omega^2(\mathrm{Hom}(\xi, \xi))$ is called the *curvature form* of (ξ, ∇). A connection ∇ is called *flat* if $F^\nabla = 0$.

Let $X, Y \in \Omega^0(\tau_M)$ be two vector fields. By evaluating a 2-form τ at (X, Y), we get an $\Omega^0(M)$-linear map

$$ \mathrm{Ev}_{X,Y} : \Omega^2(M) \to \Omega^0(M) $$

which induces a map

$$\mathrm{Ev}_{X,Y}\colon \Omega^2(\mathrm{Hom}(\xi,\xi)) \to \Omega^0(\mathrm{Hom}(\xi,\xi)).$$

We write $F^\nabla_{X,Y} = \mathrm{Ev}_{X,Y}(F^\nabla)$. As for connections, $F^\nabla_{X,Y}(p)\colon \xi_p \to \xi_p$ depends only on the values $X_p, Y_p \in T_pM$ of X, Y in p.

We can calculate F^∇ locally by using (4),

$$
\begin{aligned}
d^\nabla \circ \nabla(e_i) &= \sum dA_{ij} \otimes e_j - \sum A_{ij} \wedge \nabla(e_j) \\
&= \sum dA_{ij} \otimes e_j - \sum A_{ij} \wedge \sum A_{j\nu} \otimes e_\nu \\
&= \sum_\nu \left(dA_{i\nu} \otimes e_\nu - \left(\sum_j A_{ij} \wedge A_{j\nu} \right) \otimes e_\nu \right)
\end{aligned}
$$

so that $F^\nabla(e_i) = \sum_\nu (dA - A \wedge A)_{i\nu} \otimes e_\nu$. In matrix notation:

$$(10) \qquad\qquad F^\nabla = dA - A \wedge A$$

where A is the connection matrix for ∇. In other words, the matrix of the linear map $F^\nabla_{X_p,Y_p}\colon \xi_p \to \xi_p$ in the basis $e_1(p), \dots, e_k(p)$ is $(dA - A \wedge A)_{X_p,Y_p}$.

We next consider the $\Omega^0(M)$-bilinear product

$$\wedge\colon \Omega^i(\xi) \times \mathrm{Hom}_{\Omega^0(M)}\big(\Omega^0(\xi), \Omega^2(\xi)\big) \to \Omega^{i+2}(\xi)$$

which maps a pair $(\omega \otimes s, G)$, with

$$\omega \otimes s \in \Omega^i(M) \otimes_{\Omega^0(M)} \Omega^0(\xi), \qquad G \in \mathrm{Hom}_{\Omega^0(M)}\big(\Omega^0(\xi), \Omega^2(\xi)\big)$$

into

$$(11) \qquad\qquad (\omega \otimes s) \wedge G = \omega \wedge G(s)$$

with the right-hand side given by (7). Alternatively we can use (9) to rewrite (11) as the composition

$$\Omega^i(\xi) \otimes \Omega^2(\mathrm{Hom}\,(\xi,\xi)) \xrightarrow{\wedge} \Omega^{i+2}(\xi \otimes \mathrm{Hom}\,(\xi,\xi)) \to \Omega^{i+2}(\xi)$$

where the last map is induced from the evaluation bundle homomorphism $\xi \otimes \mathrm{Hom}\,(\xi,\xi) \to \xi$.

Lemma 17.8 *The composition* $d^\nabla \circ d^\nabla \colon \Omega^i(\xi) \to \Omega^{i+2}(\xi)$ *maps* t *to* $t \wedge F^\nabla$.

Proof. Let $\omega \otimes s \in \Omega^i(M) \otimes_{\Omega^0(M)} \Omega^0(\xi)$. By Lemma 17.6,

$$
\begin{aligned}
d^\nabla \circ d^\nabla(\omega \otimes s) &= d^\nabla\big(d\omega \otimes s + (-1)^i \omega \wedge \nabla s\big) \\
&= d \circ d(\omega) \otimes s + \omega \wedge d^\nabla \circ \nabla(s) = \omega \wedge F^\nabla(s). \qquad \square
\end{aligned}
$$

We see that the sequence (8) is a chain complex precisely when ∇ is a flat connection ($F^\nabla = 0$). However, as will be clear later, not every vector bundle admits a flat connection.

Example 17.9 Let H be the canonical complex line bundle over $\mathbb{CP}^1$ from Example 15.2. Its total space $E(H)$ consists of pairs $(L, u) \in \mathbb{CP}^1 \times \mathbb{C}^2$ with $u \in L$. Indeed, the map

$$i: S^3 \times_{S^1} \mathbb{C} \to \mathbb{CP}^1 \times \mathbb{C}^2; \quad [z_1, z_2, u] \mapsto ([z_1, z_2], uz_1, uz_2)$$

is a fiberwise monomorphism, whose image is precisely $E(H)$. It follows that a complement to H is the bundle $H^\perp$ with total space

$$E\left(H^\perp\right) = \{(L, v) \mid v \in L^\perp\}.$$

We want to explicate the projection $\pi: \mathbb{CP}^1 \times \mathbb{C}^2 \to E(H)$, which maps (L, u_1, u_2) to the pair (L, u), where u is the orthogonal projection of (u_1, u_2) onto the line L. If $L = [z_1, z_2]$ with $|z_1|^2 + |z_2|^2 = 1$, then

$$\pi(L, u_1, u_2) = (u_1, u_2) \cdot P_L$$

where P_L is the 2×2 matrix

$$P_{[z_1, z_2]} = \begin{pmatrix} \bar{z}_1 z_1 & \bar{z}_1 z_2 \\ \bar{z}_2 z_1 & \bar{z}_2 z_2 \end{pmatrix}.$$

Indeed, if L contains the unit vector $\mathbf{z} = (z_1, z_2)$, then orthogonal projection in $\mathbb{C}^2$ onto L is given by the formula

$$\pi_L(u_1, u_2) = (\bar{z}_1 u_1 + \bar{z}_2 u_2)(z_1, z_2) = (u_1, u_2) P_{[z_1, z_2]}.$$

We examine the (complex) connection from Example 17.3,

$$\nabla: \Omega^0(H) \xrightarrow{i_*} \Omega^0(\epsilon^2) \xrightarrow{\nabla_0} \Omega^1(\epsilon^2) \xrightarrow{\pi_*} \Omega^1(H), \quad \nabla_0 = (d, d),$$

by calculating the connection form A in (4) with respect to sections over the stereographic charts U_1 and U_2 of Example 15.2. Let g be the local parametrization defined as

$$g: \mathbb{R}^2 \to U_1 \subset \mathbb{CP}^1; \quad g(x, y) = [1, z]$$

with $z = x + iy$, and let us consider the section e over U_1

$$e(g(x, y)) = (g(x, y), (1, z)) \in \mathbb{CP}^1 \times \mathbb{C}^2$$

where we also use z to denote the function on U_1 whose value at $g(x, y)$ is $x + iy$. Now

$$\nabla_0(e) = (g(x, y), (0, dz)), \quad dz = dx + idy$$

and hence

$$\nabla(e) = (g(x, y), (0, dz) \cdot P_{g(x, y)}) = \left(g(x, y), \frac{1}{1 + |z|^2} (0, dz) \begin{pmatrix} 1 & z \\ \bar{z} & |z|^2 \end{pmatrix}\right)$$

$$= \left(g(x, y), \frac{1}{1 + |z|^2} (\bar{z} dz, |z|^2 dz)\right).$$

We have shown that $\nabla(e) = A \otimes e$ where A is given as

$$A_{g(x,y)} = \frac{\overline{z}}{1 + |z|^2} \, dz, \quad z(g(x,y)) = x + iy$$

or equivalently

$$g^*(A) = \frac{\overline{z}}{1 + |z|^2} \, dz, \quad z(x,y) = x + iy.$$

We use formula (10) to calculate the curvature form. First note that

$$dz \wedge dz = (dx + idy) \wedge (dx + idy) = 0$$
$$d\overline{z} \wedge dz = (dx - idy) \wedge (dx + idy) = 2idx \wedge dy$$

so that

$$dg^*(A) = \frac{\left(1 + |z|^2\right) - \overline{z} \cdot z}{\left(1 + |z|^2\right)^2} \, d\overline{z} \wedge dz = \frac{2i}{\left(1 + |z|^2\right)^2} \, dx \wedge dy.$$

Since $A \wedge A = 0$ we have the following formula in $\Omega^2(\mathrm{Hom}(g^*H, g^*H))$:

(12)
$$g^*(F^\nabla) = \frac{2i}{\left(1 + |z|^2\right)^2} \, dx \wedge dy.$$

Any complex line bundle H has trivial complex endomorphism bundle $\mathrm{Hom}(H, H)$, because it is a complex line bundle and has a section $e(p) = \mathrm{id}_{H_p}$, which is a basis in every fiber.   In particular the curvature form $F^\nabla \in \Omega^2(\mathrm{Hom}(H, H))$ is just a 2-form with complex values.

It is left for the reader to calculate $h^*(F^\nabla)$ where $h \colon \mathbb{R}^2 \to U_2$ is the parametrization

$$h(x, y) = [z, 1], \quad z = x + iy.$$

This ends the example.

We conclude this chapter by showing that the constructions $f^*(\xi)$, ξ^*, $\mathrm{Hom}(\xi, \eta)$ and $\xi \otimes \eta$ can be extended to constructions on vector bundles equipped with connections.   We begin with the pull-back construction.   Let $f \colon M' \to M$ be a smooth map and ξ a vector bundle over M with connection ∇.   The map $f^* \colon \Omega^0(\xi) \to \Omega^0(f^*(\xi))$; $f^*(s)(p) = s(f(p))$ can be tensored with $f^* \colon \Omega^1(M) \to \Omega^1(M')$, to obtain a linear map

$$f^* \colon \Omega^1(M) \otimes_{\Omega^0(M)} \Omega^0(\xi) \to \Omega^1(M') \otimes_{\Omega^0(M')} \Omega^0(f^*(\xi)).$$

Lemma 17.10 *There exists a unique connection $f^*(\nabla)$ on $f^*(\xi)$ such that the diagram below commutes:*

$$
\begin{array}{ccc}
\Omega^0(\xi) & \xrightarrow{\nabla} & \Omega^1(\xi) \\
\Big\downarrow{f^*} & & \Big\downarrow{f^*} \\
\Omega^0(f^*(\xi)) & \xrightarrow{f^*(\nabla)} & \Omega^1(f^*(\xi))
\end{array}
$$

Proof. The map $f: M' \to M$ induces a homomorphism of rings $\Omega^0(M) \to \Omega^0(M')$, so that every $\Omega^0(M')$-module becomes an $\Omega^0(M)$-module. In particular $\Omega^0(f^*(\xi))$ becomes an $\Omega^0(M)$-module, and there is a homomorphism of $\Omega^0(M)$-modules

$$
f^*: \Omega^0(\xi) \to \Omega^0(f^*(\xi)),
$$

with $f^*(s)(x') = s(f(x'))$. We can then define a homomorphism of $\Omega^0(M')$-modules

$$
\Omega^0(M') \otimes_{\Omega^0(M)} \Omega^0(\xi) \to \Omega^0(f^*(\xi))
$$

by sending $\phi' \otimes s$ into $\phi' \cdot f^*(s)$. This is an isomorphism; cf. Exercise 17.13. It follows that

$$
\Omega^k(f^*(\xi)) = \Omega^k(M') \otimes_{\Omega^0(M')} \Omega^0(f^*(\xi)) \cong \Omega^k(M') \otimes_{\Omega^0(M)} \Omega^0(\xi).
$$

Similarly, pull-back of differential forms

$$
f^*: \Omega^k(M) \to \Omega^k(M')
$$

is $\Omega^0(M)$-linear and induces a homomorphism

$$
\Omega^0(M') \otimes_{\Omega^0(M)} \Omega^k(M) \to \Omega^k(M'); \quad \phi \otimes \omega \mapsto \phi f^*(\omega).
$$

This is not an isomorphism, but applying the functor $(-) \otimes_{\Omega^0(M)} \Omega^0(\xi)$ one gets a homomorphism

$$
\rho: \Omega^0(M') \otimes_{\Omega^0(M)} \Omega^k(\xi) \to \Omega^k(M') \otimes_{\Omega^0(M)} \Omega^0(\xi).
$$

The sum of the maps

$$
d \otimes 1: \Omega^0(M') \otimes_{\Omega^0(M)} \Omega^0(\xi) \to \Omega^1(M') \otimes_{\Omega^0(M)} \Omega^0(\xi)
$$
$$
\rho(1 \otimes \nabla): \Omega^0(M') \otimes_{\Omega^0(M)} \Omega^0(\xi) \to \Omega^1(M') \otimes_{\Omega^0(M)} \Omega^0(\xi)
$$

defines the required connection

$$
f^*(\nabla): \Omega^0(f^*(\xi)) \to \Omega^1(f^*(\xi)). \qquad \square
$$

We note that if $A(\mathbf{e})$ is the connection matrix for ∇ w.r.t. a frame $\mathbf{e}$ for $\xi|_U$ then $f^*(A(\mathbf{e}))$ is the connection matrix for $f^*(\nabla)$ w.r.t. the frame $\mathbf{e} \circ f$ for $f^*(\xi)_{f^{-1}(U)}$. There is a commutative diagram corresponding to that of Lemma 17.10 where ∇ is replaced by $d^\nabla : \Omega^1(\xi) \to \Omega^2(\xi)$, and thus also a diagram

$$
\begin{array}{ccc}
\Omega^0(\xi) & \xrightarrow{\ F^\nabla\ } & \Omega^2(\xi) \\
\Big\downarrow{\scriptstyle f^*} & & \Big\downarrow{\scriptstyle f^*} \\
\Omega^0(f^*(\xi)) & \xrightarrow{\ F^{f^*(\nabla)}\ } & \Omega^2(f^*(\xi))
\end{array}
$$

Since $f^*\mathrm{Hom}(\xi,\xi) = \mathrm{Hom}(f^*(\xi), f^*(\xi))$ the above gives

(13) $$ f^*(F^\nabla) = F^{f^*(\nabla)}. $$

Consider the non-singular pairing

$$ (\, , \,) : \Omega^i(\xi) \otimes \Omega^j(\xi^*) \xrightarrow{\ \wedge\ } \Omega^{i+j}(\xi \otimes \xi^*) \to \Omega^{i+j}(M) $$

where the last map is induced from the bundle map $\xi \otimes \xi^* \to \varepsilon^1$. For $i = j = 0$,

$$ \Omega^0(\xi^*) \cong \mathrm{Hom}_{\Omega^0(M)}\big(\Omega^0(\xi), \Omega^0(M)\big) $$

by Theorem 16.13.(iii), and the above pairing corresponds to the evaluation

$$ \langle\, , \,\rangle : \Omega^0(\xi) \otimes \mathrm{Hom}_{\Omega^0(M)}\big(\Omega^0(\xi), \Omega^0(M)\big) \to \Omega^0(M). $$

For general i and j,

$$ (\omega \otimes s, \tau \otimes s^*) = (\omega \wedge \tau) \otimes \langle s, s^* \rangle $$

where $\omega \in \Omega^i(M)$, $\tau \in \Omega^j(M)$ and $s \in \Omega^0(\xi)$, $s^* \in \Omega^0(\xi)$.

Given a connection ∇_ξ on ξ, we define the connection ∇_{ξ^*} on ξ^* by requiring

(14) $$ d(s, s^*) = \big(\nabla_\xi(s), s^*\big) + \big(s, \nabla_{\xi^*}(s^*)\big). $$

This specifies ∇_{ξ^*} uniquely because the pairing $(\, , \,)$ is non-singular. The desired connection on the tensor product is defined analogously. Indeed the product from (7) induces a $\Omega^0(M)$-linear map

$$ \wedge : \Omega^i(\xi) \otimes_{\Omega^0(M)} \Omega^j(\eta) \to \Omega^{i+j}(\xi \otimes \eta), $$

which for $i = j = 0$ is the isomorphism

$$ \Omega^0(\xi) \otimes_{\Omega^0(M)} \Omega^0(\eta) \cong \Omega^0(\xi \otimes \eta) $$

from Theorem 16.13.(ii). Define

(15) $$ \nabla_{\xi \otimes \eta}(s \otimes t) = \nabla_\xi(s) \wedge t + s \wedge \nabla_\eta(t). $$

Finally we can combine (14) and (15) to define

$$ \nabla_{\xi^* \otimes \eta}(s \otimes t) = \nabla_{\xi^*}(s) \wedge t + s \wedge \nabla_\xi(t). $$

Since $\xi^* \otimes \eta \cong \mathrm{Hom}(\xi, \eta)$, this defines a connection on $\mathrm{Hom}(\xi, \eta)$. Alternatively one can apply the evaluation $\Omega^0(\mathrm{Hom}(\xi, \eta)) \times \Omega^0(\xi) \to \Omega^0(\eta)$ and the induced $\Omega^0(M)$-bilinear product $(\, , \,) : \Omega^i(\xi) \times \Omega^j(\mathrm{Hom}(\xi, \eta)) \to \Omega^{i+j}(\eta)$ and define $\nabla_{\mathrm{Hom}(\xi,\eta)}$ by the formula

(16) $$ \nabla_\eta((s, \phi)) = \big(\nabla_\xi(s), \phi\big) + \big(s, \nabla_{\mathrm{Hom}(\xi,\eta)}(\phi)\big). $$

Lemma 17.11 *Under the identification* $\alpha \colon \xi^* \otimes \eta \xrightarrow{\cong} \mathrm{Hom}\,(\xi, \eta)$, $\nabla_{\xi^* \otimes \eta} = \nabla_{\mathrm{Hom}(\xi,\eta)}$.

Proof. There is a commutative diagram of vector bundles over M

$$
\begin{array}{ccc}
\xi \otimes \xi^* \otimes \eta & \xrightarrow{\mathrm{id}\otimes\alpha} & \xi \otimes \mathrm{Hom}(\xi, \eta) \\
\downarrow{\scriptstyle (\;,\;)\otimes\mathrm{id}_\eta} & & \downarrow{\scriptstyle (\;,\;)} \\
\varepsilon_M^1 \otimes \eta & \xrightarrow{\mathrm{mult}} & \eta
\end{array}
$$

and a corresponding diagram of sections. Let $s \in \Omega^0(\xi)$, $t \in \Omega^0(\eta)$ and $s^* \in \Omega^0(\xi^*)$. Then

$$
\begin{aligned}
\nabla_\eta((s, \alpha(s^* \otimes t))) &= \big(\nabla_\xi(s), \alpha(s^* \otimes t)\big) + \big(s, \nabla_{\mathrm{Hom}(\xi,\eta)}(\alpha(s^* \otimes t))\big) \\
d((s, s^*)) &= \big(\nabla_\xi(s), s^*\big) + \big(s, \nabla_{\xi^*}(s^*)\big) \\
\nabla_{\xi^* \otimes \eta}(s^* \otimes t) &= \nabla_{\xi^*}(s^*) \wedge t + s^* \wedge \nabla_\eta(t).
\end{aligned}
$$

From the diagram we get that $(s, \alpha(s^* \otimes t)) = (s, s^*)t$, and hence

$$
\begin{aligned}
\big(\nabla_\xi(s), \alpha(s^* \otimes t)\big) &= \big(\nabla_\xi(s), s^*\big) \wedge t \\
\big(s, \nabla_{\xi^* \otimes \eta} Im(s^* \otimes t)\big) &= \big(s, \nabla_{\xi^*}(s^*)\big) \wedge t + (s, s^*) \nabla_\eta (t).
\end{aligned}
$$

On the other hand, using these formulas we have

$$
\begin{aligned}
\big(s, \nabla_{\mathrm{Hom}(\xi,\eta)}(\alpha(s^* \otimes t))\big) &= d((s, s^*)t) - \big(\nabla_\xi(s), s^*\big) \wedge t \\
&= d((s, s^*)) \wedge t + (s, s^*) \nabla_\eta (t) - \big(\nabla_\xi(s), s^*\big) \wedge t \\
&= \big(s, \nabla_{\xi^*}(s^*)\big) \wedge t + (s, s^*) \nabla_\eta (t),
\end{aligned}
$$

and the assertion follows. $\square$

Each of the connections from (14), (15) and (16) can be extended to linear maps

$$
\Omega^i(\xi^*) \xrightarrow{d^\nabla} \Omega^{i+1}(\xi^*)
$$
$$
\Omega^i(\xi \otimes \eta) \xrightarrow{d^\nabla} \Omega^{i+1}(\xi \otimes \eta)
$$
$$
\Omega^i(\mathrm{Hom}(\xi, \eta)) \xrightarrow{d^\nabla} \Omega^{i+1}(\mathrm{Hom}(\xi, \eta))
$$

and the defining formulas generalize to the following lemma, whose proof is left to the reader.

Lemma 17.12 *Let* $s \in \Omega^i(\xi)$, $s^* \in \Omega^i(\xi^*)$, $t \in \Omega^j(\eta)$ *and* $\phi \in \Omega^j(\mathrm{Hom}(\xi, \eta))$. *We have*

(i) $d((s, s^*)) = (d^\nabla(s), s^*) + (-1)^i (s, d^\nabla(s^*))$

(ii) $d^\nabla(s \otimes t) = d^\nabla s \otimes t + (-1)^i s \otimes d^\nabla(t)$

(iii) $d^\nabla((s, \phi)) = (d^\nabla(s), \phi) + (-1)^i (s, d^\nabla \phi)$

where d^∇ corresponds to $\nabla = \nabla_\xi, \nabla_\eta, \nabla_{\xi^*}$ and $\nabla_{\mathrm{Hom}(\xi,\xi^*)}$, respectively.   □

The definitions above may appear somewhat abstract, so let us state in local coordinates the case of most importance for our later use. Let $\nabla = \nabla_\xi$ be a connection on ξ, and let $e = (e_1, \ldots, e_k)$ be a frame over U. This defines isomorphisms

$$\xi|_U \cong U \times \mathbb{R}^k, \qquad \mathrm{Hom}(\xi, \xi)|_U \cong U \times M_k(\mathbb{R})$$

and induces

$$\Omega^n(\xi|_U) \cong \Omega^n(U)^{\oplus k}, \qquad \Omega^n(\mathrm{Hom}(\xi, \xi)|_U) \cong M_k(\Omega^n(U)).$$

The connections $\nabla = \nabla_\xi$ and $\hat{\nabla} = \nabla_{\mathrm{Hom}(\xi,\xi)}$ and the induced d^∇ and $d^{\hat{\nabla}}$ then become

$$d^\nabla : \Omega^n(U)^{\oplus k} \to \Omega^{n+1}(U)^{\oplus k}, \qquad d^{\hat{\nabla}} : M_k(\Omega^n(U)) \to M_k(\Omega^{n+1}(U)).$$

They are given as

(17)
$$d^\nabla(s_1, \ldots, s_k) = (ds_1, \ldots, ds_k) + (s_1, \ldots, s_k) \wedge A$$
$$d^{\hat{\nabla}}(\Phi) = d\Phi - (A \wedge \Phi - (-1)^n \Phi \wedge A)$$

where $A = A(e)$ is the connection matrix. The first formula follows from (5); the second is proved quite similarly.

If e' is another frame for $\xi|_U$ then $e' = G \cdot e$ with $G \in \mathrm{GL}_k(\Omega^0(U))$, and the connection matrices $A = A(e)$, $A' = A(e)$ and the curvature forms $F^\nabla(e)$, $F^\nabla(e')$ are related by

(18)
$$A' = (dG)G^{-1} + GAG^{-1}$$
$$F^\nabla(e') = GF^\nabla(e)G^{-1}$$

cf. Exercise 17.8. The first formula follows from (5) applied to the equation $(s_1, \ldots, s_k) = (s'_1, \ldots, s'_k)G$. The second formula follows from the first one and (10).

Theorem 17.13 (Bianchi's identity) *We have $d^\nabla F^\nabla = 0$, where d^∇ is associated to the connection $\nabla = \nabla_{\mathrm{Hom}(\xi,\xi)}$.*

Proof. Use the local forms (10) and (17) to get

$$F^\nabla = dA - A \wedge A$$
$$d^\nabla F^\nabla = -d(A \wedge A) + F^\nabla \wedge A - A \wedge F^\nabla$$
$$= -d(A \wedge A) + dA \wedge A - A \wedge dA = 0.$$ □

The product in $\Omega^0(\mathrm{Hom}(\xi,\xi))$ associated to fiberwise composition induces a product

$$\wedge : \Omega^i(\mathrm{Hom}(\xi,\xi)) \otimes \Omega^j(\mathrm{Hom}(\xi,\xi)) \to \Omega^{i+j}(\mathrm{Hom}(\xi,\xi)).$$

It is not hard to show that d^∇ is a derivation with respect to this product, i.e. that

(19) $$d^\nabla(R_1 \wedge R_2) = d^\nabla(R_1) \wedge R_2 + (-1)^i R_1 \wedge d^\nabla(R_2).$$

The trace homomorphism

$$\mathrm{Tr}\colon \mathrm{Hom}\,(V,V) \to \mathbb{R}$$

can be defined without reference to choice of basis as the composition

$$\mathrm{Hom}(V,V) \overset{\cong}{\to} V^* \otimes V \overset{\mathrm{ev}}{\to} \mathbb{R}$$

where $\mathrm{ev}(f \otimes v) = f(v)$. It induces a trace

$$\mathrm{Tr}\colon \mathrm{Hom}(\xi,\xi) \to \varepsilon^1$$

of vector bundles, and hence in turn a trace

$$\mathrm{Tr}\colon \Omega^i(\mathrm{Hom}\,(\xi,\xi)) \to \Omega^i(M)$$

into the i-forms on M, and we have:

Theorem 17.14 *For $\phi \in \Omega^i(\mathrm{Hom}(\xi,\xi))$,*

$$d\,\mathrm{Tr}\,(\phi) = \mathrm{Tr}(d^\nabla \phi)$$

where d^∇ is associated to $\nabla = \nabla_{\mathrm{Hom}(\xi,\xi)}$.

Proof. Let $s \in \Omega^0(\xi)$, $s^* \in \Omega^0(\xi^*)$, $\omega \in \Omega^i(M)$ and suppose

$$\phi = \omega \otimes s \otimes s^* \in \Omega^i(M) \otimes_{\Omega^0(M)} \Omega^0(\xi) \otimes_{\Omega^0(M)} \Omega^0(\xi^*) \cong \Omega^i(\mathrm{Hom}(\xi,\xi)).$$

Then

$$\begin{aligned} d^\nabla \phi &= d\omega \otimes (s \otimes s^*) + (-1)^i \omega \otimes \nabla(s \otimes s^*) \\ &= d\omega \otimes (s \otimes s^*) + (-1)^i \omega \otimes \left(\nabla_\xi(s) \otimes s^* + s \otimes \nabla_{\xi^*}(s^*)\right) \end{aligned}$$

and we get

$$\begin{aligned} \mathrm{Tr}\,d^\nabla \phi &= (d\omega)(s,s^*) + (-1)^i \omega \wedge \left(\left(\nabla_\xi(s), s^*\right) + \left(s, \nabla_{\xi^*}(s^*)\right)\right) \\ &= (s,s^*)d\omega + (-1)^i \omega \wedge d((s,s^*)) \\ &= (s,s^*)d\omega + d((s,s^*)) \wedge \omega = d((s,s^*)\omega) = d\,\mathrm{Tr}\phi. \quad \square \end{aligned}$$

We have mostly formulated the above theorems for real smooth vector bundles, but there are of course completely analogous results for complex vector bundles upon starting with a complex connection ∇. One simply replaces $\Omega^i(M)$ by $\Omega^i(M; \mathbb{C})$, and $\otimes_{\mathbb{R}}$ by $\otimes_{\mathbb{C}}$ and $\mathrm{Hom}_{\mathbb{R}}$ by $\mathrm{Hom}_{\mathbb{C}}$ throughout, and requires maps to be $\mathbb{C}$-linear rather than $\mathbb{R}$-linear. This complex version will be used below in Chapter 18.

Let ξ be a complex vector bundle with complex connection ∇. Combining Theorems 17.13 and 17.14 we see that the 2-form $\mathrm{Tr}(F^\nabla) \in \Omega^2(M; \mathbb{C})$ is closed, and thus defines a cohomology class in complex de Rham cohomology: $[\mathrm{Tr}(F^\nabla)] \in H^2(M; \mathbb{C})$ More generally, it follows from (19) that the trace of

$$F^\nabla \wedge \ldots \wedge F^\nabla \in \Omega^{2k}(\mathrm{Hom}_{\mathbb{C}}(\xi, \xi))$$

is a closed form in $\Omega^{2k}(M)$.

Definition 17.15 The k-th *Chern character class* of (ξ, ∇) is the cohomology class

$$\mathrm{ch}_k(\xi, \nabla) = \frac{(-1)^k}{\left(2\pi\sqrt{-1}\right)^k k!} \left[\mathrm{Tr}\left(F^\nabla \wedge \ldots \wedge F^\nabla\right)\right] \in H^{2k}(M; \mathbb{C}).$$

Here $H^*(M; \mathbb{C}) = H^*(M) \otimes_{\mathbb{R}} \mathbb{C}$ is the cohomology of the complexified de Rham complex. The normalizing factor in Definition 17.15 is chosen so that the cohomology class is actually real, i.e. a class in $H^{2k}(M)$. This will be proved in the following chapters, where we also show that the cohomology class is independent of the choice of connection.

A real vector bundle ξ can be complexified $\xi_{\mathbb{C}} = \xi \otimes_{\mathbb{R}} \varepsilon^1_{\mathbb{C}}$, and a real connection ∇ on ξ induces a complex connection $\nabla_{\mathbb{C}}$ on $\xi_{\mathbb{C}}$.

Definition 17.16 The k-th Pontryagin character class $\mathrm{ph}_k(\xi, \nabla)$ is the class $\mathrm{ph}_k(\xi, \nabla) = \mathrm{ch}_{2k}(\xi_{\mathbb{C}}, \nabla_{\mathbb{C}}) \in H^{4k}(M; \mathbb{C})$.

18. CHARACTERISTIC CLASSES OF COMPLEX VECTOR BUNDLES

In this chapter ξ will be an n-dimensional complex smooth vector bundle over a smooth compact manifold M, and $\Omega^*(M;\mathbb{C})$ will denote the de Rham complex with *complex* coefficients. Connections in this chapter will always be complex. Let $P(A) = P(\ldots, A_{ij}, \ldots)$ be a homogeneous invariant polynomial of n^2 variables displayed as an $n \times n$ matrix; cf. Appendix B. The most important examples are

$$(1) \qquad P(A) = \sigma_k(A) \quad \text{and} \quad P(A) = s_k(A) = \mathrm{Tr}(A^k)$$

where $\sigma_k(A)$ is the coefficient of t^k in the characteristic polynomial

$$\det(I + tA) = \sum_{k=0}^{n} \sigma_k(A) t^k.$$

They both have degree k. Since the wedge product is commutative on even-dimensional forms, we can replace the variables $A = (A_{ij})$ by differential 2-forms, $A_{ij} \in \Omega^2(M;\mathbb{C})$, and thus obtain a $2k$-form $P(A) \in \Omega^{2k}(M;\mathbb{C})$. More generally we define a map

$$(2) \qquad P:\Omega^2(\mathrm{Hom}_{\mathbb{C}}(\xi,\xi)) \to \Omega^{2k}(M;\mathbb{C}).$$

Let $\mathbf{e}$ be a frame of $\xi_{|U}$; it induces an isomorphism onto the trivial bundle,

$$\mathrm{Hom}_{\mathbb{C}}(\xi,\xi)\big|_U \to U \times M_n(\mathbb{C})$$

and hence an isomorphism

$$\Omega^2\left(\mathrm{Hom}_{\mathbb{C}}(\xi,\xi)_{|U}\right) \to \Omega^2(U; M_n(\mathbb{C})) \overset{\cong}{\to} M_n\left(\Omega^2(U;\mathbb{C})\right).$$

Thus a 2-form R of $\mathrm{Hom}_{\mathbb{C}}(\xi,\xi)$ gives a matrix of 2-forms $R(\mathbf{e}) = (R_{ij})$. Apply P and get an element $P(R(\mathbf{e})) \in \Omega^{2k}(U;\mathbb{C})$. Since P is invariant, and since for any other choice of frame $R(\mathbf{e}) = gR(\mathbf{e}')g^{-1}$ with $g \in \Omega^0(U; M_n(\mathbb{C}))$ invertible, we have

$$P(R(\mathbf{e})) = P\left(R(\mathbf{e}')\right).$$

It follows that we have defined a global $2k$-form on M which we denote $P(R)$. For example we have (locally)

$$(3) \quad s_1(R) = \sum R_{ii}, \quad s_2(R) = \sum R_{ij} \wedge R_{ji}, \quad \sigma_2(R) = \frac{1}{2}\left(s_1(R)^2 - s_2(R)\right).$$

Choose a complex connection ∇ on ξ and apply the above with $R = F^\nabla$ to get a $2k$-form

$$(4) \qquad P\left(F^\nabla\right) \in \Omega^{2k}(M;\mathbb{C}).$$

Here are two fundamental lemmas:

Lemma 18.1 *For each invariant polynomial and connection ∇, $P(F^\nabla)$ is a closed form.*

Lemma 18.2 *The cohomology class $[P(F^\nabla)]$ in $H^*(M; \mathbb{C})$ is independent of the choice of connection.*

The first lemma follows from Theorems 17.13 and 17.14 and results of Appendix B, but there is also the following attractive alternative proof ([Milnor–Stasheff]).

Proof of Lemma 18.1. Choose a frame for ξ over U, and let ∇ have the connection matrix $A = (A_{ij})$, so that

$$F^\nabla = dA - A \wedge A = (F_{ij}).$$

In local terms Bianchi's identity is $dF^\nabla = A \wedge F^\nabla - F^\nabla \wedge A$, cf. (17.17), so

$$(5) \qquad dP(F^\nabla) = \sum \frac{\partial P}{\partial A_{ij}}(F^\nabla) \wedge dF_{ij} = \mathrm{Tr}\big(P'(F^\nabla) \wedge dF^\nabla\big)$$

where $P'(A)$ is the transpose of the matrix of partial derivatives

$$P'(A) = \left(\frac{\partial P}{\partial A_{ij}}\right)^t.$$

For an invariant polynomial P one has

$$(6) \qquad\qquad\qquad P'(A)A = AP'(A).$$

This is seen by applying the operator $\frac{d}{dt}$ to the equation

$$P((I + tE_{ij})A) = P(A(I + tE_{ij}))$$

where E_{ij} is the basic matrix with 1 in the (i, j)-th entry and zero elsewhere. Now (6) yields the relation

$$(7) \qquad\qquad\qquad P'(F^\nabla) \wedge F^\nabla = F^\nabla \wedge P'(F^\nabla),$$

and using (5) and the Bianchi identity we get

$$-dP(F^\nabla) = \mathrm{Tr}\big(P'(F^\nabla) \wedge F^\nabla \wedge A - P'(F^\nabla) \wedge A \wedge F^\nabla\big)$$
$$= \mathrm{Tr}\big(F^\nabla \wedge (P'(F^\nabla) \wedge A) - (P'(F^\nabla) \wedge A) \wedge F^\nabla\big) = 0. \qquad \square$$

Proof of Lemma 18.2. Let ∇_0, ∇_1 be two connections on ξ, and $\pi\colon M \times \mathbb{R} \to M$ the projection onto the first factor. Let $\widehat{\nabla}_\nu = \pi^*(\nabla_\nu)$ be the induced connections on $\pi^*(\xi)$; cf. Lemma 17.10. Define a new connection on $\pi^*(\xi)$ by

$$\widehat{\nabla}(s)(p,t) = (1-t)\widehat{\nabla}_0(s)(p,t) + t\widehat{\nabla}_1(s)(p,t)$$

where $(p,t) \in M \times \mathbb{R}$. Apply Lemma 17.10 to see that

$$i_0^*\big(\widehat{\nabla}\big) = \nabla_0, \quad i_1^*\big(\widehat{\nabla}\big) = \nabla_1$$

where $i_\nu\colon M \to M \times \mathbb{R}$ are the two inclusions in respectively top and bottom. From (17.21) it follows that

$$i_0^*(F^{\widehat{\nabla}}) = F^{\nabla_0}, \quad i_1^*(F^{\widehat{\nabla}}) = F^{\nabla_1}$$

and hence $i_\nu^*(P(F^{\widehat{\nabla}})) = P\big(F^{\nabla_\nu}\big)$, $\nu = 0,1$. Since $i_0 \simeq i_1$ and $P(F^{\widehat{\nabla}})$ is closed, we have that $i_0^*([P(F^{\widehat{\nabla}})]) = i_1^*([P(F^{\widehat{\nabla}})])$. $\square$

Note that isomorphic vector bundles define identical cohomology classes $[P(F^\nabla)]$, since a smooth fiberwise isomorphism $\hat{f}\colon \xi \to \xi'$ induces isomorphisms between section spaces, and since we can choose connections to make the diagram

$$
\begin{array}{ccc}
\Omega^0(\xi) & \xrightarrow{\nabla} & \Omega^1(\xi) \\
\downarrow{\scriptstyle \hat{f}_*} & & \downarrow{\scriptstyle \hat{f}_*} \\
\Omega^0(\xi') & \xrightarrow{\nabla'} & \Omega^1(\xi')
\end{array}
$$

commute.   Thus the matrices for F^∇ and $F^{\nabla'}$ are identical with respect to corresponding frames for ξ and ξ', and $P(F^\nabla) = P(F^{\nabla'})$.

In particular, if ξ is a trivial vector bundle, then $[P(\xi)] = [P(\varepsilon_\mathbb{C}^n)] = 0$. Indeed, we just use the flat connection ∇_0 on $\varepsilon_\mathbb{C}^n$.

Definition 18.3

(i) The k-*th Chern class* of the complex vector bundle ξ is

$$c_k(\xi) = \left[\sigma_k\left(\frac{-1}{2\pi\sqrt{-1}} F^\nabla\right)\right] \in H^{2k}(M;\mathbb{C}).$$

(ii) The k-*th Chern character class* is

$$\mathrm{ch}_k(\xi) = \frac{1}{k!}\left[s_k\left(\frac{-1}{2\pi\sqrt{-1}} F^\nabla\right)\right] \in H^{2k}(M;\mathbb{C}).$$

Here ∇ is any complex connection on ξ. If $k = 0$ then $c_0(\xi) = 1$ and $\mathrm{ch}_0(\xi) = \dim \xi$.

The reader may check that Definition 18.3.(ii) agrees with Definition 17.15. We shall prove some properties of these classes. First note that they determine each other, since

$$\mathrm{ch}_k(\xi) = \frac{1}{k!} s_k(\xi), \quad s_k(\xi) = Q_k(c_1(\xi), \ldots, c_k(\xi)), \quad c_k(\xi) = P_k(s_1(\xi), \ldots, s_k(\xi))$$

for certain polynomials P_k and Q_k; cf. Appendix B. For example we have

$$\mathrm{ch}_1(\xi) = c_1(\xi) \quad \text{and} \quad \mathrm{ch}_2(\xi) = \tfrac{1}{2} c_1(\xi)^2 - c_2(\xi)$$

The integration homomorphism

$$I : H^2(\mathbb{CP}^1; \mathbb{C}) \to \mathbb{C}$$

is an isomorphism by Corollary 10.14, and the inclusion $j : \mathbb{CP}^1 \subset \mathbb{CP}^n$ induces an isomorphism

$$j^* : H^2(\mathbb{CP}^n; \mathbb{C}) \to H^2(\mathbb{CP}^1; \mathbb{C})$$

by Theorem 14.3. We now chose $c \in H^2(\mathbb{CP}^n; \mathbb{C})$ once and for all with the property that

$$(8) \qquad\qquad\qquad I(j^*(c)) = -1.$$

It follows from Example 14.10 that $-\pi c$ is the cohomology class of the volume form of $\mathbb{CP}^1$ with the Fubini–Study metric (cf. Theorem 14.8) and if we identify S^2 with $\mathbb{CP}^1$ via Ψ then $-4\pi c$ corresponds to the volume form of S^2 in its natural metric as the unit sphere and with its complex orientation.

Let H_n be the canonical line bundle on $\mathbb{CP}^n$ with total space

$$E(H_n) = \{(L, u) \in \mathbb{CP}^n \times \mathbb{C}^{n+1} \mid u \in L\}.$$

Then $j^*(H_n) = H_1$ is the canonical line bundle of Example 15.2.

Theorem 18.4 *The integration homomorphism maps $c_1(H_1)$ to -1.*

Proof. Apply the two positively oriented stereographic charts ψ_- and $\overline{\psi}_+$ on $S^2 = \mathbb{CP}^1$ from Example 14.1. In Example 17.9 we calculated the pre-image of the curvature form F^∇ under $g = (\psi_-)^{-1}$ to be

$$g^*(F^\nabla) = \frac{2i}{\left(1 + |z|^2\right)} \, dx \wedge dy.$$

We integrate this form by changing to polar coordinates $(x, y) = (r \cos \theta, r \sin \theta)$. Since

$$dx = \cos \theta dr - r \sin \theta d\theta \quad \text{and} \quad dy = \sin \theta dr + r \cos \theta d\theta$$

we see that $dx \wedge dy = r \, dr \wedge d\theta$, and

$$\int_{\mathbb{R}^2} g^* \left(F^\nabla \right) = 2i \int_0^{2\pi} \int_0^\infty \frac{r dr \wedge d\theta}{(1 + r^2)^2} = 4\pi i \int_0^\infty \frac{r \, dr}{(1 + r^2)^2}$$

$$= 2\pi i \int_0^\infty \frac{ds}{(1 + s)^2} = -2\pi i \left[\frac{1}{1 + s} \right]_{s=0}^\infty = 2\pi i.$$

This calculation implies that

$$\int_{\mathbb{CP}^1} F^\nabla = 2\pi i.$$

Indeed we can apply a partition of unity $1 = \rho_0 + \rho_1$ with suppρ_0 an arbitrarily large ϵ^{-1}-sphere in the chart g and suppρ_1 a correspondingly small ϵ-sphere in the other chart. In the limit $\epsilon \to 0$ the integral of $g^*(F^\nabla)$ (over all of $\mathbb{R}^2$) is equal to the integral of F^∇ over $\mathbb{CP}^1$. □

Theorem 18.5 *Let $f: N \to M$ be a smooth map and ξ a complex vector bundle on M. For every invariant polynomial we have $f^*[P(\xi)] = [P(f^*(\xi))]$.*

Proof. We give $f^*(\xi)$ the connection $f^*(\nabla)$ of Lemma 17.10. By formula (17.13), $f^*(F^\nabla) = F^{f^*(\nabla)}$. Hence $f^*(P(F^\nabla)) = P(F^{f^*(\nabla)})$. □

For a line bundle L, $\Omega^2(\text{Hom}(L, L)) = \Omega^2(M; \mathbb{C})$ so that $F^\nabla \in \Omega^2(M; \mathbb{C})$, and

$$s_k(F^\nabla) = F^\nabla \wedge \ldots \wedge F^\nabla.$$

This gives

(9) $$\text{ch}_k(L) = \frac{1}{k!} \text{ch}_1(L)^k = \frac{1}{k!} c_1(L)^k$$

so that $\text{ch}_k(L)$ becomes the k-th term in the power series $\exp(c_1(L))$.

Theorem 18.6 *For a sum of complex vector bundles,*

(i) $\text{ch}_k(\xi_0 \oplus \xi_1) = \text{ch}_k(\xi_0) + \text{ch}_k(\xi_1)$

(ii) $c_k(\xi_0 \oplus \xi_1) = \sum_{\nu=0}^k c_\nu(\xi_0) c_{k-\nu}(\xi_1).$

Proof. Choose complex connections ∇_ν on ξ_ν. We identify $\Omega^i(\xi_0 \oplus \xi_1)$ with $\Omega^i(\xi_0) \oplus \Omega^i(\xi_1)$; then

$$\nabla_0 \oplus \nabla_1 \colon \Omega^0(\xi_0 \oplus \xi_1) \to \Omega^1(\xi_0 \oplus \xi_1)$$

is a connection on $\xi_0 \oplus \xi_1$ with curvature

$$F^{\nabla_0} \oplus F^{\nabla_1} \in \Omega^2(\mathrm{Hom}(\xi_0 \oplus \xi_1, \xi_0 \oplus \xi_1)).$$

For direct sum of matrices

$$A_0 \oplus A_1 = \begin{pmatrix} A_0 & 0 \\ 0 & A_1 \end{pmatrix} \in M_{n+m}(\mathbb{C})$$

formula (3) of Appendix B gives the equations

$$s_k(A_0 \oplus A_1) = s_k(A_0) + s_k(A_1) \quad \text{and} \quad \sigma_k(A_0 \oplus A_1) = \sum_{\nu=0}^{k} \sigma_\nu(A_0) \cdot \sigma_{k-\nu}(A_1),$$

which prove the assertions. $\qquad\qquad\qquad\qquad\qquad\qquad\qquad\qquad\qquad\qquad\quad$ $\square$

Theorem 18.7 *For a tensor product of complex vector bundles,*

$$\mathrm{ch}_k(\xi_0 \otimes \xi_1) = \sum_{\nu=0}^{k} \mathrm{ch}_\nu(\xi_0)\mathrm{ch}_{k-\nu}(\xi_1)$$

where $\mathrm{ch}_0(\xi_\nu) = \dim_{\mathbb{C}} \xi_\nu$.

Proof. The tensor product of linear maps, applied fiberwise, defines a map of vector bundles

$$\mathrm{Hom}(\xi_0, \xi_0) \otimes \mathrm{Hom}(\xi_1, \xi_1) \to \mathrm{Hom}(\xi_0 \otimes \xi_1, \xi_0 \otimes \xi_1)$$

and thus a product

$$\wedge \colon \Omega^i(\mathrm{Hom}(\xi_0, \xi_0)) \otimes \Omega^j(\mathrm{Hom}(\xi_1, \xi_1)) \to \Omega^{i+j}(\mathrm{Hom}(\xi_0 \otimes \xi_1, \xi_0 \otimes \xi_1)).$$

For connections ∇_0, ∇_1 on ξ_0, ξ_1, we have the connection ∇ on $\xi_0 \otimes \xi_1$ from (17.15):

$$\nabla(s_0 \otimes s_1) = \nabla_0(s_0) \wedge s_1 + s_0 \wedge \nabla_1(s_1).$$

The corresponding curvature form becomes $F^\nabla = F^{\nabla_0} \wedge \mathrm{id} + \mathrm{id} \wedge F^{\nabla_1}$ where $\mathrm{id} \in \Omega^0(\mathrm{Hom}(\xi_\nu, \xi_\nu))$ is the section that maps $p \in M$ to $\mathrm{id} \colon \xi_p \to \xi_p$. It follows that

(10) $$F^\nabla \wedge \ldots \wedge F^\nabla = \sum_{i=1}^{k} \binom{k}{i} (F^{\nabla_0})^{\wedge i} \wedge (F^{\nabla_1})^{\wedge(k-i)}.$$

There is a commutative diagram:

$$\Omega^i(\mathrm{Hom}(\xi_0,\xi_0)) \otimes \Omega^j(\mathrm{Hom}(\xi_1,\xi_1)) \xrightarrow{\;\wedge\;} \Omega^{i+j}(\mathrm{Hom}(\xi_0\otimes\xi_1,\xi_0\otimes\xi_1))$$

(11)
$$\downarrow \mathrm{Tr}\otimes\mathrm{Tr} \qquad\qquad\qquad\qquad\qquad \downarrow \mathrm{Tr}$$

$$\Omega^i(M;\mathbb{C})\otimes\Omega^j(M;\mathbb{C}) \xrightarrow{\;\wedge\;} \Omega^{i+j}(M;\mathbb{C})$$

From (10) and (11) we get

$$s_k(F^\nabla) = \sum_{i=0}^{k} \binom{k}{i} s_i(F^{\nabla_0}) s_{k-i}(F^{\nabla_1})$$

which is equivalent to the statement of the theorem. □

Let $H^{2*}(M;\mathbb{C})$ denote the graded algebra

$$H^{2*}(M;\mathbb{C}) = \bigoplus_{i\geq 0} H^{2i}(M;\mathbb{C}).$$

For a complex smooth vector bundle ξ, we define the *Chern character* by

$$\mathrm{ch}(\xi) = \sum \mathrm{ch}_i(\xi) \in H^{2*}(M;\mathbb{C}).$$

This defines a homomorphism $\mathrm{ch}\colon \mathrm{Vect}^{\mathbb{C}}(M) \to H^{2*}(M;\mathbb{C})$, which by Theorem 18.6.(ii) and the universal property of the Grothendieck construction can be extended to a homomorphism

$$\mathrm{ch}\colon K(M) \to H^{2*}(M;\mathbb{C}).$$

An application of Theorem 18.7 shows that ch is a multiplicative map, when the product in $K(M)$ is defined by

$$([\xi_0] - [\eta_0])([\xi_1] - [\eta_1]) = [\xi_0\otimes\xi_1] + [\eta_0\otimes\eta_1] - [\xi_0\otimes\eta_1] - [\eta_0\otimes\xi_1].$$

Without proof we state:

Theorem 18.8 *The Chern character induces an isomorphism of algebras*

$$\mathrm{ch}\colon K(M)\otimes_{\mathbb{Z}}\mathbb{C} \to H^{2*}(M;\mathbb{C}). \qquad \square$$

Theorem 18.9 *There exists precisely one set of cohomology classes* $c_k(\xi) \in H^{2k}(M;\mathbb{C})$, *depending only on the isomorphism class of* ξ, *and such that*

(i) $I(c_1(H_1)) = -1$, $c_k(H_n) = 0$ *when* $k > 1$, *and* $c_0(H_n) = 1$
(ii) $f^*c_k(\xi) = c_k(f^*(\xi))$
(iii) $c_k(\xi_0\oplus\xi_1) = \sum_{i=0}^{k} c_i(\xi_0)c_{k-i}(\xi_1).$

The uniqueness part of Theorem 18.9 rests on the so-called *splitting principle*, whose proof is deferred to Chapter 20.

Theorem 18.10 (Splitting principle) *For any complex vector bundle ξ on M there exists a manifold $T = T(\xi)$ and a proper smooth map $f\colon T \to M$ such that*

(i) $f^*\colon H^k(M) \to H^k(T)$ *is injective*

(ii) $f^*(\xi) \cong \gamma_1 \oplus \ldots \oplus \gamma_n$

for certain complex line bundles $\gamma_1, \ldots, \gamma_n$.

Proof of Theorem 18.9. The Chern classes of Definition 18.3 satisfy the three conditions, so it remains to consider the uniqueness part.

From (i) it follows that $c_1(H_1) = c$ in the notation of (8). Let L be an arbitrary line bundle and $L^\perp$ a complement to L, with

$$L \oplus L^\perp = M \times \mathbb{C}^{n+1}.$$

We can define

$$\pi\colon M \to \mathbb{C}\mathrm{P}^n; \quad x \mapsto \mathrm{proj}_2(L_x)$$

where $\mathrm{proj}_2\colon M \times \mathbb{C}^{n+1} \to \mathbb{C}^{n+1}$. There is an obvious diagram

$$
\begin{array}{ccc}
L & \xrightarrow{\hat{\pi}} & H_n \\
\downarrow & & \downarrow \\
M & \xrightarrow{\pi} & \mathbb{C}\mathrm{P}^n
\end{array}
$$

with $\hat{\pi}_p$ an isomorphim for every $p \in M$. Hence $\pi^*(H_n) \cong L$. From (ii) it follows that

$$c_1(L) = \pi^*(c).$$

Since $c_k(H_n) = 0$ when $k > 1$, the same holds for any line bundle. Therefore (i) and (ii) determine the Chern classes of an arbitrary line bundle. Inductive application of (iii) shows that for a sum of line bundles, $c_k(L_1 \oplus \ldots \oplus L_n)$ is determined by $c_1(L_1), \ldots, c_1(L_n)$. Finally we can apply Theorem 18.10 to see that $c_k(\xi)$ is uniquely determined for every complex vector bundle. □

The graded class, called the *total* Chern class,

(12) $$c(\xi) = 1 + c_1(\xi) + c_2(\xi) + \cdots \in H^*(M; \mathbb{C})$$

is exponential by Theorem 18.9.(iii), and $c(L) = 1 + c_1(L)$ for a line bundle. Hence

$$c(L_1 \oplus \ldots \oplus L_k) = \prod_{\nu=1}^{k} (1 + c_1(L_\nu)) = \sum \sigma_i(c_1(L_1), \ldots, c_1(L_k))$$

and it follows that $c_i(L_1 \oplus \ldots \oplus L_k) = \sigma_i(c_1(L_1), \ldots, c_1(L_k))$. We have additional calculational rules for Chern classes:

Properties 18.11

(a) $c_k(\xi) = 0$ if $k > \dim \xi$

(b) $c_k(\xi^*) = (-1)^k c_k(\xi)$, $\quad \mathrm{ch}_k(\xi^*) = (-1)^k \mathrm{ch}_k(\xi)$

(c) $c_{2k+1}(\eta_\mathbb{C}) = 0$ and $\mathrm{ch}_{2k+1}(\eta_\mathbb{C}) = 0$ for a real vector bundle η.

Proof. For a line bundle, (a) follows from assertions (i) and (ii) of Theorem 18.9, because every line bundle ξ has the form $\pi^*(H_n)$. If $\xi = \gamma_1 \oplus \ldots \oplus \gamma_n$ is a sum of line bundles, then

$$c(\xi) = \prod (1 + c_1(\gamma_j))$$

and it follows that $c_k(\xi) = 0$ when $k > n$. For an arbitrary ξ we can apply Theorem 18.10. The proof of (b) is analogous: if $\dim_\mathbb{C} \xi = 1$ then $\xi^* \otimes \xi = \mathrm{Hom}(\xi, \xi)$ is trivial and Theorem 18.7 gives that $\mathrm{ch}_1(\xi^*) + \mathrm{ch}_1(\xi) = 0$, hence $c_1(\xi^*) = -c_1(\xi)$. For a sum of line bundles,

$$(13) \qquad c(\xi^*) = \prod \left(1 + c_1(\gamma_j^*)\right) = \prod (1 - c_1(\gamma_j)).$$

This shows that $c_k(\xi^*) = (-1)^k c_k(\xi)$, and the splitting principle implies (b) in general. For a real vector bundle η, $\eta^* \cong \eta$, as we can choose a metric $\langle , \rangle$ on η and use the isomorphism

$$a: \eta \to \mathrm{Hom}(\eta, \mathbb{R}); \quad a(v) = \langle u, - \rangle.$$

Then $(\eta_\mathbb{C})^* = (\eta^*)_\mathbb{C}$ so that $c_k\left(\eta_\mathbb{C}^*\right) = c_k(\eta_\mathbb{C})$. Now (c) follows from (b).

Note that (c) implies that $\mathrm{ch}(\eta_\mathbb{C})$ is a graded cohomology class, which can only be non-zero in the dimensions $4k$.

One defines *Pontryagin classes* and *Pontryagin character classes* for real vector bundles by the equations:

$$(14) \qquad p_k(\eta) = (-1)^k c_{2k}(\eta_\mathbb{C}), \qquad \mathrm{ph}_k(\eta) = \mathrm{ch}_{2k}(\eta_\mathbb{C}).$$

We leave to the reader to check that the total Pontryagin class $p(\eta) = 1 + p_1(\eta) + \cdots$ is exponential.

Remark 18.12 Definition 18.3 gives cohomology classes in $H^*(M; \mathbb{C})$, but actually all classes lie in real cohomology. This follows from Theorem 18.9.(i) for H_n, and for a sum of line bundles from (ii) and (iii). The general case is a consequence of Theorem 18.10. Theorem 18.8 actually gives isomorphisms

$$\mathrm{ch}: K(M) \otimes_\mathbb{Z} \mathbb{R} \xrightarrow{\cong} H^{2*}(M)$$

$$\mathrm{ph}: KO(M) \otimes_\mathbb{Z} \mathbb{R} \xrightarrow{\cong} H^{4*}(M).$$

Example 18.13 Given a line $L \subset \mathbb{C}^{n+1}$, consider the map

$$g_L : \mathrm{Hom}(L, L^{\perp}) \to \mathbb{CP}^n$$

which maps an element $\phi \in \mathrm{Hom}(L, L^{\perp})$ into the graph of ϕ. Its image is the open set $U_L \subseteq \mathbb{CP}^n$ of lines not orthogonal to L. The functions h_j^{-1} of (14.2) are equal to g_{L_j} where L_j is the line that contains the basis vector $e_j = (0, \ldots, 1, 0, \ldots, 0)$. Each (U_L, g_L^{-1}) is a holomorphic coordinate chart on $\mathbb{CP}^n$.

Let $H^{\perp}$ be the n-plane bundle over $\mathbb{CP}^n$ with total space

$$E(H^{\perp}) = \{ (L, u) \in \mathbb{CP}^n \times \mathbb{C}^{n+1} \mid u \in L^{\perp} \}.$$

Then $H \oplus H^{\perp}$ is the trivial $(n+1)$-dimensional vector bundle where $H = H_n$, and

(15) $$\mathrm{Hom}(H, H^{\perp}) \cong \tau_{\mathbb{CP}^n}.$$

Indeed, the fiber of $\mathrm{Hom}(H, H^{\perp})$ at $L \in \mathbb{CP}^n$ is the vector space $\mathrm{Hom}(L, L^{\perp})$, and the differential

$$(Dg_L)_0 : \mathrm{Hom}(L, L^{\perp}) \to T_L \mathbb{CP}^n$$

defines the required fiberwise isomorphism.

One can use (15) to evaluate the Chern classes of the complex n-plane bundle $\tau_{\mathbb{CP}^n}$. Indeed, $\mathrm{Hom}(H, H) \cong \varepsilon_{\mathbb{C}}^1$, so

$$\tau_{\mathbb{CP}^n} \oplus \varepsilon_{\mathbb{C}}^1 \cong \mathrm{Hom}(H, H^{\perp}) \oplus \mathrm{Hom}(H, H) \cong \mathrm{Hom}(H, H \oplus H^{\perp}) = (n+1) H^*.$$

Hence the total Chern class can be calculated from Theorem 18.9 and Properties 18.11,

$$c(\tau_{\mathbb{CP}^n}) = c(\tau_{\mathbb{CP}^n} \oplus \varepsilon_{\mathbb{C}}^1) = c(H^*)^{n+1} = (1 - c_1(H))^{n+1},$$

and the binomial formula gives

(16) $$c_k(\tau_{\mathbb{CP}^n}) = (-1)^k \binom{n+1}{k} c_1(H)^k.$$

The class $c_1(H) \in H^2(\mathbb{CP}^n)$ is a generator, and Theorem 14.3 shows that $c_k(\tau_{\mathbb{CP}^n})$ is non-zero for all $k \leq n$.

Example 18.14 One of the main applications of characteristic classes is to the question of whether a given closed manifold is (diffeomorphic to) the boundary of a compact manifold. We refer the reader to [Milnor–Stasheff] for the general theory and just present an example. We show that $\mathbb{CP}^{2n}$ is not the boundary of any $4n + 1$-dimensional manifold R^{4n+1}. Indeed, suppose this was the case. By Stokes's theorem,

(17) $$\int_{\partial R} \omega = \int_R d\omega = 0$$

for any closed form $\omega \in \Omega^{4n}(R)$. But we can exhibit a closed $4n$-dimensional form on R which contradicts this as follows. The tangent bundle of ∂R satisfies the equation $\tau_{\partial R} \oplus \varepsilon_{\mathbf{R}}^1 = i^*(\tau_R)$, so when $\partial R = \mathbb{CP}^{2n}$ we get after complexification

$$(\tau_{\mathbb{CP}^{2n}})_{\mathbf{RC}} \oplus \varepsilon_{\mathbf{C}}^1 = i^*(\tau_R \otimes \mathbb{C})$$

and from the above together with Lemma 16.10,

$$(\tau_{\mathbb{CP}^{2n}})_{\mathbf{RC}} \oplus \varepsilon_{\mathbf{C}}^2 = (2n+1)H_{2n}^* \oplus (2n+1)H_{2n}.$$

The total Chern class of the right hand side is

$$c = (1 - c_1(H_{2n}))^{2n+1}(1 + c_1(H_{2n}))^{2n+1} = \left(1 - c_1(H)^2\right)^{2n+1}$$

so that

$$c_{2k}(\tau_{\mathbb{CP}^{2n}})_{\mathbf{RC}} = (-1)^k \binom{2n+1}{k} c_1(H)^{2k}.$$

Now take ω in (17) to be the $2n$-th Chern form of $\tau_R \otimes \mathbb{C}$.

19. THE EULER CLASS

Let ξ be a smooth *real* $2k$-dimensional vector bundle over M with inner product $\langle\,,\rangle$. The inner product induces a pairing

$$\langle\,,\rangle\colon \Omega^i(\xi)\otimes\Omega^j(\xi)\to\Omega^{i+j}(M);$$
$$\langle\omega_1\otimes s_1,\omega_2\otimes s_2\rangle = \omega_1\wedge\omega_2\otimes\langle s_1,s_2\rangle$$

where $\langle s_1,s_2\rangle$ is the function that maps $p\in M$ to $\langle s_1(p),s_2(p)\rangle$ and $\omega_1,\omega_2\in\Omega^*(M)$.

Definition 19.1 A connection ∇ on $(\xi,\langle\,,\rangle)$ is said to be *metric* or orthogonal if

$$d\langle s_1,s_2\rangle = \langle\nabla s_1,s_2\rangle + \langle s_1,\nabla s_2\rangle.$$

We express this condition locally in terms of the connection form A associated to an *orthonormal frame*. Let $e_1,\ldots,e_k\in\Omega^0(\xi)$ be sections over U, so that $e_1(p),\ldots,e_k(p)$ forms an orthonormal basis of ξ for $p\in U$. Let A be the associated connection form,

$$\nabla(e_i)=\sum A_{ij}\otimes e_j.$$

For every pair (i,k) we have $\langle e_i,e_k\rangle = \delta_{ik}$ (on U), so $d\langle e_i,e_k\rangle = 0$. If ∇ is metric one gets

$$0 = \langle\Sigma A_{ij}\otimes e_j,e_k\rangle + \langle e_i,\Sigma A_{kj}\otimes e_j\rangle$$
$$= \Sigma_j\,A_{ij}\langle e_j,e_k\rangle + \Sigma_j\,A_{kj}\langle e_i,e_j\rangle = A_{ik}+A_{ki}.$$

Thus the connection matrix with respect to an orthonormal frame is *skew-symmetric*. If conversely A is skew-symmetric with respect to an orthonormal frame, then ∇ is metric.

Let $F^\nabla\in\Omega^2(\mathrm{Hom}\,(\xi,\xi))$ be the curvature form associated to a metric connection. After choice of an orthonormal frame for $\xi_{|U}$,

$$\Omega^2\big(\mathrm{Hom}\,(\xi,\xi)_{|U}\big)\cong M_{2k}\big(\Omega^2(U)\big).$$

In (17.10) the corresponding matrix of 2-forms $F^\nabla(\mathbf{e})$ was calculated to be

$$F^\nabla(\mathbf{e})=dA-A\wedge A$$

where A is the connection form associated to $\mathbf{e}$. In particular, $F^\nabla(\mathbf{e})$ is skew-symmetric, and we can apply the Pfaffian polynomial from Appendix B to $F^\nabla(\mathbf{e})$ to get

$$\text{(1)} \qquad\qquad \mathrm{Pf}\,(F^\nabla(\mathbf{e}))\in\Omega^{2k}(U).$$

In another orthonormal frame e' over U

(2)
$$F^\nabla(e')_p = B_p F^\nabla(e) B_p^{-1}$$

where B_p is the orthogonal transisition matrix between $e(p)$ and $e'(p)$.

Now suppose further that the vector bundle ξ is oriented, and that $e(p)$ and $e'(p)$ are oriented orthonormal bases for ξ_p, $p \in U$. Then $B_p \in SO_{2k}$, and by Theorem B.5,

(3)
$$Pf(F^\nabla(e)) = Pf\left(F^\nabla(e')\right).$$

It follows that $Pf(F^\nabla)$ becomes a well-defined global $2k$-form on M. The proof of Lemma 18.1 shows that $Pf(F^\nabla)$ is a closed $2k$-form.

We must verify that its cohomology class is independent of the choice of metric on ξ and of the metric connection. First note that connections can be glued together by a partition of unity: if $(\nabla_\alpha)_{\alpha \in A}$ is a family of connections on ξ and $(\rho_\alpha)_{\alpha \in A}$ is a smooth partition of unity on M, then $\nabla s = \sum \rho_\alpha \nabla_\alpha s$ defines a connection on ξ. Furthermore, if each ∇_α is a metric for $g = \langle\,,\rangle$ then $\nabla = \sum \rho_\alpha \nabla_\alpha$ is also metric. Indeed, if

(4)
$$d\langle s_1, s_2\rangle = \langle \nabla_\alpha s_1, s_2\rangle + \langle s_1, \nabla_\alpha s_2\rangle$$

then

$$\langle \nabla s_1, s_2\rangle + \langle s_1, \nabla s_2\rangle = \sum \langle \rho_\alpha \nabla_\alpha s_1, s_2\rangle + \sum \langle s_1, \rho_\alpha \nabla_\alpha s_2\rangle$$
$$= \sum \rho_\alpha(\langle \nabla_\alpha s_1, s_2\rangle + \langle s_1, \nabla_\alpha s_2\rangle)$$
$$= \sum \rho_\alpha d\langle s_1, s_2\rangle = d\langle s_1, s_2\rangle.$$

In this calculation we have only used (4) over open sets that contain $\mathrm{supp}_M(\rho_\alpha)$, and not neccesarily on all of M. This will be used in the proof of Lemma 19.2 below.

Consider the maps

$$M \underset{i_1}{\overset{i_0}{\rightrightarrows}} M \times \mathbf{R} \overset{\pi}{\to} M$$

with $i_\nu(x) = (x, \nu)$ and $\pi(x, t) = x$, and let $\widetilde{\xi} = \pi^*(\xi)$ over $M \times \mathbf{R}$. Then $i_\nu^*(\widetilde{\xi}) = \xi$ for $\nu = 0, 1$ and we have:

Lemma 19.2 *For any choice of inner products and metric connections g_ν, ∇_ν ($\nu = 0, 1$) on the smooth real vector bundle ξ over M, there is an inner product $\widetilde{g}$ on $\widetilde{\xi}$ and a metric connection $\widetilde{\nabla}$ compatible with $\widetilde{g}$ such that $i_\nu^*(\widetilde{g}) = g_\nu$ and $i_\nu^*(\widetilde{\nabla}) = \nabla_\nu$.*

Proof. We can pull back by π^* the metric g_ν and the metric connections ∇_ν to $\tilde{\xi}$. Let $\{\rho_0, \rho_1\}$ be a partition of unity on $M \times \mathbb{R}$ subordinate to the cover $M \times (-\infty, 3/4)$ and $M \times (1/4, \infty)$. Then $\tilde{g} = \rho_0 \pi^*(g_0) + \rho_1 \pi^*(g_1)$ is a metric on $\tilde{\xi}$ which agrees with $\pi^*(g_0)$ over $M \times (-\infty, 1/4)$ and with $\pi^*(g_1)$ on $M \times (3/4, \infty)$. In particular $i_\nu^*(\tilde{g}) = g_\nu$.

Let $\tilde{\nabla}$ be any metric connection on $\tilde{\xi}$ compatible with $\tilde{g}$. We have connections $\pi^*(\nabla_0)$, $\tilde{\nabla}$ and $\pi^*(\nabla_1)$ compatible with $\tilde{g}$ over $M \times (-\infty \times 1/4)$, $M \times (1/8, 7/8)$ and $M \times (3/4, \infty)$ respectively. We use a partition of unity, subordinate to this cover, to glue together the three connections to construct a connection $\tilde{\nabla}$ over $M \times \mathbb{R}$. This is metric w.r.t. $\tilde{g}$, and by construction, $i_\nu^* \tilde{\nabla} = \nabla_\nu$. $\quad\square$

Corollary 19.3 *The cohomology class* $\left[\mathrm{Pf}(F^\nabla)\right] \in H^{2k}(M)$ *is independent of the metric and the compatible metric connection.*

Proof. Let (g_0, ∇_0) and (g_1, ∇_1) be two different choices and let $(\tilde{g}, \tilde{\nabla})$ be the metric and connection of the previous lemma. Then $i_\nu^*(F^{\tilde{\nabla}}) = F^{\nabla_\nu}$, and hence $i_\nu^* \mathrm{Pf}(F^{\tilde{\nabla}}) = \mathrm{Pf}(F^{\nabla_\nu})$. The maps i_0 and i_1 are homotopic, so $i_0^* = i_1^* \colon H^n(M \times \mathbb{R}) \to H^n(M)$. Thus the cohomology classes of $\mathrm{Pf}(F^{\nabla_0})$ and $\mathrm{Pf}(F^{\nabla_1})$ agree. $\quad\square$

Definition 19.4 The cohomology class

$$e(\xi) = \left[\mathrm{Pf}\left(\frac{-F^\nabla}{2\pi}\right)\right] \in H^{2k}(M)$$

is called the Euler class of the oriented real $2k$-dimensional vector bundle ξ.

Example 19.5 Suppose M is an oriented surface with Riemannian metric and that $\xi = \tau^* \cong \tau_M$ is the cotangent bundle. Let e_1, e_2 be an oriented orthonormal frame for $\Omega^0(\tau_{|U}^*) = \Omega^1(U)$, such that $e_1 \wedge e_2 = \mathrm{vol}$ on U. Let a_1, a_2 be the smooth functions on U determined by

$$de_1 = a_1(e_1 \wedge e_2), \qquad de_2 = a_2(e_1 \wedge e_2)$$

and let $A_{12} = a_1 e_1 + a_2 e_2$. We give $\tau_{|U}^*$ the connection with connection form

$$A = \begin{pmatrix} 0 & A_{12} \\ -A_{12} & 0 \end{pmatrix}$$

so that $\nabla(e_1) = A_{12} \otimes e_2$ and $\nabla(e_2) = -A_{12} \otimes e_1$. This is the so-called Levi-Civita connection; cf. Exercise 19.6. By (17.10)

$$F^\nabla = dA - A \wedge A = \begin{pmatrix} 0 & dA_{12} \\ -dA_{12} & 0 \end{pmatrix}$$

since $A_{12} \wedge A_{12} = 0$. In this case $\mathrm{Pf}(F^\nabla) = dA_{12}$ is called the Gauss–Bonnet form, and the Gaussian curvature $\kappa \in \Omega^0(M)$ is defined by the formula

$$-\kappa \, \mathrm{vol} = \mathrm{Pf}(F^\nabla).$$

This definition is compatible with Example 12.18; cf. Exercise 19.6.

There is also a concept of *metric* or *hermitian* connections for complex vector bundles equipped with a hermitian metric. Indeed hermitian connections are defined as above, Definition 19.1, with the sole change that $\langle \, , \, \rangle$ now indicates a hermitian inner product on the complex vector bundle in question.

The connection form A of a hermitian connection with respect to a local orthonormal frame is skew-hermitian rather than skew-symmetric: $A_{ik} + \overline{A}_{ki} = 0$ or in matrix terms

(5) $$A^* + A = 0.$$

Given a hermitian smooth vector bundle $(\zeta, \langle \, , \, \rangle_\mathbb{C})$ of complex dimension k with a hermitian connection, the underlying real vector bundle $\zeta_\mathbb{R}$ is naturally oriented, and inherits an inner product $\langle \, , \, \rangle_\mathbb{R}$, namely the real part of $\langle \, , \, \rangle_\mathbb{C}$, and an orthogonal connection.

If A is the skew-hermitian connection form of $(\zeta, \langle \, , \, \rangle_\mathbb{C})$ with respect to an orthonormal frame $\mathbf{e}$, then the connection form associated with the underlying real situation is $A_\mathbb{R}$, the matrix of 1-forms given by the usual embedding of $M_k(\mathbb{C})$ into $M_{2k}(\mathbb{R})$. This embedding sends skew-hermitian matrices into skew-symmetric matrices, and

(6) $$\mathrm{Pf}(F^\nabla(\mathbf{e})_\mathbb{R}) = (-i)^k \det(F^\nabla(\mathbf{e}))$$

by Theorem B.6.

For a complex vector bundle ζ we write $e(\zeta)$ instead of $e(\zeta_\mathbb{R})$. Then we have

Theorem 19.6

(i) For a complex k-dimensional vector bundle ζ, $e(\zeta) = c_k(\zeta)$.
(ii) For oriented real vector bundles ξ_1 and ξ_2, $e(\xi_1 \oplus \xi_2) = e(\xi_1)e(\xi_2)$.
(iii) $e(f^*(\xi)) = f^*e(\xi)$.

Proof. The first assertion follows from (6) upon comparing with Definition 18.3. Indeed, $\sigma_k \colon M_k(\mathbb{C}) \to \mathbb{C}$ is precisely the determinant, so by (6)

$$\mathrm{Pf}(-F_\mathbb{R}^\nabla/2\pi) = (-1)^k/(2\pi)^k \mathrm{Pf}(F_\mathbb{R}^\nabla)$$
$$= i^k/(2\pi)^k \, \sigma_k(F^\nabla)$$

when F^∇ is the curvature of a hermitian connection on $(\zeta, \langle\ ,\ \rangle_C)$. Thus

$$\mathrm{Pf}\left(-F^\nabla_\mathbb{R}/2\pi\right) = \sigma_k\left(iF^\nabla/2\pi\right).$$

This proves (i).

The second assertion is similar to Theorem 18.6. With the direct sum connection on $\zeta_1 \oplus \zeta_2$,

$$F^\nabla = F^{\nabla_1} \oplus F^{\nabla_2}$$

and for matrices A and B,

$$\mathrm{Pf}\,(A \oplus B) = \mathrm{Pf}(A)\mathrm{Pf}\,(B).$$

Finally assertion (iii) follows from (17.13).   □

In order to prove uniqueness of Euler classes we need a version of the splitting principle for real oriented vector bundles, namely

Theorem 19.7 (Real splitting principle) *For any oriented real vector bundle ζ over M there exists a manifold $T(\zeta)$ and a smooth proper map $f: T(\zeta) \to M$ such that*

(i) $f^*: H^*(M) \to H^*(T)$ *is injective.*

(ii) $f^*(\zeta) = \gamma_1 \oplus \ldots \oplus \gamma_n$ *when* $dim\,\zeta = 2n$, *and* $f^*(\zeta) = \gamma_1 \oplus \ldots \oplus \gamma_n \oplus \varepsilon^1$ *when* $dim\,\zeta = 2n + 1$, *where* $\gamma_1, \ldots \gamma_n$ *are oriented 2-plane bundles, and ε^1 is the trivial line bundle.*

The proof of this theorem will be postponed to the next chapter.

Theorem 19.8 *Suppose that to each oriented isomorphism class of $2n$-dimensional oriented real vector bundles ζ^{2n} over M we have associated a class $\hat{e}(\zeta^{2n}) \in H^{2n}(M)$ that satisfies*

(i) $f^*(\hat{e}(\zeta)) = \hat{e}(f^*(\zeta))$ *for a smooth map $f: N \to M$*

(ii) $\hat{e}(\zeta_1 \oplus \zeta_2) = \hat{e}(\zeta_1)\hat{e}(\zeta_2)$ *for oriented even-dimensional vector bundles over the same base space.*

Then there exists a real constant $a \in \mathbb{R}$ such that $\hat{e}(\zeta^{2n}) = a^n e(\zeta^{2n})$.

Proof. Given a complex line bundle L over M, we can define $c(L) = \hat{e}(L_\mathbb{R})$. Then $f^*c(L) = c(f^*L)$, and the argument used at the beginning of the proof of Theorem 18.9 shows that $c(L) = ac_1(L)$. Thus $\hat{e}(\gamma) = ae(\gamma)$ for each oriented 2-plane bundle γ. Indeed, an oriented 2-plane bundle is of the form $L_\mathbb{R}$ for a complex line bundle which is uniquely determined up to isomorphism. One simply defines multiplication by $\sqrt{-1}$ to be a positive rotation by $\pi/2$.

If $\zeta^{2n} = \gamma_1 \oplus \ldots \oplus \gamma_n$ is a sum of oriented 2-plane bundles then we can use (ii) and Theorem 19.6.(ii) to see that $\hat{e}(\zeta^{2n}) = a^n e(\zeta^{2n})$. Finally Theorem 19.7 implies the result in general.   □

In (18.14) we defined the Pontryagin classes $p_k(\zeta)$ of a real vector bundle by $p_k(\zeta) = (-1)^k c_{2k}(\zeta_C)$. The total Pontryagin class

$$(7) \qquad\qquad p(\zeta) = 1 + p_1(\zeta) + p_2(\zeta) + \cdots$$

is exponential: $p(\zeta_1 \oplus \zeta_2) = p(\zeta_1)p(\zeta_2)$. Indeed, this follows from the exponential property of the total Chern class together with the fact (Properties 18.11) that the odd Chern classes of a complexified bundle are trivial.

Proposition 19.9 *For an oriented $2k$-dimensional real vector bundle ζ, $p_k(\zeta) = e(\zeta)^2$.*

Proof. We give ζ a metric $\langle \, , \, \rangle$ and chose a compatible metric connection $\triangledown$. Then $e(\zeta)$ is represented locally by $(-1)^k/(2\pi)^k \operatorname{Pf}(F^\triangledown(e))$ where e is an orthonormal frame. If on the complexified bundle ζ_C we use the complexified metric then e is still an orthonormal frame, and the connection $\triangledown$ becomes a hermitian connection on $(\zeta_C, \langle \, , \, \rangle)$. It follows that $F^\triangledown(e)$ is the curvature form for ζ_C, and $c_{2k}(\zeta_C)$ is represented by $i^{2k}/(2\pi)^{2k} \det(F^\triangledown(e))$. The result now follows from Theorem B.5.(i) of Appendix B.   □

20. COHOMOLOGY OF PROJECTIVE AND GRASSMANNIAN BUNDLES

In this chapter we calculate the cohomology of the total space of certain smooth fiber bundles, associated to vector bundles, as a module over the cohomology of the base manifold. As corollaries we obtain the splitting principles for complex and oriented vector bundles used in Chapters 18 and 19.

Let $\pi: E \to M$ be a smooth fiber bundle over M with fiber F. There is a product

$$H^i(M) \otimes H^j(E) \to H^{i+j}(E)$$

given by the formula

(1) $$a.e = \pi^*(a) \wedge e \quad \text{for } a \in H^i(M), \ e \in H^j(E).$$

Thus $H^*(E)$ becomes a (graded) module over the (graded) algebra $H^*(M)$. We shall examine this module structure in the particular simple case where we suppose given classes $e_\alpha \in H^{n_\alpha}(E)$ for $\alpha \in A$ with the property that for every $p \in M$,

(2) $$\{i_p^*(e_\alpha) \mid \alpha \in A\} \quad \text{is a basis for the vector space } H^*(F_p).$$

Here $F_p = \pi^{-1}(p)$ is the fiber over p and i_p is the inclusion of F_p into E.

Theorem 20.1 *In the above situation $H^*(E)$ is a free $H^*(M)$-module with basis $\{e_\alpha \mid \alpha \in A\}$.*

Proof. The proof follows the pattern used to prove Poincaré duality in Chapter 13. Let $\mathcal{V}$ be the cover consisting of open sets $V \subset M$, such that E is trivial over V. Let $\mathcal{U}$ be the cover of M by open sets, so that the theorem is satisfied with M replaced by $U \in \mathcal{U}$ and E replaced by $\pi^*(U)$. We must verify the conditions of Theorem 13.9. We leave conditions (i), (ii) and (iv) to the reader and prove condition (iii). So suppose

$$U = U_1 \cup U_2, \quad U_{12} = U_1 \cap U_2$$

and let E_1, E_2 and E_{12} denote the restriction of the bundle E over U_1, U_2 and U_{12}. The classes $e_\alpha \in H^{n_\alpha}(E)$ restrict to classes which again satisfy condition (2), and we denote the restricted classes by the same letters. We suppose that the theorem is true for $H^*(E_1)$, $H^*(E_2)$, $H^*(E_{12})$, and want to conclude it is true for $H^*(E_U)$. This employs the two Mayer–Vietoris sequences

$$\cdots \xrightarrow{J^*} H^{*-1}(E_{12}) \xrightarrow{\delta^*} H^*(E) \xrightarrow{I^*} H^*(E_1) \oplus H^*(E_2) \xrightarrow{J^*} \cdots$$
$$\cdots \xrightarrow{J^*} H^{*-1}(U_{12}) \xrightarrow{\delta^*} H^*(U) \xrightarrow{I^*} H^*(U_1) \oplus H^*(U_2) \xrightarrow{J^*} \cdots$$

where we write E instead of E_U. We must show that every element $e \in H^*(E)$ has a unique representation of the form $e = \sum m_\alpha e_\alpha$ with $m_\alpha \in H^*(U)$. We give the existence proof and leave uniqueness to the reader. By assumption we know that

$$i_\nu^*(e) = \sum m_\alpha^{(\nu)}.e_\alpha, \quad \nu = 1, 2$$

where $i_\nu : E_\nu \to E$ is the inclusion. Since $J^*I^* = 0$,

$$\sum j_1^*(m_\alpha^{(1)})e_\alpha = \sum j_2^*(m_\alpha^{(2)})e_\alpha$$

in $H^*(E_{12})$, where $j_\nu : E_{12} \to E_\nu$ is the inclusion.

Uniqueness of representations for $H^*(E_{12})$ shows that $j_1^*(m_\alpha^{(1)}) = j_2^*(m_\alpha^{(2)})$ for each $\alpha \in A$, and the Mayer–Vietoris sequence for the base spaces implies elements $m_\alpha \in H^*(U)$ with $I^*(m_\alpha) = (m_\alpha^{(1)}, m_\alpha^{(2)})$, so that

$$I^*(e - \sum m_\alpha e_\alpha) = 0.$$

It thus suffices to argue that every element of $\operatorname{Ker} I^* = \operatorname{Im} \delta^*$ has a representation as asserted. This in turn is an easy consequence of the theorem for $H^*(E_{12})$ and the formula

(3) $$\delta^*(m.i_{12}^*(e)) = \delta^*(m).e,$$

valid for any $m \in H^*(U)$ and $e \in H^*(E)$, with $i_{12} : E_{12} \to E$ the inclusion. We leave the proof of (3) as an exercise. $\qquad\square$

We are now ready to prove the complex splitting principle, as stated in Theorem 18.10. Let ξ be a complex vector bundle over M with $\dim_{\mathbb{C}} \xi = n + 1$. We form an associated fiber bundle $P(\xi)$ over M with total fiber space

$$E(P(\xi)) = \{(p, L) \mid p \in M, \ L \in P(\xi_p)\}.$$

Here $P(\xi_p)$ denotes the projective space of complex lines in the vector space ξ_p. Projection onto the first factor

$$\pi : E(P(\xi)) \to M$$

makes $P(\xi)$ into a fiber bundle over M. We leave the reader to show that $P(\xi)$ is a smooth manifold and that π is a proper smooth map. There is a complex line bundle $H(\xi)$ over $P(\xi)$ with total space

$$E(H(\xi)) = \{(p, L, v) \mid (p, L) \in P(\xi), \ v \in L\}.$$

If M consists of a single point then $P(\xi)$ is the complex projective n-space $\mathbb{CP}^n$ and $H(\xi)$ is the canonical line bundle of Example 15.2. If more generally

$\xi = M \times \mathbb{C}^{n+1}$ is the trivial bundle then $P(\xi) = M \times \mathbb{C}\mathbb{P}^n$ and $H(\xi) = \mathrm{pr}_2^*(H_n)$. Let us give ξ an inner product. Then $\pi^*(\xi)$ has an inner product, and we can form the fiberwise orthogonal complement $H(\xi)^\perp$ of the subbundle $H(\xi) \subset \pi^*(\xi)$, i.e.

$$H(\xi)^\perp = \{(p, L, v) \mid (p, L) \in P(\xi), \ v \in L^\perp\}$$

where the orthogonal complement $L^\perp$ is calculated in the fiber ξ_p. Clearly $L \oplus L^\perp = \xi_p = \pi^*(\xi)_p$. So that we have an isomorphism of vector bundles

$$(4) \qquad\qquad \pi^*(\xi) = H(\xi) \oplus H(\xi)^\perp.$$

Let e be the first Chern class of $H(\xi)$, $e = c_1(H(\xi))$. We want to apply Theorem 20.1 to the classes

$$(5) \qquad\qquad 1, e, e^2, \ldots, e^n \in H^*(P(\xi)).$$

Property (2) is satisfied because the fiber of $\pi\colon P(\xi) \to M$ over $p \in M$ is the projective space $P(\xi_p) = \mathbb{C}\mathbb{P}^n$, and because the restriction of $H(\xi)$ to $P(\xi_p)$ is the canonical line bundle H_n over $\mathbb{C}\mathbb{P}^n$. Now $i_p^*(e) = c_1(H_n) \neq 0$ and the powers e^i restrict by i_p^* to $c_1(H_n)^i$ which are non-zero in $H^{2i}(\mathbb{C}\mathbb{P}^n)$, and hence a basis as long as $i \leq n$, by Theorem 14.3.

In the situation of Theorem 20.1 one has in particular that $\pi^*\colon H^*(M) \to H^*(E)$ is injective. Indeed $\pi^*(m) = m.1$, and 1 is an $\mathbb{R}$-linear combination of basis elements e_α. We have proved:

Theorem 20.2 *For any complex n-dimensional vector bundle ξ over M, $H^*(P(\xi))$ is a free $H^*(M)$-module with basis*

$$1, c_1(H(\xi)), \ldots, c_1(H(\xi))^{n-1}.$$

In particular, $\pi^*\colon H^*(M) \to H^*(P(\xi))$ is injective. $\qquad\qquad\square$

We may now prove the splitting principle for complex vector bundles.

Proof of Theorem 18.10. Starting with ξ over M with $\dim_{\mathbb{C}}\xi = n + 1$, we consider the composition

$$P(\xi_{n-1}) \overset{\pi_{n-1}}{\to} \cdots \to P(\xi_1) \overset{\pi_1}{\to} P(\xi) \overset{\pi_0}{\to} M$$

where $\xi_1 = H(\xi)^\perp$ was defined above and where ξ_2, $H(\xi_1)$ are the corresponding bundles over $P(\xi_1)$, i.e. $\xi_2 \oplus H(\xi_1) = \pi_1^*(\xi_1)$ etc. Thus if we let $f = \pi_0 \circ \ldots \circ \pi_{n-1}$, $f^*(\xi)$ is the sum of the pull-backs of the line bundles $H(\xi_i)$ over $P(\xi_i)$, and $f^* = \pi_{n-1}^* \circ \ldots \circ \pi_0^*$ is injective by Theorem 20.2. $\qquad\qquad\square$

The above discussion contains no statement about the class $e^{n+1} = c_1(H(\xi))^{n+1}$ in $H^{2n+2}(P(\xi))$ except of course that

$$c_1(H(\xi))^{n+1} = \lambda_0(\xi).1 + \lambda_1(\xi).e + \cdots + \lambda_n(\xi).e^n$$

for some uniquely determined classes

$$\lambda_i(\xi) \in H^{2n+2-2i}(M).$$

We assert that

(6) $$\lambda_i(\xi) = (-1)^{i+1} c_{n+1-i}(\xi).$$

To see this we use that $\pi_1^*(\xi) = H(\xi) \oplus \xi_1$ and the exponential property of the total Chern class so that $c(H(\xi))c(\xi_1) = c(\pi_1^*(\xi))$. Hence

$$c(\xi_1) = \pi^*(c(\xi)) \wedge c(H(\xi))^{-1} = c(\xi).(1 + c_1(H(\xi)))^{-1}.$$

In $H^{2n+2}(P(\xi))$ we get the formula

$$c_{n+1}(\xi_1) = \sum_{i=0}^{n+1} (-1)^i c_{n+1-i}(\xi).c_1(H(\xi))^i$$

which is equivalent to (6), because $\dim_{\mathbb{C}} \xi_1 = n$ and thus $c_{n+1}(\xi_1) = 0$.

Remark 20.3 One can turn the above argument upside down and use (6) to define the Chern classes, once $c_1(L)$ is defined for a line bundle. One then must show that the Chern classes so defined satisfy the two last conditions of Theorem 18.10. This treatment of Chern classes is due to A. Grothendieck. It is useful in numerous situations and gives for example Chern classes in singular cohomology, in K-theory and in étale cohomology.

The rest of this chapter is about the splitting principle for oriented real vector bundles. The construction is similar in spirit to the case of complex bundles, but the details are somewhat harder. The projectivized bundle $P(\xi)$ is replaced by the bundle $\widetilde{G}_2(\zeta)$ of fiberwise oriented 2-planes in the oriented real vector bundle ζ over M, and the canonical line bundle $H(\xi)$ over $P(\xi)$ is replaced by the oriented 2-plane bundle $\gamma_2 = \gamma_2(\zeta)$ over $\widetilde{G}_2(\zeta)$ whose fiber over an oriented plane in ζ_p is that plane itself. If ζ has an inner product then $\pi^*(\zeta) = \gamma_2(\zeta) \oplus \gamma_2(\zeta)^\perp$ as oriented bundles, so that the procedure may be iterated. The analogue of (5) is a set of classes in $H^*(\widetilde{G}_2(\zeta))$, namely the classes

(7) $$1, e(\gamma_2), e(\gamma_2)^2, \ldots, e(\gamma_2)^{2n-2}, e(\gamma_2^\perp) \in H^*(\widetilde{G}_2(\zeta))$$

where $\gamma_2 = \gamma_2(\zeta)$, $\dim_{\mathbb{R}} \zeta = 2n$ and where $e(-)$ is the Euler class of the previous chapter. In order to apply Theorem 20.1 we must show that the classes in (7) are a basis for $H^*\big(\widetilde{G}_2(\mathbb{R}^{2n})\big)$. We now give the details, starting with a proper definition of $\widetilde{G}_2(\mathbb{R}^{2n})$ and then proceeding with the somewhat cumbersome calculation of its cohomology.

Let $V_2(\mathbb{R}^m)$ denote the set of orthonormal pairs (x, y) of vectors in $\mathbb{R}^m$. We view $x \in S^{m-1}$ and y as a unit tangent vector in $T_x S^{m-1}$. Thus $V_2(\mathbb{R}^m)$ becomes the unit vectors in the tangent bundle TS^{m-1}. It is a smooth submanifold of $\mathbb{R}^{2m}$ via the embedding

$$V_2(\mathbb{R}^m) \subset TS^{m-1} \subset T(\mathbb{R}^m) = \mathbb{R}^{2m}.$$

It is better for our own purpose however to consider the embedding

(8) $$\varphi : V_2(\mathbb{R}^m) \to S^{2m-1} \subset \mathbb{C}^m, \quad \varphi(x, y) = \tfrac{1}{\sqrt{2}}(x - iy).$$

The manifold $V_2(\mathbb{R}^m)$ is called the Stiefel manifold of (orthogonal) 2-frames in $\mathbb{R}^m$; it is evidently compact. The group $SO(2)$ of rotation matrices

$$R_\theta = \begin{pmatrix} \cos\theta & -\sin\theta \\ \sin\theta & \cos\theta \end{pmatrix}$$

acts (smoothly) on $V_2(\mathbb{R}^m)$ by

$$(x, y).R_\theta = ((\cos\theta)x + (\sin\theta)y, -(\sin\theta)x + (\cos\theta)y).$$

The orbit space $V_2(\mathbb{R}^m)/SO(2)$ is identified with the space $\widetilde{G}_2(\mathbb{R}^m)$ of oriented 2-dimensional linear subspaces of $\mathbb{R}^m$ by associating to $(x, y) \in V_2(\mathbb{R}^m)$ the subspace they span, oriented so as to make (x, y) a positively oriented orthonormal basis. We leave it to the reader as an exercise to specify a smooth manifold structure on $\widetilde{G}_2(\mathbb{R}^m)$. The resulting manifold is the *Grassmann manifold* of oriented 2-dimensional subspaces of $\mathbb{R}^m$.

It is clear from (8) that

$$\varphi((x, y)R_\theta) = (\cos\theta + i\sin\theta)\varphi(x, y)$$

so when we identify $S^1 \subset \mathbb{C}$ with $SO(2)$, the action of $SO(2)$ on $V_2(\mathbb{R}^m)$ and the action of S^1 on S^{2m-1} correspond under φ. This gives a commutative diagram

(9)
$$
\begin{array}{ccccc}
V_2(\mathbb{R}^m) & \xrightarrow{\ \varphi\ } & S^{2m-1} & \longrightarrow & \mathbb{C}^m - \{0\} \\
\downarrow{\scriptstyle \pi_0} & & \downarrow{\scriptstyle \pi_1} & & \downarrow{\scriptstyle \pi} \\
\widetilde{G}_2(\mathbb{R}^m) & \xrightarrow{\ \tilde{\varphi}\ } & \mathbb{CP}^{m-1} & \xrightarrow{\ \mathrm{id}\ } & \mathbb{CP}^{m-1}
\end{array}
$$

where $\pi_0(x, y) = \mathrm{span}_{\mathbf{R}}(x, y)$ and π_1 is the restriction of the canonical map

$$\pi : \mathbb{C}^m - \{0\} \to \mathbb{CP}^{m-1}.$$

If we use homogeneous coordinates on $\mathbb{CP}^{m-1}$ then

(10) $$\widetilde{\varphi}(\pi_0(x,y)) = [x - iy].$$

It is not difficult to see that $\widetilde{\varphi}$ is injective and that its image is a smooth submanifold of $\mathbb{CP}^{m-1}$ of (real) dimension $2m-4$; cf. Exercise 20.3. We note that π_0 is a fiber bundle; cf. Example 15.2. Complex conjugation on the homogeneous coordinates for $\mathbb{CP}^{m-1}$ is an involution, whose fixed set is the real projective space $\mathbb{RP}^{m-1}$, all of whose homogeneous coordinates are real. Since in (10), x and y are linearly independent, $\widetilde{\varphi}(\pi_0(x,y))$ is never fixed under conjugation, so

(11) $$\widetilde{\varphi} \colon \widetilde{G}_2(\mathbb{R}^m) \to \mathbb{CP}^{m-1} - \mathbb{RP}^{m-1} = W_m.$$

We show below that this map is a homotopy equivalence, so that

$$H^*\big(\widetilde{G}_2(\mathbb{R}^m)\big) \cong H^*\big(\mathbb{CP}^{m-1} - \mathbb{RP}^{m-1}\big),$$

and use this to calculate $H^*\big(\widetilde{G}_2(\mathbb{R}^m)\big)$.

We begin with a discussion of the group $GL_2^+(\mathbb{R})$ of real $2{\times}2$ matrices with positive determinant. The action of the multiplicative group $\mathbb{C}^*$ on $\mathbb{C} = \mathbb{R}^2$ by complex multiplication identifies $\mathbb{C}^*$ with a subgroup of $GL_2^+(\mathbb{R})$, where $a + ib \in \mathbb{C}^*$ corresponds to

(12) $$\begin{pmatrix} a & -b \\ b & a \end{pmatrix} \in GL_2^+(\mathbb{R}).$$

Let $Q \subset GL_2^+(\mathbb{R})$ be the subset of positive definite symmetric matrices with determinant 1. The subgroup $SO(2) \subset GL_2^+(\mathbb{R})$ acts on Q by conjugation.

Lemma 20.4 *The map*

$$\psi \colon Q \times \mathbb{C}^* \to GL_2^+(\mathbb{R}); \quad (A, a + ib) \mapsto A.\begin{pmatrix} a & -b \\ b & a \end{pmatrix}$$

is a homeomorphism.

Proof. Let $B \in GL_2^+(\mathbb{R})$ with transpose B^*. Then BB^* is positive definite, and by the spectral theorem it has a unique square root $(BB^*)^{1/2}$ which commutes with BB^*. This gives the polar decomposition

$$B = (BB^*)^{1/2}.R, \quad R = (BB^*)^{-1/2}.B,$$

and $RR^* = I$, so $R = R_\theta \in SO(2)$. For $d = \det B = \det(BB^*)^{1/2}$ and $A = d^{-1}(BB^*)^{1/2} \in Q$, we obtain $B = \psi(A, de^{i\theta})$. The polar decomposition is unique. Indeed if $B_1 R_1 = B_2 R_2$ with B_i symmetric and $R_i \in SO(2)$ then $B_1^2 = B_2^2$, and square roots are unique. Hence ψ is a bijection. Its inverse is continuous, since A, d and R depend continuously on B. $\qquad\square$

For symmetric, positive definite matrices one can form powers A^t for any $t \in \mathbb{R}$, and we have

Lemma 20.5 *The space Q is contractible by the homotopy*

$$F: Q \times [0,1] \to Q; \quad F(A,t) = A^t.$$

Proof. This is again a consequence of the spectral decomposition. A matrix $A \in Q$ has positive eigenvalues λ, λ^{-1} with say $\lambda \geq 1$. Then λ depends continuously on A. The case $\lambda = 1$ occurs only for the identity matrix I. If $A \neq I$ we can write

$$A = \lambda P + \lambda^{-1}(I - P) = \lambda^{-1}I + (\lambda - \lambda^{-1})P,$$

where P is the orthogonal projection on the (1-dimensional) λ-eigenspace of A. Here P depends continuously on $A \in Q - \{I\}$. We define F on $(Q - \{I\}) \times [0,1]$ by

$$F(A,t) = \lambda^t P + \lambda^{-t}(I - P) = \lambda^{-t}I + (\lambda^t - \lambda^{-t})P.$$

Observing that each matrix entry in $(\lambda^t - \lambda^{-t})P$ has numerical value at most $\lambda - \lambda^{-1}$, we see that F extends continuously to $Q \times [0,1]$ by $F(I,t) = I$. $\square$

Proposition 20.6 *The map*

$$\widetilde{\varphi}: \widetilde{G}_2(\mathbb{R}^m) \to \mathbb{CP}^{m-1} - \mathbb{RP}^{m-1}$$

from (11) is a homotopy equivalence.

Proof. We write $W_m = \mathbb{CP}^{m-1} - \mathbb{RP}^{m-1}$, and consider the smooth map (cf. (?))

$$\Phi: V_2(\mathbb{R}^m) \times GL_2^+(\mathbb{R}) \to \pi^{-1}(W_m);$$

$$\Phi\left(x, y, \begin{pmatrix} a & c \\ b & a \end{pmatrix}\right) = (ax + by) - i(cx + dy).$$

A point in $\pi^{-1}(W_m)$ has the form $z = v - iw$, where v and w are linearly independent vectors in $\mathbb{R}^m$. The fiber $\Phi^{-1}(z)$ is in 1-1 correspondence with the orthonormal bases (x, y) of $\mathrm{span}_{\mathbb{R}}(v, w)$ which determine the same orientation as (v, w). There is a global smooth section S of Φ, constructed using the Gram–Schmidt orthonormalization process. The fibers of Φ are the orbits of the $SO(2)$-action,

$$(x, y, B).R_\theta = ((x, y).R_\theta, R_\theta^{-1}B),$$

so Φ induces a bijection from the set of orbits

$$V_2(\mathbb{R}^m) \times_{SO(2)} GL_2^+(\mathbb{R}) \to \pi^{-1}(W_m).$$

This bijection commutes with the $\mathbb{C}^*$-actions, if we let $\mathbb{C}^*$ act on the domain by right multiplication in $GL_2^+(\mathbb{R})$ and on $\pi^{-1}(W_m) \subset \mathbb{C}^m$ by scalar multiplication.

This gives a commutative diagram where the horizontal maps are bijections

$$\begin{array}{ccc} V_2(\mathbf{R}^m) \times_{SO(2)} GL_2^+(\mathbf{R}) & \longrightarrow & \pi^{-1}(W_m) \\ \downarrow & & \downarrow \\ \left(V_2(\mathbf{R}^m) \times_{SO(2)} GL_2^+(\mathbf{R})\right)/\mathbf{C}^* & \longrightarrow & W_m \end{array}$$

Lemma 20.4 gives a bijection $Q \cong GL_2^+(\mathbf{R})/\mathbf{C}^*$, so we may identify the lower left-hand corner of the diagram with the quotient space

$$X = V_2(\mathbf{R}^m) \times_{SO(2)} Q$$

for the $SO(2)$-action on $V_2(\mathbf{R}^m) \times Q$ defined by

$$(x, y, A).R_\theta = \left((x, y).R_\theta, R_\theta^{-1} A R_\theta\right).$$

Altogether we obtain a bijection $\widetilde{\Phi} \colon X \to W_m$ given by

$$(13) \qquad \widetilde{\Phi}\left(\left[(x, y), \begin{pmatrix} \alpha & \beta \\ \beta & \gamma \end{pmatrix}\right]\right) = [(\alpha x + \beta y) - i(\beta x + \gamma y)].$$

Evidently $\widetilde{\Phi}$ is continuous in the quotient topology on X, in fact it is a homeomorphism. In a neighborhood V of any given point in W_m the inverse $\widetilde{\Phi}^{-1}$ can be written as the composition

$$V \xrightarrow{s} \pi^{-1}(W_m) \xrightarrow{S} V_2(\mathbf{R}^m) \times GL_2^+(\mathbf{R}) \xrightarrow{\mathrm{id} \times \psi^{-1}} V_2(\mathbf{R}^m) \times Q \times \mathbf{C}^* \xrightarrow{\rho^{\mathrm{opr}}} X$$

where s is a local section given by Lemma 14.4, S the global section of Φ mentioned above, and ψ^{-1} given by Lemma 20.4. Each map in the sequence is continuous, so $\widetilde{\Phi}^{-1}$ is continuous. As a byproduct we find that the canonical map $\rho \colon V_2(\mathbf{R}^m) \times Q \to X$ has continuous local sections defined on open sets covering X.

The conjugation action of $SO(2)$ on Q fixes the identity matrix $I \in Q$, so one has the subspace

$$(14) \qquad \widetilde{G}_2(\mathbf{R}^m) \overset{\pi_0}{\underset{\cong}{\leftarrow}} V_2(\mathbf{R}^m) \times_{SO(2)} \{I\} \subset X$$

(cf. (9)). Comparing (10) and (13), we see that $\widetilde{\Phi} \circ \pi_0^{-1}$ is precisely equal to $\widetilde{\varphi}$. It remains to be shown that the inclusion map i_0 in (14) is a homotopy equivalence. There is an obvious retraction induced by the constant map $Q \to \{I\}$

$$r \colon X \to V_2(\mathbf{R}^m) \times_{SO(2)} \{I\} \subseteq X.$$

Finally the required homotopy $H \colon X \times [0, 1] \to X$ between $i_0 \circ r$ and id_X is induced by

$$\mathrm{id}_{V_2(\mathbf{R}^m)} \times F \colon V_2(\mathbf{R}^m) \times Q \times [0, 1] \to V_2(\mathbf{R}^m) \times Q,$$

where F is defined in Lemma 20.5. Observe that H is well-defined because $F(R_\theta^{-1} A R_\theta, t) = R_\theta^{-1} F(A, t) R_\theta$. Continuity of H can be shown with use of local sections of ρ. $\qquad\square$

Remark 20.7 We can offer the following more conceptual explanation of the construction in the previous proof. Consider the set $G_2^{\mathbb{C}}(\mathbb{R}^m)$ of pairs (V, J) where $V \subseteq \mathbb{R}^m$ is a 2-dimensional real oriented linear subspace and J a complex structure on V compatible with the orientation, i.e. an $\mathbb{R}$-linear map $J \colon V \to V$ with $J^2 = -\mathrm{id}$ and such that (x, Jx) for $x \in V$, $x \neq 0$ is a positively oriented basis for V. One forms the complexifications $V_{\mathbb{C}} = V \otimes_{\mathbb{R}} \mathbb{C} \subset \mathbb{C}^m$ and $J_{\mathbb{C}} = J \otimes_{\mathbb{R}} \mathrm{id}_{\mathbb{C}} \colon V_{\mathbb{C}} \to V_{\mathbb{C}}$. Then $V_{\mathbb{C}} = V_+ \oplus V_-$, where $V_{\pm}$ are the $(\pm i)$-eigenspaces of $J_{\mathbb{C}}$. These are 1-dimensional over $\mathbb{C}$. In fact for $x \in V - \{0\}$,

$$V_+ = \mathrm{span}_{\mathbb{C}}(x - iJx), \quad V_- = \mathrm{span}_{\mathbb{C}}(x + iJx)$$

The pair (V, J) may be recovered from V_+ since V_- is complex conjugate to V_+, $V = (V_+ + V_-) \cap \mathbb{R}^m$ and since J is the restriction of the $\mathbb{C}$-linear endomorphism of $V_+ \oplus V_-$ which acts on $V_{\pm}$ by multiplication with $\pm i$. This gives an identification of $G_2^{\mathbb{C}}(\mathbb{R}^m)$ with W_m in which (V, J) corresponds to V_+ considered as a point in $W_m \subseteq \mathbb{CP}^{m-1}$.

The space $Q \cong GL_2^+(\mathbb{R})/\mathbb{C}^*$ parametrizes the complex structures on $\mathbb{R}^2$ compatible with the standard orientation (cf. Exercise 20.5) so $X = V_2(\mathbb{R}^m) \times_{SO(2)} Q$ is another model for $G_2^{\mathbb{C}}(\mathbb{R}^m)$. The homeomorphism $\tilde{\Phi}$ in (13) is the natural identification of the two models.

Each $V \in \tilde{G}_2(\mathbb{R}^m)$ has a canonical complex structure J_0 such that any unit vector $x \in V$ leads to a positively oriented orthonormal basis $(x, J_0 x)$ for V. The inclusion $\tilde{G}_2(\mathbb{R}^m) \to G_2^{\mathbb{C}}(\mathbb{R}^m)$, which sends V to (V, J_0), corresponds exactly to the embedding $\tilde{\varphi}$ in (10).

We shall now use Proposition 20.6 to calculate the cohomology of $\tilde{G}_2(\mathbb{R}^m)$. First we apply Poincaré duality to evaluate $H^*(W_m)$. Consider the inclusions

$$i \colon W_m \to \mathbb{CP}^{m-1}, \quad j \colon \mathbb{RP}^{m-1} \to \mathbb{CP}^{m-1}.$$

When $m = 2$, $S^2 \overset{\cong}{\to} \mathbb{CP}^1$ by Example 14.1, and $\mathbb{RP}^1 \subset \mathbb{CP}^1$ becomes identified with a great circle in S^2 so that W_2 becomes the disjoint union of two open hemispheres in S^2.

Lemma 20.8 *The map* $i_* \colon H_c^q(W_m) \to H^q(\mathbb{CP}^{m-1})$ *is an isomorphism except in possibly two cases: for $q = 0$, and, if m is even, for $q = m$. In fact $H_c^0(W_m) = 0$, and for m even there is a short exact sequence*

$$0 \to \mathbb{R} \to H_c^m(W_m) \overset{i_*}{\to} H^m(\mathbb{CP}^{m-1}) \to 0.$$

Proof. The exact sequence of Proposition 13.11 for the pair $(\mathbb{CP}^{m-1}, \mathbb{RP}^{m-1})$ takes the form

$$\cdots \overset{j^*}{\to} H^{q-1}(\mathbb{RP}^{m-1}) \overset{\delta}{\to} H_c^q(W_m) \overset{i_*}{\to} H^q(\mathbb{CP}^{m-1}) \overset{j^*}{\to} H^q(\mathbb{RP}^{m-1}) \overset{\delta}{\to} \cdots$$

The terms involving $\mathbb{RP}^{m-1}$ and $\mathbb{CP}^{m-1}$ have been calculated in Example 9.31 and Theorem 14.2, respectively. Note that $H^0(\mathbb{CP}^{m-1}) \cong \mathbb{R}$ maps isomorphically to $H^0(\mathbb{RP}^{m-1}) \cong \mathbb{R}$ under j^*, whereas $j^* = 0$ in other degrees. $\square$

Proposition 20.9 *The cohomology* $H^{2p-1}(W_m) = 0$ *for all p, and*

$$H^{2p}(W_m) \cong \begin{cases} \mathbb{R}^2 & \text{if } 2p = m - 2 \\ \mathbb{R} & \text{if } 2p \neq m - 2 \text{ and } 0 \leq 2p \leq 2m - 4 \\ 0 & \text{if } 2p \geq 2m - 2. \end{cases}$$

Moreover $H^q(i) \colon H^q(\mathbb{CP}^{m-1}) \to H^q(W_m)$ *is a monomorphism if* $q \neq 2m - 2$, *and is zero if* $q = 2m - 2$.

Proof. We apply Poincaré duality to the oriented $(2m - 2)$-dimensional manifold $\mathbb{CP}^{m-1}$ and to W_m. Lemma 13.6 gives a commutative diagram

$$\begin{array}{ccc} H^p(\mathbb{CP}^{m-1}) & \xrightarrow{H^p(i)} & H^p(W_m) \\ \Big\downarrow{\cong} & & \Big\downarrow{\cong} \\ H^{2m-2-p}(\mathbb{CP}^{m-1})^* & \xrightarrow{i^!} & H_c^{2m-2-p}(W_m)^* \end{array}$$

The previous lemma implies that $H^p(i)$ is an isomorphism except if $p = 2m - 2$, or if $p = m - 2$ and m is even, because $i^!$ is the vector space dual of i_*. In the first case

$$H^{2m-2}(W_m) \cong H_c^0(W_m)^* = 0$$

but $H^{2m-2}(\mathbb{CP}^{m-1}) \cong \mathbb{R}$. In the second case, the exact sequence of Lemma 20.8 dualizes to the exact sequence

$$0 \to H^{m-2}(\mathbb{CP}^{m-1}) \xrightarrow{H^{m-2}(i)} H^{m-2}(W_m) \to \mathbb{R} \to 0,$$

and the proposition follows from Theorem 14.2. $\square$

We have the two canonical bundles γ_2 and $\gamma_2^\perp$ over $\widetilde{G}_2(\mathbb{R}^m)$ with total spaces

$$E(\gamma_2) = \{(V, v) \in \widetilde{G}_2(\mathbb{R}^m) \times \mathbb{R}^m \mid v \in V\}$$
$$E(\gamma_2^\perp) = \{(V, v) \in \widetilde{G}_2(\mathbb{R}^m) \times \mathbb{R}^m \mid v \in V^\perp\}$$

and with $\gamma_2 \oplus \gamma_2^\perp = \varepsilon^m$. Alternatively $E(\gamma_2) = V_2(\mathbb{R}^m) \times_{SO(2)} \mathbb{R}^2$, the orbit space of the $SO(2)$-action given by $(x, y, v)R_\theta = ((x, y)R_\theta, R_\theta^{-1}v)$. The embedding of $V_2(\mathbb{R}^m) \times_{SO(2)} \mathbb{R}^2$ into $\widetilde{G}_2(\mathbb{R}^m) \times \mathbb{R}^m$ sends $[x, y, v]$ into $(\pi_0(x, y), xv_1 + yv_2)$ where $v = (v_1, v_2)$.

There is a bundle map over $\tilde{\varphi}$ from γ_2 to the underlying real bundle $H_{\mathbb{R}}$ of the canonical line bundle over $\mathbb{CP}^{m-1}$,

$$V_2(\mathbb{R}^m) \times_{SO(2)} \mathbb{R}^2 \xrightarrow{\varphi \times \mathrm{id}} S^{2m-1} \times_{S^1} \mathbb{C}$$

$$\downarrow \qquad\qquad\qquad\qquad \downarrow$$

$$\tilde{G}_2(\mathbb{R}^m) \qquad \xrightarrow{\ \tilde{\varphi}\ } \qquad \mathbb{CP}^{m-1}$$

Since $c_1(H) = e(H_{\mathbb{R}})$ by Theorem 19.6.(ii), naturality of the Euler class gives $e(\gamma_2) = \tilde{\varphi}^*(c_1(H))$.

Let $c \in H^2(\tilde{G}_2(\mathbb{R}^m))$ be the Euler class of γ_2, $c = e(\gamma_2)$. If m is even, we let $e = e(\gamma_2^{\perp})$ be the Euler class in $H^{m-2}(\tilde{G}_2(\mathbb{R}^m))$.

Theorem 20.10 *With the notation above*

(i) *For m odd and $m \geq 3$, $H^*(\tilde{G}_2(\mathbb{R}^m)) = \mathbb{R}[c]/(c^{m-1})$*

(ii) *For m even and $m \geq 4$,*

$$H^*(\tilde{G}_2(\mathbb{R}^m)) = \mathbb{R}[c, e]/(c^{m-1}, ce, e^2 + (-1)^{m/2} c^{m-2})$$

with $\deg c = 2$ and $\deg e = m - 2$.

Proof. We already know the additive structure by Propositions 20.6 and 20.9, and also that

$$H^*(\tilde{\varphi}): H^*(\mathbb{CP}^{m-1}) \to H^*(\tilde{G}_2(\mathbb{R}^m))$$

has kernel $H^{2m-2}(\mathbb{CP}^{m-1})$ and is onto for odd m. Since

$$H^*(\mathbb{CP}^{m-1}) = \mathbb{R}[c_1(H)]/(c_1(H)^m)$$

and $c = H^*(\tilde{\varphi})(c_1(H))$, this proves (i).

Suppose now $m = 2n \geq 4$. We first establish the relations:

(15) $$c^{m-1} = 0, \quad ce = 0, \quad e^2 + (-1)^n c^{m-2} = 0.$$

Indeed, the first one follows from Proposition 20.9, since c and hence any power of c is in the image of

$$H^*(\mathbb{CP}^{m-1}) \xrightarrow{H^*(i)} H^*(W_m) \xrightarrow{H^*(\tilde{\varphi})} H^*(\tilde{G}_2(\mathbb{R}^m)).$$

The second relation is a consequence of Theorem 19.6.(ii),

$$ec = e(\gamma_2)e(\gamma_2^{\perp}) = e(\gamma_2 \oplus \gamma_2^{\perp}) = e(\varepsilon^m) = 0.$$

For the third relation we use that the total Pontryagin class is exponential (cf. (18.14)) so that

$$(1 + p_1(\gamma_2))\big(1 + p_1(\gamma_2^{\perp}) + \ldots + p_{n-1}(\gamma_2^{\perp})\big) = 1$$

and hence $p_j(\gamma_2^{\perp}) = (-1)^j p_1(\gamma_2)^j$ for $j \leq n-1$. Moreover by Proposition 19.9

$$e^2 = e(\gamma_2^{\perp})^2 = p_{n-1}(\gamma_2^{\perp}) = (-1)^{n-1} p_1(\gamma_2)^{n-1}.$$

Since $\gamma_2 = \tilde{\phi}^*(H_{\mathbf{R}})$ and $H_{\mathbf{RC}} = H \oplus H^*$,

$$p_1(\gamma_2) = -\tilde{\phi}^*(c_2(H \oplus H^*)) = \tilde{\phi}^*\big(c_1(H)^2\big) = c^2$$

so $e^2 = (-1)^{n-1} c^{m-2}$, which is the last equation of (15). From Proposition 20.9 and the non-triviality of $c_1(H)^{m-2}$ we know that $c^{m-2} \neq 0$, so by (15) also that $e \neq 0$. The vector space $H^{m-2}(\widetilde{G}_2(\mathbf{R}^m))$ is 2-dimensional, and $c^{n-1} = \lambda e$, $\lambda \in \mathbf{R} - \{0\}$ gives $c^n = \lambda ec = 0$, which contradicts that $m - 2 \geq n$ and $c^{m-2} \neq 0$. We have proved that the set $\{1, c, \ldots, c^{m-2}, e\}$ is a vector space basis for $H^*(\widetilde{G}_2(\mathbf{R}^m))$, but this is also a basis for the ring $\mathbf{R}[c, e]/\big(c^{m-1}, ce, e^2 + (-1)^n c^{m-2}\big)$. $\qquad\square$

Let ζ be a smooth m-dimensional oriented real vector bundle over M, and suppose ζ has an inner product. Consider the associated smooth fiber bundle $\widetilde{G}_2(\zeta)$ over M with total space

$$E(\widetilde{G}_2(\zeta)) = \{(p, V) \mid p \in M, \ V \in \widetilde{G}_2(\zeta_p)\}$$

and with $\pi \colon \widetilde{G}_2(\zeta) \to M$ being the projection onto the first factor. There are two oriented vector bundles over $\widetilde{G}_2(\zeta)$ with total spaces

$$E(\gamma_2(\zeta)) = \{(p, V, v) \mid (p, V) \in \widetilde{G}_2(\zeta), \ v \in V\}$$
$$E(\gamma_2^{\perp}(\zeta)) = \{(p, V, v) \mid (p, V) \in \widetilde{G}_2(\zeta), \ v \in V^{\perp}\}$$

and $\gamma_2(\zeta) \oplus \gamma_2^{\perp}(\zeta) \cong \pi^*(\zeta)$. The orientation of $\gamma_2^{\perp}(\zeta_p)$ is such that $\gamma_2(\zeta_p) \oplus \gamma_2^{\perp}(\zeta_p)$ has the same orientation as ζ_p. An application of Theorem 20.1 gives

Theorem 20.11 *For any oriented m-dimensional real vector bundle ζ, $H^*(\widetilde{G}_2(\zeta))$ is a free $H^*(M)$ module with basis*

$$\begin{cases} 1, e(\gamma_2(\zeta)), \ldots, e(\gamma_2(\zeta))^{m-2}, e(\gamma_2^{\perp}(\zeta)) & \text{if } m = 2n \geq 4 \\ 1, e(\gamma_2(\zeta)), \ldots, e(\gamma_2(\zeta))^{m-2} & \text{if } m = 2n+1 \geq 3 \end{cases}$$

In particular $\pi^* \colon H^*(M) \to H^*(\widetilde{G}_2(\zeta))$ is injective. $\qquad\square$

Given this result, the proof of the real splitting principle, Theorem 19.7, is precisely analogous to the proof of the complex splitting principle, treated earlier in this chapter.

21. THOM ISOMORPHISM AND THE GENERAL GAUSS–BONNET FORMULA

Let ξ be an oriented vector bundle over M with total space $E = E(\xi)$. The Thom isomorphism theorem calculates the compactly supported cohomology $H_c^*(E)$ in terms of $H^*(M)$, namely $H^q(M) \cong H_c^{q+m}(E)$ where $m = \dim \xi$. Assuming ξ to be smooth (cf. Exercise 15.8) and M to be oriented the Thom isomorphism theorem is a consequence of Poincaré duality. Indeed,

$$H_c^{m+q}(E) \cong H^{n-q}(E)^* \cong H^{n-q}(M)^* \cong H_c^q(M)$$

where $n = \dim M$. The second isomorphism is induced from the homotopy equivalence $E \simeq M$. For M compact we give below a more direct proof of Thom isomorphism.

Suppose ξ is smooth and has an inner product, and let $\xi \oplus 1$ denote the (orthogonal) direct sum of ξ and the trivial line bundle over M. We write $S(\xi \oplus 1)$ for its unit sphere bundle over M, or what amounts to the same thing, for the fiberwise one-point compactification of E. Let

$$\pi \colon S(\xi \oplus 1) \to M, \quad s_\infty \colon M \to S(\xi \oplus 1)$$

be the bundle projection and the "section at infinity", respectively, that is $s_\infty(p) = (0,1) \in \xi_p \oplus \mathbb{R}$. This makes $S(\xi \oplus 1)$ into a smooth fiber bundle over M, and hence $H^*(S(\xi \oplus 1))$ into a $H^*(M)$-module (cf. Theorem 20.1). The fiber of π is the m-sphere $S^m \cong \pi^{-1}(p)$, and the orientation of ξ_p induces an orientation of each fiber S^m. In particular the integration homomorphism of Chapter 10 induces a fixed isomorphism

$$(1) \qquad\qquad I \colon H^m(\pi^{-1}(p)) \xrightarrow{\cong} \mathbb{R}.$$

Stereographic projection from $s_\infty(p)$ identifies $\pi^{-1}(p) - \{s_\infty(p)\}$ with ξ_p, and globally it identifies the total space E with $S(\xi \oplus 1) - s_\infty(M)$.

Definition 21.1 An orientation class for ξ is a cohomology class $u \in H^m(S(\xi \oplus 1))$ that satisfies

(a) $s_\infty^*(u) = 0$

(b) For each $p \in M$, the restriction of u to $\pi^{-1}(p)$ has integral 1.

Theorem 21.2 *Each oriented vector bundle ξ admits a unique orientation class u, and $H^*(S(\xi \oplus 1))$ is a free $H^*(M)$-module with basis $\{1, u\}$*

Proof. The second part of the theorem follows from Theorem 20.1, since $H^*\left(\pi^{-1}(p)\right)$ has basis 1 and $i_p^*(u)$ according to Example 9.29 and Corollary 10.14. Here i_p denotes the inclusion of the fiber into the total space $S(\xi \oplus 1)$. In particular, it suffices to find a class v that satisfies Definition 21.1.(b). Indeed, if v is such a class then any other class in $H^m(S(\xi \oplus 1))$ has the form

$$u = \pi^*(x) + \pi^*(a) \wedge v$$

for some $x \in H^m(M)$ and $a \in H^0(M)$. The restriction of $\pi^*(x)$ to $\pi^{-1}(p)$ vanishes for all p, so the locally constant function a must have value 1 at each $p \in M$ if u is to satisfy Definition 21.1.(b). But then

$$s_\infty^*(u) = x + s_\infty^*(v),$$

and $s_\infty^*(u) = 0$ if and only if $x = -s_\infty^*(v)$.

The existence of a class $u \in H^m(S(\xi \oplus 1))$ that satisfies condition (b) is based upon Theorem 13.9. Write $S_U = \pi^{-1}(U)$ where $U \subseteq M$ is open, and let $\mathcal{U}$ be the collection of open sets for which the restriction $\xi_U = \xi_{|U}$ satisfies the conclusion of the theorem. The preceding discussion shows that $U \in \mathcal{U}$ if and only if there exists a class in $H^m(S_U)$ with integral equal to 1 over each fiber $\pi^{-1}(p)$ for $p \in U$. Let $\mathcal{V} = (V_\beta)$ in Theorem 13.9 be the cover with V_β the open sets in M such that $\xi_{|V_\beta}$ is a trivial bundle ($\xi_{|V_\beta} \cong \varepsilon_{V_\beta}^m$). We must verify the four conditions of Theorem 13.9.

The first condition is trivial. If $U \subseteq V_\beta$ then we may trivialize ξ_U (compatible with the Riemannian metric) so as to identify S_U with $U \times S^m$. Let

$$S_U \cong U \times S^m \xrightarrow{\text{pr}} S^m$$

denote the resulting projection, and let $u \in H^m(S^m)$ have integral equal to 1. Then $\text{pr}^*(u)$ restricts to a class with integral 1 on each fiber, and condition (ii) of Theorem 13.9 is satisfied. Next we verify condition (iii). So suppose U_1, U_2 and $U_1 \cap U_2$ belong to $\mathcal{U}$. The orientation classes $u_\nu \in H^m(S_{U_\nu})$, $\nu = 1, 2$ restrict to classes in $H^m(S_{U_1 \cap U_2})$ that satisfy both condition (a) and (b) for $\xi_{U_1 \cap U_2}$. Uniqueness applied to $\xi_{U_1 \cap U_2}$ shows that u_1 and u_2 have the same restriction to $S_{U_1 \cap U_2}$. In the Mayer–Vietoris sequence

$$H^m(S_{U_1 \cup U_2}) \xrightarrow{I^*} H^m(S_{U_1}) \oplus H^m(S_{U_2}) \xrightarrow{J^*} H^m(S_{U_1 \cap U_2}),$$

$(u_1, u_2) \in \text{Ker} J^* = \text{Im} I^*$, so we can find a class $u \in H^m(S_{U_1 \cup U_2})$ with restriction u_ν to S_{U_ν}. This class has integral 1 over all fibers $\pi^{-1}(p) \subseteq S_{U_1 \cup U_2}$ and $U_1 \cup U_2$ belongs to $\mathcal{U}$.

Finally consider a sequence $U_1, U_2, \ldots$ of disjoint open sets in $\mathcal{U}$ with union $U = \bigcup_i U_i$. We have the isomorphism from Proposition 13.4,

$$H^m(S_U) \to \prod_i H^m(S_{U_i}).$$

The family of orientation classes $u_i \in H^m(S_{U_i})$ is the image of some $u \in H^m(S_U)$ with integral 1 over all fibers $\pi^{-1}(p) \subseteq S_U$. Hence $U \in \mathcal{U}$. Now Theorem 13.9 applies to show that $M \in \mathcal{U}$. $\qquad\square$

The above does not require M to be compact, but if it is, we may apply Proposition 13.11 to the compact manifold pair $(S(\xi \oplus 1), s_\infty(M))$. Since $S(\xi \oplus 1) - s_\infty(M) \cong E$ and $s_\infty(M) \cong M$ we obtain a long exact sequence

$$\cdots \xrightarrow{\delta} H^q_c(E) \xrightarrow{i_*} H^q(S(\xi \oplus 1)) \xrightarrow{s^*_\infty} H^q(M) \xrightarrow{\delta} H^{q+1}_c(E) \to \cdots.$$

Now $s^*_\infty \circ \pi^* = (\pi \circ s_\infty)^* = \mathrm{id}$, so that s^*_∞ is an epimorphism and $\delta = 0$. By exactness i_* is a monomorphism, leading to the short exact sequence

$$(2) \qquad\qquad 0 \to H^*_c(E) \xrightarrow{i_*} H^*(S(\xi \oplus 1)) \xrightarrow{s^*_\infty} H^*(M) \to 0.$$

Theorem 21.3 (Thom isomorphism) *Let ξ be an oriented m-dimensional real vector bundle over a compact manifold M. There is a unique class $U \in H^m_c(E)$ with integral 1 over each fiber ξ_p, and the map*

$$\Phi \colon H^q(M) \xrightarrow{\cong} H^{m+q}_c(E); \quad \Phi(x) = (\pi^* x) \wedge U$$

is an isomorphism. The class $U = \Phi(1)$ is called the Thom class.

Proof. The exact sequence (2) shows that the orientation class $u \in H^m(S(\xi \oplus 1))$ has the form $u = i_*(U)$ for a uniquely determined $U \in H^m_c(E)$. The first statement now follows from Theorem 21.2. The homomorphisms i_* and s^*_∞ in (2) are $H^*(M)$-linear and the last part of Theorem 21.2 shows that $H^*_c(E)$ is a free $H^*(M)$-module generated by U. Thus Φ is an isomorphism. $\qquad\square$

Definition 21.4 With the notation of the previous theorem, let $\hat{e}(\xi) \in H^m(M)$ be the class with $\Phi(\hat{e}(\xi)) = U \wedge U$.

The product in $H^*_c(E)$ is anti-commutative (since this is the case for the representing differential forms), so $U \wedge U = 0$ if m is odd. Thus $\hat{e}(\xi) = 0$ for odd-dimensional oriented vector bundles.

Lemma 21.5 *Let $s: M \to E$ be an arbitrary smooth section of E. Then $\hat{e}(\xi) = s^*(U)$.*

Proof. Since s is a proper map, it induces a homomorphism

$$H_c^*(E) \to H_c^*(M) = H^*(M),$$

cf. Chapter 13. A closed form $\omega \in \Omega_c^m(E)$ that represents U also defines a class $[\omega] \in H^m(E)$. Now $s \circ \pi: E \to E$ is homotopic to id_E (use the linear structure in the fibers), so $[\omega] = [\pi^* s^*(\omega)]$. But then

$$\Phi(s^*(U)) = (\pi^* s^*(U)) \wedge U = [\pi^* s^*(\omega)] \wedge U$$
$$= [\omega] \wedge U = U \wedge U. \qquad \square$$

The next two lemmas show that $\hat{e}(\xi)$ satisfies the conditions of Theorem 19.8, and hence that

(3) $$\hat{e}(\xi^m) = a^{m/2} e(\xi^m)$$

for even-dimensional oriented vector bundles over compact manifolds. We shall see later that $a = 1$.

Lemma 21.6 *Let $f: N \to M$ be a smooth map of compact manifolds and ξ an oriented vector bundle over M. Then $\hat{e}(f^*\xi) = f^*\hat{e}(\xi)$.*

Proof. We have the pull-back diagram

$$
\begin{array}{ccc}
E' & \xrightarrow{\hat{f}} & E \\
\downarrow & & \downarrow \\
N & \xrightarrow{f} & M
\end{array}
$$

where E and E' are the total spaces of ξ and $f^*\xi$ respectively. The map $\hat{f}$ is proper, so the class $U \in H_c^m(E)$ of Theorem 21.3 pulls back to $U' = \hat{f}^*(U) \in H_c^m(E')$. Since $\hat{f}$ maps a fiber $f^*\xi_p$ to $\xi_{f(p)}$ by a linear oriented isomorphism, U' will have integral 1 over fibers. There is another commutative diagram

$$
\begin{array}{ccc}
E' & \xrightarrow{\hat{f}} & E \\
\uparrow{\scriptstyle s'} & & \uparrow{\scriptstyle s} \\
N & \xrightarrow{f} & M
\end{array}
$$

where s and s' are the zero sections, and by Lemma 21.5 we find

$$\hat{e}(f^*\xi) = (s')^*(U') = (s')^* \circ \hat{f}^*(U)$$
$$= (\hat{f} \circ s')^*(U) = (s \circ f)^*(U)$$
$$= f^* \circ s^*(U) = f^*(\hat{e}(\xi)). \qquad \square$$

Lemma 21.7 *If ξ_1 and ξ_2 are oriented real vector bundles over the compact manifold M, then $\hat{e}(\xi_1 \oplus \xi_2) = \hat{e}(\xi_1)\hat{e}(\xi_2)$.*

Proof. Let ξ_ν have total space E_ν, bundle projection $\pi_\nu \colon E_\nu \to M$ and fiber dimension m_ν, and let $U_\nu \in H_c^{m_\nu}(E_\nu)$ be given by Theorem 21.3. The product map

$$\pi_1 \times \pi_2 \colon E_1 \times E_2 \to M \times M$$

is the projection of an oriented $(m_1 + m_2)$-dimensional vector bundle ξ over $M \times M$ with $\Delta^*(\xi) = \xi_1 \oplus \xi_2$, where $\Delta \colon M \to M \times M$ is the diagonal map. If $\mathrm{pr}_\nu \colon E_1 \times E_2 \to E_\nu$ is the projection and $\omega_\nu \in \Omega_c^{m_\nu}(E_\nu)$ is a closed form representing U_ν, then we can form $U \in H_c^{m_1 + m_2}(E_1 \times E_2)$ represented by

$$\omega = \mathrm{pr}_1^*(\omega_1) \wedge \mathrm{pr}_2^*(\omega_2).$$

It follows from Fubini's theorem that ω has integral 1 over each fiber in ξ. Let $s_\nu \colon M \to E_\nu$ be the zero section. Then

$$\begin{aligned}
\hat{e}(\xi) &= (s_1 \times s_2)^*(U) = \left[(s_1 \times s_2)^*(\omega) \right] \\
&= \left[(s_1 \times s_2)^* \circ \mathrm{pr}_1^*(\omega_1) \right] \left[(s_1 \times s_2)^* \circ \mathrm{pr}_2^*(\omega_2) \right].
\end{aligned}$$

If $p_\nu \colon M \times M \to M$ denotes the projections, then

$$\left[(s_1 \times s_2)^* \circ \mathrm{pr}_\nu^*(\omega_\nu) \right] = \left[p_\nu^* \circ s_\nu^*(\omega_\nu) \right] = p_\nu^*(\hat{e}(\xi_\nu))$$

so $\hat{e}(\xi) = p_1^*(\hat{e}(\xi_1)) \wedge p_2^*(\hat{e}(\xi_2))$. Finally Lemma 21.6 yields

$$\begin{aligned}
\hat{e}(\xi_1 \oplus \xi_2) &= \hat{e}(\Delta^*\xi) = \Delta^*(\hat{e}(\xi)) \\
&= \Delta^* \circ p_1^*(\hat{e}(\xi_1)).\Delta^* \circ p_2^*(\hat{e}(\xi_2)) \\
&= (p_1 \circ \Delta)^*(\hat{e}(\xi_1)).(p_2 \circ \Delta)^*(\hat{e}(\xi_2)) \\
&= \hat{e}(\xi_1)\hat{e}(\xi_2).
\end{aligned}$$

$\square$

In Chapter 11 we defined the local index of a tangent vector field with isolated singularities, and in Chapter 12 we proved the Poincaré–Hopf theorem, that the sum of the local indices is the Euler characteristic of the manifold. We shall now extend these notions to sections of an arbitrary oriented vector bundle ξ over a compact, oriented smooth manifold M, provided $\dim \xi = \dim M = m$. Let $E = E(\xi)$ be the total space and $\pi \colon E \to M$ the bundle projection. We let $s_0 \colon M \to E$ be the *zero section* of ξ that to each $p \in M$ associates the origin in the fiber ξ_p. Let $s \colon M \to E$ be a second smooth section. The differentials $D_p s$ and $D_p s_0$ from $T_p M$ to $T_q E$ are monomorphisms since s and s_0 have one-sided inverses.

Definition 21.8 Let $p \in M$ be a zero (singularity) for s, $s(p) = s_0(p)$. Then s is called transversal to s_0 at p if

$$(4) \qquad\qquad D_p s(T_p M) \cap D_p s_0(T_p M) = 0$$

and s is called transversal to s_0 if this holds for all zeros of s.

The tangent space $T_{s_0(p)} E$ is the direct sum of the tangent space $D_p s_0(T_p M)$ to the zero section $s_0(M)$ and the tangent space at $s_0(p)$ to the fiber ξ_p which is naturally identified with ξ_p itself. In other words:

$$T_{s_0(p)} E \cong T_p M \oplus \xi_p.$$

With this identification, (4) is equivalent to the statement that $D_p s(T_p M)$ is the graph of a linear isomorphism $A : T_p M \to \xi_p$. Both vector spaces are oriented by assumption; we define the *local index* $\iota(s; p)$ to be $+1$, if A preserves the orientations, and -1 if not. In the special case where $\xi = \tau_M$ is the tangent bundle this is in agreement with Definition 11.16 (cf. Lemma 11.20).

Given an oriented local trivialization of ξ over U, $\xi_{|U} \cong U \times \mathbb{R}^m$, we can identify the restriction of s to U with a smooth map $F : U \to \mathbb{R}^m$. Then $A : T_p M \to \xi_p$ corresponds to $D_p F : T_p M \to \mathbb{R}^m$, and (4) becomes the statement that $D_p F$ is an isomorphism. Hence $\iota(s; p) = \pm 1$ depending on the orientation behavior of $D_p F$. The inverse function theorem implies that F is a local diffeomorphism at p. In particular (4) forces p to be an isolated zero of s. If $s : M \to E$ is transversal to the zero section, then the number of zeros of s is finite, since M was assumed compact.

Theorem 21.9 *In the situation above, if s is transverse to the zero section, then*

$$(5) \qquad\qquad I(\hat{e}(\xi)) = \sum_p \iota(s; p)$$

where we sum over the zeros of s, and where $I : H^m(M) \to \mathbb{R}$ is the integration homomorphism.

Example 21.10 Let H be the canonical complex line bundle over $\mathbb{CP}^1$, H^* its dual bundle, and $\xi = (H^*)_{\mathbb{R}}$ the underlying oriented real bundle. The bundle H is a subbundle of the trivial 2-dimensional bundle

$$E(H) = \{ (L; \mathbf{z}) \mid L \in \mathbb{CP}^1, \mathbf{z} \in L \} \subset \mathbb{CP}^2 \times \mathbb{C}^2$$

(cf. Example 17.2). Dually there is an epimorphism from the dual product bundle onto H^*, and we let $s : \mathbb{CP}^1 \to E(H^*)$ be the section that is the image of the constant section in the dual product bundle given by the linear form

$$\sigma : \mathbb{C}^2 \to \mathbb{C}; \quad \sigma(w_0, w_1) = w_1.$$

In terms of homogeneous coordinates on $\mathbb{CP}^1$, $s([z_0, z_1])$ is the restriction of σ to the fiber $\text{span}_{\mathbb{C}}(z_0, z_1)$ in H. The only zero of s is $p_0 = [1, 0]$. Over the coordinate chart $U_0 = \{[1, z] \mid z \in \mathbb{C}\}$ in $\mathbb{CP}^1$ we have a trivialization of H defined by

$$U_0 \times \mathbb{C} \to H; \quad ([1, z], a) \mapsto ([1, z], (a, az)).$$

In the dual trivialization of H^* we find that $s_{|U_0}$ corresponds to the function $U_0 \to \mathbb{C}$, which maps $[1, z]$ into z; thus in terms of local coordinates s is the identity. It follows that s is transversal to the zero section at p_0, and that $\iota(s; p_0) = 1$. From Theorem 21.9 we conclude that $I(\hat{e}(\xi)) = 1$.

Theorem 21.11 *For oriented vector bundles over compact manifolds, $\hat{e}(\xi) = e(\xi)$.*

Proof. We have already seen in (3) above that $\hat{e}(\xi^{2m}) = a^m e(\xi^{2m})$ for some constant a; it remains to be shown that $a = 1$. For $\xi^2 = (H^*)_{\mathbb{R}}$, the previous example shows that $I(\hat{e}(\xi^2)) = 1$. On the other hand, $e(\xi^2) = c_1(H^*)$ by Theorem 19.6.(ii), and $I(c_1(H^*)) = 1$ by Theorem 18.9.(i) and Property 18.11.(b). Since I is injective $e(\xi^2) = \hat{e}(\xi^2)$ in this case, so $a = 1$. $\square$

If the dimension m is odd then we saw in the discussion following Definition 21.4 that $\hat{e}(\xi) = 0$, and consequently the index sum in (5) will always vanish. If ξ is even-dimensional and admits a section s without zeros, then the index sum is zero, and $\hat{e}(\xi) = 0$. One often expresses this by saying that $\hat{e}(\xi)$ is the *obstruction* for ξ to admit a non-zero section. Note that a non-zero section is equivalent to a splitting $\xi \cong \xi' \oplus \varepsilon_M^1$. Indeed, we may choose an inner product on ξ and define ξ' to be the orthogonal complement to the trivial subbundle of ξ consisting of lines generated by s.

Theorem 21.12 *For any oriented compact smooth manifold M,*

$$I(e(\tau_M)) = \chi(M).$$

Proof. We simply apply Theorem 21.9 to $\xi = \tau_M$, taking for s a gradient-like vector field X w.r.t. some Morse function; cf. Definition 12.7. The proof of Lemma 12.8 shows that X is transversal to the zero section, and that the sum in (5) is equal to $\text{Index}(X)$. The Poincaré–Hopf theorem finishes the proof. $\square$

We can combine the two previous theorems to give a generalization of the classical Gauss–Bonnet theorem to even-dimensional compact, oriented smooth manifolds.

Theorem 21.13 (Generalized Gauss-Bonnet formula)

$$\int_M \text{Pf}\left(\frac{-F^\nabla}{2\pi}\right) = \chi(M^{2n}),$$

where F^∇ is the curvature associated to any metric connection on the tangent bundle of M^{2n}.   □

The rest of this chapter is devoted to a proof of Theorem 21.9. Let $p_1, \ldots, p_k$ be the zeros of s. First we construct a closed form $\omega \in \Omega_c^m(E)$ which represents $U \in H_c^m(E)$, and local trivializations of ξ over disjoint open neighborhoods $V_1, \ldots, V_k$ of the zeros. Let E_{V_i} be the inverse image of V_i in E, and define $f_i \colon E_{V_i} \to \mathbf{R}^m$ to be the composition of a trivialization $E_{V_i} \overset{\cong}{\to} V_i \times \mathbf{R}^m$ with the projection on $\mathbf{R}^m$.

Lemma 21.14 *In the above, V_i and f_i may be chosen so that*

$$\omega_{|E_{V_i}} = f_i^*(\omega_i), \quad 1 \le i \le k$$

where $\omega_i \in \Omega_c^m(\mathbf{R}^m)$ are forms with $\int_{\mathbf{R}^m} \omega_i = 1$.

Proof. Let $h \colon M \to M$ be a map such that

(α) h is smoothly homotopic to id_M.

(β) h is constant with value p_i on some open neighborhood V_i of p_i, $i = 1, \ldots, k$.

It follows from (α) and Theorem 15.21 that $\xi' = h^*(\xi)$ is isomorphic to ξ, so it suffices to construct the required trivializations and forms for ξ' instead of ξ. Consider the diagram

$$
\begin{array}{ccc}
E' & \overset{\tilde{h}}{\longrightarrow} & E \\
\downarrow{\scriptstyle \pi'} & & \downarrow{\scriptstyle \pi} \\
M & \overset{h}{\longrightarrow} & M
\end{array}
$$

where $E = E(\xi)$, $E' = E(\xi')$. Pick a closed form $\omega \in \Omega_c^m(E)$ with integral 1 over each fiber in ξ. Then $\omega' = \tilde{h}^*(\omega) \in \Omega_c^m(E')$ has integral 1 over each fiber of ξ'. Since $h(V_i) = p_i$, $\tilde{h} \colon E'_{V_i} \to \xi_{p_i}$. We pick an oriented linear isomorphism $\xi_{p_i} \cong \mathbf{R}^m$, and let $\omega_i \in \Omega_c^m(\mathbf{R}^m)$ correspond to $\omega_{|\xi_{p_i}} \in \Omega_c^m(\xi_p)$, and f_i to $\tilde{h}$ under this isomorphism. Then

$$\omega'_{|E'_{V_i}} = f_i^*(\omega_i), \quad \int_{\mathbf{R}^m} \omega_i = 1, \quad 1 \le i \le k$$

as required. We have left to construct h with the stated properties. To this end we use a smooth map $G \colon \mathbf{R}^m \times \mathbf{R} \to \mathbf{R}^m$ which satisfies:

$$G(x, 0) = 0 \quad \text{for } x \in \tfrac{1}{2} D^m$$
$$G(x, 1) = x \quad \text{for } x \in \mathbf{R}^m$$
$$G(x, t) = x \quad \text{for } t \in \mathbf{R} \text{ and } x \in \mathbf{R}^m - D^m.$$

For example we can take

$$(6) \qquad\qquad G(x,t) = ((1-t)\rho(\|x\|) + t)x$$

where $\rho \colon \mathbb{R} \to \mathbb{R}$ is a smooth function with

$$\rho(y) = \begin{cases} 0 & \text{for } y \leq \frac{1}{2} \\ 1 & \text{for } y \geq 1. \end{cases}$$

Now construct a homotopy $F \colon M \times \mathbb{R} \to M$ as follows. Choose disjoint charts W_i, $1 \leq i \leq k$, and diffeomorphisms $\varphi_i \colon W_i \to \mathbb{R}^m$ such that $\varphi_i(p_i) = 0$, and define

$$F(p,t) = \begin{cases} \varphi_i^{-1} \circ F(\varphi_i(p), t) & \text{if } p \in W_i \\ p & \text{if } p \notin \cup_i W_i. \end{cases}$$

Then (α) holds for $h(p) = F(p,0)$ and (β) holds for $V_i = \varphi_i^{-1}\left(\frac{1}{2}\mathring{D}^m\right)$. $\square$

Proof of Theorem 21.9 Pick data as in Lemma 21.14. Replacing V_i with a smaller open neighborhood of p_i, we may assume that $f_i \circ (s_{|V_i})$ maps V_i diffeomorphically onto an open set in $\mathbb{R}^m$. Pick an open neighborhood W_i of p_i with closure $\overline{W_i} \subseteq V_i$. Since $M - \bigcup_i W_i$ and $\operatorname{supp}_E(\omega)$ are compact, we can find a constant $c > 0$ such that the scaled section $\tilde{s} = cs$ satisfies

$$\operatorname{supp}_{\mathbb{R}^m}(\omega_i) \subset f_i\big(\tilde{s}(V_i)\big), \quad 1 \leq i \leq k$$
$$\tilde{s}\Big(M - \bigcup_i W_i\Big) \cap \operatorname{supp}_E(\omega) = \emptyset.$$

Now $e(\xi) \in H^m(M)$ is represented by the m-form $\tilde{s}^*(\omega)$, which is identically zero on $M - \bigcup_i \overline{W}_i$. By Lemma 21.14

$$\tilde{s}^*(\omega)_{|V_i} = \big(f_i \circ (\tilde{s}_{|V_i})\big)^*(\omega_i)$$

where $f_i \circ (\tilde{s}_{|V_i})$ is a diffeomorphism from V_i to an open set in $\mathbb{R}^m$ containing $\operatorname{supp}_{\mathbb{R}^m}(\omega_i)$. This diffeomorphism preserves or reverses orientations depending on the value of $\iota(\tilde{s}; p_i) = \iota(s; p_i) = \pm 1$. Formula (5) follows by the computation:

$$(7) \qquad \begin{aligned} I(e(\xi)) &= \int_M \tilde{s}^*(\omega) = \sum_{i=1}^k \int_{V_i} \big(f_i \circ (\tilde{s}_{|V_i})\big)^*(\omega_i) \\ &= \sum_{i=1}^k \iota(s; p_i) \int_{\mathbb{R}^m} \omega_i = \sum_{i=1}^k \iota(s; p_i). \end{aligned}$$

$\square$

A. SMOOTH PARTITION OF UNITY

The following technical theorem is a much used tool when working with smooth maps and smooth manifolds.

For a function $f: U \to \mathbb{R}$ with domain $U \subseteq \mathbb{R}^n$ the *support* of f *in* U is the set

$$\text{supp}_U(f) = \overline{\{x \in U \mid f(x) \neq 0\}},$$

where the bar denotes the closure of the set in the *induced topology* on U. If U is open in $\mathbb{R}^n$ then $U - \text{supp}_U(f)$ is the largest open subset of U on which f vanishes.

Theorem A.1 *If $U \subseteq \mathbb{R}^n$ is open and $\mathcal{V} = (V_i)_{i \in I}$ is a cover of U by open sets V_i, then there exist smooth functions $\phi_i: U \to [0,1]$ $(i \in I)$, satisfying*

(i) $\text{supp}_U(\phi_i) \subseteq V_i$ *for all* $i \in I$.
(ii) *Every* $x \in U$ *has a neighborhood W on which only finitely many ϕ_i do not vanish.*
(iii) *For every* $x \in U$ *we have* $\sum_{i \in I} \phi_i(x) = 1$

We say that $(\phi_i)_{i \in I}$ is a (smooth) *partition of unity*, which is subordinate to the cover $\mathcal{V}$.

A family of functions $\phi_i: U \to \mathbb{R}$ that satisfy (ii) is called *locally finite*. Note that the sum $\sum_{i \in I} \phi_i$ in this case becomes a well-defined function $U \to \mathbb{R}$. Moreover, it is smooth when all the ϕ_i are smooth. The proof of Theorem A.1 requires some preparations.

Lemma A.2 *The function $\omega: \mathbb{R} \to \mathbb{R}$ defined by*

$$\omega(t) = \begin{cases} 0 & \text{if } t \leq 0 \\ \exp(-1/t) & \text{if } t > 0 \end{cases}$$

is smooth.

Proof. It is only smoothness at $t = 0$ which causes difficulties. It is sufficient to see that

$$\lim_{t \to 0+} \frac{\omega^{(n-1)}(t)}{t} = 0$$

for all $n \geq 1$. By induction there exist polynomials $p_0, p_1, p_2, \ldots$, such that

$$\omega^{(n)}(t) = p_n(1/t) \exp(-1/t),$$

for $t > 0$ and $n \geq 0$. The result now follows because

$$\lim_{t \to 0+} \left((1/t)^k \exp(-1/t) \right) = \lim_{x \to \infty} \frac{x^k}{\exp(x)} = 0$$

for $k \geq 0$. $\square$

Corollary A.3 *For real numbers $a < b$ there exists a smooth function $\psi: \mathbb{R} \to [0,1]$ such that $\psi(t) = 0$ for $t \leq a$ and $\psi(t) = 1$ for $t \geq b$.*

Proof. Set $\psi(t) = \omega(t - a)/(\omega(t - a) + \omega(b - t))$. □

For $x \in \mathbb{R}$ and $\epsilon > 0$ let $D_\epsilon(x) = \{y \in \mathbb{R}^n \mid \|y - x\| < \epsilon\}$.

Corollary A.4 *For $x \in \mathbb{R}^n$ and $\epsilon > 0$ there exists a smooth function $\phi: \mathbb{R}^n \to [0, \infty)$, such that $D_\epsilon(x) = \phi^{-1}((0, \infty))$.*

Proof. Define ϕ by the formula

$$\phi(y) = \omega(\epsilon^2 - \|y - x\|^2) = \omega\left(\epsilon^2 - \sum_{j=1}^{n}(y_j - x_j)^2\right).$$ □

Note that the support of ϕ is the corresponding closed disc

$$\overline{D}_\epsilon(x) = \{y \in \mathbb{R}^n \mid \|y - x\| \leq \epsilon\}.$$

Lemma A.5 *An arbitrary open set $U \subseteq \mathbb{R}^n$ can be written in the form $U = \bigcup_{m=1}^{\infty} K_m$, where the sets K_m are compact and $K_m \subseteq \mathring{K}_{m+1}$ (the interior of K_{m+1}) for $m \geq 1$.*

Proof. The conditions hold for $K_m = \overline{D}_{2^m}(0) - \bigcup_{x \in \mathbb{R}^n - U} D_{1/2^m}(x)$. □

Proposition A.6 *For an arbitrary open set $U \subseteq \mathbb{R}^n$ and a cover $\mathcal{V} = (V_i)_{i \in I}$ of U by open sets, we can find a sequence (x_j) in U and a sequence (ϵ_j) of positive real numbers that satisfy the following conditions:*

  (i) $U = \bigcup_{j=1}^{\infty} D_{\epsilon_j}(x_j)$.
  (ii) *For every j there exists $i(j) \in I$ with $D_{2\epsilon_j}(x_j) \subseteq V_{i(j)}$.*
  (iii) *Every $x \in U$ has a neighborhood W that intersects only finitely many of the balls $D_{2\epsilon_j}(x_j)$.*

Proof. We choose $K_m (m \geq 1)$ as in Lemma A.5. Additionally we set $K_0 = K_{-1} = \emptyset$. For $m \geq 1$ we introduce the sets

$$B_m = K_m - \mathring{K}_{m-1}, \qquad U_m = \mathring{K}_{m+1} - K_{m-2}.$$

Here B_m is compact, U_m is open, $B_m \subseteq U_m$ and $U = \bigcup_{m=1}^{\infty} B_m$. For $x \in B_m$ we can find $\epsilon(x) > 0$ such that $D_{2\epsilon(x)}(x)$ is contained in both U_m and at least one of the sets V_i. The Heine–Borel property for B_m ensures the existence of $x_{m,j} \in B_m$ and $\epsilon_{m,j} > 0$ $(1 \leq j \leq d_m)$ such that

  (α) $B_m \subseteq \bigcup_{j=1}^{d_m} D_{\epsilon_{m,j}}(x_{m,j})$.
  (β) Every $D_{2\epsilon_{m,j}}(x_{m,j})$ is contained in U_m and in at least one of the sets V_i.

The desired result is achieved by re-indexing the families $(x_{m,j})$ and $(\epsilon_{m,j})$, where $m \geq 1$ and $1 \leq j \leq d_m$. From (α) and (β) follows

$$U = \bigcup_{m=1}^{\infty} B_m \subseteq \bigcup_{m=1}^{\infty} \bigcup_{j=1}^{r_m} D_{\epsilon_{m,j}}(x_{m,j}) \subseteq \bigcup_{m=1}^{\infty} U_m \subseteq U,$$

which yields (i). One obtains (ii) directly from (β). For $x \in U$ we choose $m_0 \geq 1$ with $x \in U_{m_0}$. Since $U_{m_0} \cap U_m = \emptyset$ when $m \geq m_0 + 3$, we see that U_{m_0} can intersect $D_{2\epsilon}(x_{m,j})$ only when $m \leq m_0 + 2$. This proves (iii). $\square$

Proof of Theorem A.1. Choose $(x_j), (\epsilon_j)$ and $i(j) \in I$ as in Proposition A.6. Apply Corollary A.4 to find smooth functions $\psi_j : \mathbb{R}^n \to [0, \infty)$ with $D_{\epsilon_j}(x_j) = \psi_j^{-1}((0, \infty))$. Condition (iii) ensures that the function $\psi : U \to \mathbb{R}$ given by

$$\psi(x) = \sum_j \psi_j(x)$$

is smooth, because the sum is finite on a neighborhood of an arbitrary $x \in U$. From (i) it follows that $\psi(x) > 0$ for all $x \in U$.

We introduce the modified functions $\tilde{\psi}_j : U \to [0, \infty)$ given by $\tilde{\psi}_j(x) = \psi_j(x) \psi(x)^{-1}$. They are smooth with $D_{\epsilon_j}(x_j) = \tilde{\psi}_j^{-1}((0, \infty))$ and $\sum \tilde{\psi}_j(x) = 1$ for all $x \in U$. Set $\phi_i = \sum \tilde{\psi}_j$, for $i \in I$, summed over the set J_i of indices j for which $i(j) = i$ (in particular $\phi_i = 0$ when $J_i = \emptyset$). By Proposition A.6.(iii) it follows that ϕ_i is smooth on U. Moreover, these functions satisfy (ii) and (iii) in the theorem. Any $x \in \mathrm{supp}_U(\phi_i)$ has a neighborhood $W \subseteq U$ that satisfies Proposition A.6.(iii). Then the restriction $\phi_{i|W}$ becomes a sum of finitely many $\tilde{\psi}_{j_\nu|W}$, with $j_\nu \in J_i$, and there is at least one $j_\nu \in J_i$ with $x \in \mathrm{supp}_U(\tilde{\psi}_{j_\nu})$. Since

$$\mathrm{supp}_U(\tilde{\psi}_{j_\nu}) = \overline{D}_{\epsilon_{j_\nu}}(x_{j_\nu}) \subseteq D_{2\epsilon_{j_\nu}}(x_{j_\nu}) \subseteq V_i,$$

we get that $x \in V_i$. Hence part (i) of the theorem is satisfied. $\square$

Corollary A.3 has the following generalization to several variables:

Lemma A.7 *If $A \subseteq \mathbb{R}^n$ is closed and $U \subseteq \mathbb{R}^n$ is open with $A \subseteq U$, then there exists a smooth function $\psi : \mathbb{R}^n \to [0, 1]$ with $\mathrm{supp}_{\mathbb{R}^n}(\psi) \subseteq U$ and $\psi(x) = 1$ for all $x \in A$.*

Proof. Apply Theorem A.1 to the cover of $\mathbb{R}^n$ consisting of the open sets $V_1 = U$ and $V_2 = \mathbb{R}^n - A$. Now $\psi = \phi_1$ has the desired properties. $\square$

Proposition A.8 (Whitney) *For an arbitrary closed set $A \subseteq \mathbb{R}^n$ there exists a smooth function $\phi: \mathbb{R}^n \to [0, \infty)$ with $A = \phi^{-1}(0)$.*

Proof. Apply Lemma A.5 to the open set $U = \mathbb{R}^n - A$, and use Lemma A.7 to find smooth functions $\psi_m: \mathbb{R}^n \to [0,1]$, $m \geq 1$, with $\mathrm{Supp}_{\mathbb{R}^n}(\psi_m) \subseteq \mathring{K}_{m+1}$ and $\psi_m(x) = 1$ for all $x \in K_m$. Define

$$\phi = \sum_{m=1}^{\infty} c_m \psi_m$$

for a suitably chosen sequence (c_m) of positive numbers. Let

$$D_{\mathbf{i}} = \frac{\partial^{|\mathbf{i}|}}{\partial x_1^{i_1} \partial x_2^{i_2} \ldots \partial x_n^{i_n}},$$

where $\mathbf{i} = (i_1, \ldots, i_n)$ and $|\mathbf{i}| = \sum i_\nu$. We show that the series

(1)
$$\sum_{m=1}^{\infty} c_m D_{\mathbf{i}} \psi_m$$

converges uniformly on $\mathbb{R}^n$. Since ψ_m has compact support, we can find $b_m \in [1, \infty)$ with $\sup_{x \in \mathbb{R}^n} |D_{\mathbf{i}} \psi_m(x)| \leq b_m$ for all $\mathbf{i}$ with $|\mathbf{i}| \leq m$. If we set $c_m = (2^m b_m)^{-1}$ then $\sum_{m=1}^{\infty} 2^{-m}$ is a comparison series for formula (1) from the $|\mathbf{i}|$-th term onwards. This implies uniform convergence of the series (1). Hence ϕ is smooth and $D_{\mathbf{i}} \phi$ is given by (1). For all $x \in K_m$ we get $\phi(x) \geq c_m \psi_m(x) = c_m > 0$, and thus $A = \phi^{-1}(0)$. $\square$

Lemma A.9 *Suppose that $A \subseteq U_0 \subseteq U \subseteq \mathbb{R}^n$, where U_0 and U are open in $\mathbb{R}^n$ and A is closed in U (in the induced topology from $\mathbb{R}^n$). Let $h: U \to W$ be a continuous map to an open set $W \subseteq \mathbb{R}^m$ with smooth restriction to U_0. For any continuous function $\epsilon: U \to (0, \infty)$ there exists a smooth map $f: U \to W$ that satisfies*

(i) $\|f(x) - h(x)\| \leq \epsilon(x)$ *for all $x \in U$.*

(ii) $f(x) = h(x)$ *for all $x \in A$.*

Proof. If $W \neq \mathbb{R}^m$ then $\epsilon(x)$ can be replaced by

$$\epsilon_1(x) = \min\left(\epsilon(x), \tfrac{1}{2} d(h(x), \mathbb{R}^n - W)\right)$$

where $d(y, \mathbb{R}^n - W) = \inf\{\|y - z\| \mid z \in \mathbb{R}^m - W\}$. If $f: U \to \mathbb{R}^m$ satisfies (i) with ϵ_1 instead of ϵ, we will automatically get $f(U) \subseteq W$. Hence, without loss of generality, we may assume that $W = \mathbb{R}^m$.

Using continuity of h and ϵ, we can find for each point $p \in U - A$ an open set U_p with $p \in U_p \subseteq U - A$, such that $\|h(x) - h(p)\| \leq \epsilon(x)$ for all $x \in U_p$. Apply

Theorem A.1 to the open cover of U consisting of the sets U_0 and U_p, $p \in U - A$. This yields smooth functions ϕ_0 and ϕ_p from U into $[0, 1]$, which satisfy Theorem A.1.(i), (ii) and (iii). By local finiteness, smoothness of h on U_0 and the property $\mathrm{supp}_U(\phi_0) \subseteq U_0$, we can define a smooth function $f: U \to \mathbb{R}^m$ by

$$f(x) = \phi_0(x)h(x) + \sum_{p \in U-A} \phi_p(x)h(p).$$

From Theorem A.1.(iii) one obtains $h(x) = \phi_0(x)h(x) + \sum_{p \in U-A} \phi_p(x)h(x)$ and thus

$$f(x) - h(x) = \sum_{p \in U-A} \phi_p(x)(h(p) - h(x)).$$

Now (ii) of the lemma follows because $\mathrm{Supp}_U(\phi_p) \subseteq U_p \subseteq U - A$, and (i) follows from the calculation

$$\|f(x) - h(x)\| \leq \sum_{p \in U-A} \phi_p(x)\|h(p) - h(x)\| = \sum_{p \in U-A, x \in U_p} \phi_p(x)\|h(p) - h(x)\|$$

$$\leq \sum \phi_p(x)\epsilon(x) = \left(\sum \phi_p(x)\right) \cdot \epsilon(x) \leq \epsilon(x). \qquad \square$$

B. INVARIANT POLYNOMIALS

In this appendix we shall consider polynomials in n^2 variables. We may arrange the n^2 variables A_{ij} as a matrix $A = (A_{ij})$ and write $P(A)$. We shall only consider *homogeneous* polynomials, i.e. a sum of monomials of same degree, which will be called the degree of P. The polynomial $P(A)$ is said to be *invariant* when

$$P(gAg^{-1}) = P(A)$$

for all $g \in GL_n(\mathbb{C})$. Every polynomial $P(A)$ determines a function

$$P \colon M_n(\mathbb{C}) \to \mathbb{C}$$

and this function uniquely determines the polynomial. Moreover, an invariant polynomial defines a function

$$P \colon \mathrm{Hom}(V, V) \to \mathbb{C}$$

for every n-dimensional complex vector space V independent of choice of basis. Let us consider the characteristic polynomial

$$\sigma(t) = \det(I + tA) = \sum_{i=0}^{n} \sigma_i(A) t^i, \qquad \sigma_0(A) = 1.$$

Each of the functions $\sigma_i(A)$ is an invariant polynomial. Also consider

$$(1)' \qquad s(t) = -t \frac{d}{dt} \log(\det(I - tA)) = \sum_{k=0}^{\infty} s_k(A) t^k$$

where $\log$ denotes the power series

$$\log(1 + x) = \sum_{k=1}^{\infty} \frac{(-1)^{k-1}}{k} x^k$$

and $\log(\det(I - tA))$ means that we substitute the polynomial $\det(I - tA)$ in the power series. Differentiation in (1) is performed formally,

$$\frac{d}{dt} \left(\sum_{i=0}^{\infty} a_i t^i \right) = \sum_{i=1}^{\infty} i a_i t^{i-1}.$$

Conversely

$$\sigma(t) = \sum_{i=1}^{n} \sigma_i t^i = \exp\left(\int \frac{s(-t)}{t} dt \right),$$

calculated formally too.

Lemma B.1 *For every $A \in M_n(\mathbb{C})$ we have $s_k(A) = \operatorname{tr}(A^k)$.*

Proof. Let us assume that A is a diagonal matrix, $A = \operatorname{diag}(\lambda_1, \ldots, \lambda_n)$, such that

$$\det(I - tA) = \prod_{i=1}^{n}(1 - t\lambda_i).$$

This yields the following equation of power series

$$-t\frac{d}{dt}\log\det(I - tA) = \sum_{i=1}^{n}\frac{t\lambda_i}{1 - t\lambda_i} = \sum_{i=1}^{n}\sum_{k=1}^{\infty}\lambda_i^k t^k = \sum_{k=1}^{\infty}\left(\sum_{i=1}^{n}\lambda_i^k\right)t^k.$$

Hence

$$s_k(\operatorname{diag}(\lambda_1, \ldots, \lambda_n)) = \lambda_1^k + \cdots + \lambda_n^k = \operatorname{tr}\left(\operatorname{diag}(\lambda_1, \ldots, \lambda_n)^k\right).$$

Since both $s_k(A)$ and $\operatorname{tr}(A^k)$ are unaltered when we replace A by gAg^{-1}, and since the diagonalizable matrices are dense in the vector space of all matrices, the assertion follows. $\qquad\square$

Lemma B.2 *For polynomials in n^2 variables we have that*

$$s_k(A) - s_{k-1}(A)\sigma_1(A) + s_{k-2}(A)\sigma_2(A) - \cdots + (-1)^k k\sigma_k(A) = 0.$$

Proof. It suffices to prove the equation for a diagonal matrix $A = \operatorname{diag}(\lambda_1, \ldots, \lambda_n)$ where

$$\overline{\sigma}(t) = \prod_{i=1}^{n}(1 - t\lambda_i) = \sum_{k=0}^{n}(-1)^k\sigma_k(A)t^k \quad \text{and} \quad s(t) = \sum_{i=1}^{n}\frac{t\lambda_i}{1 - t\lambda_i} = \sum_{k=1}^{\infty}s_k(A)t^k.$$

Now

$$\overline{\sigma}(t)s(t) = \left(\sum_{i=1}^{n}\frac{t\lambda_i}{1 - t\lambda_i}\right)\prod_{i=1}^{n}(1 - t\lambda_i) = \sum_{i=1}^{n}t\lambda_i\prod_{j\neq i}(1 - t\lambda_j) = -t\frac{d\overline{\sigma}}{dt}$$

$$= \sum_{k=1}^{n}(-1)^{k-1}k\sigma_k(A)t^k.$$

The coefficient of t^k in this equation yields the desired identity. $\qquad\square$

Note that the equation of Lemma B.2 inductively determines $s_k(A)$ as a polynomial with integer coefficients in the variables $\sigma_1(A), \ldots, \sigma_k(A)$. Conversely $\sigma_k(A)$ is a polynomial with rational coefficients in $s_1(A), \ldots, s_k(A)$. We write

(2)
$$s_k(A) = Q_k(\sigma_1(A), \ldots, \sigma_k(A))$$
$$\sigma_k(A) = P_k(s_1(A), \ldots, s_k(A)).$$

For instance $s_1(A) = \sigma_1(A)$ and $s_2(A) = \sigma_1(A)^2 - 2\sigma_2(A)$. We have the following identities in $n_1^2 + n_2^2$ variables

$$\sigma_k(A_1 \oplus A_2) = \sum_{i=0}^{k} \sigma_i(A_1)\sigma_{k-i}(A_2)$$

(3)

$$s_k(A_1 \oplus A_2) = s_k(A_1) + s_k(A_2)$$

$$s_k(A_1 \otimes A_2) = s_k(A_1) \cdot s_k(A_2),$$

where $A_1 \oplus A_2$ is the matrix

$$A_1 \oplus A_2 = \begin{pmatrix} A_1 & 0 \\ 0 & A_2 \end{pmatrix},$$

and $A_1 \otimes A_2$ the matrix of the tensor product of the two linear maps. For the first equation it is sufficient to show that both sides of both formulas define the same functions on $M_{n_1}(\mathbb{C}) \times M_{n_2}(\mathbb{C})$, but this is obvious from the definitions:

$$\det(I + t(A_1 \oplus A_2)) = \det(I + tA_1) \cdot \det(I + tA_2)$$

giving the first equation in (3) upon considering the coefficient of t^k. The other relations are similar, and left to the reader.

Let $\sigma_i(\lambda_1, \ldots, \lambda_n)$, $i = 1, \ldots, n$ be the polynomials defined by

$$\prod_{i=1}^{n} (1 + t\lambda_i) = \sum_{i=0}^{n} \sigma_i(\lambda_1, \ldots, \lambda_n)t^i.$$

They are the so-called *elementary symmetrical polynomials* in the variables $(\lambda_1, \ldots, \lambda_n)$.

Theorem B.3 *Every polynomial $P(\lambda_1, \ldots, \lambda_n)$ which is invariant under permutation of coordinates can be written in the form $P(\lambda_1, \ldots, \lambda_n) = p(\sigma_1, \ldots, \sigma_n)$, where σ_i are the elementary symmetrical polynomials and p is a polynomial.*

Proof. See [Lang] Chapter V.9. □

Theorem B.4 *Every invariant polynomial $P : M_n(\mathbb{C}) \to \mathbb{C}$ can be written in the form*

$$P(A) = p(\sigma_1(A), \ldots, \sigma_k(A))$$

where p is a polynomial.

Proof. Let $D_n \subset M_n(\mathbb{C})$ be the diagonal matrices. Since P is invariant and the set

$$\bigcup_{g \in GL_n(\mathbb{C})} gD_ng^{-1} \subset M_n(\mathbb{C})$$

is everywhere dense, $P\colon M_n(\mathbb{C}) \to \mathbb{C}$ is determined by its restriction to D_n. A permutation $\pi \in \Sigma_n$ of n elements induces an endomorphism $\mathbb{C}^n \to \mathbb{C}^n$ by permuting the coordinates, i.e. an element in $M_n(\mathbb{C})$, again denoted by π. If $d(\lambda_1, \ldots, \lambda_n) \in M_n(\mathbb{C})$ is the diagonal matrix with elements $\lambda_1, \ldots, \lambda_n$ then

$$\pi d(\lambda_1, \ldots, \lambda_n)\pi^{-1} = d\big(\lambda_{\pi(1)}, \ldots, \lambda_{\pi(n)}\big).$$

In particular we have

$$P(d(\lambda_1, \ldots, \lambda_n)) = P\big(d\big(\lambda_{\pi(1)}, \ldots, \lambda_{\pi(n)}\big)\big)$$

for all $\pi \in \Sigma_n$. Theorem B.3 gives that

$$P(d(\lambda_1, \ldots, \lambda_n)) = p(\sigma_1, \ldots, \sigma_n),$$

where $\sigma_i = \sigma_i(\lambda_1, \ldots, \lambda_n)$, and p is a certain polynomial. Hence

$$P(A) = p(\sigma_1(A), \ldots, \sigma_n(A))$$

for all $A \in D_n$, and hence for all $A \in M_n(\mathbb{C})$. $\square$

We finally consider the Pfaffian $\mathrm{Pf}(A)$. This is a homogeneous polynomial in $n(2n-1)$ real variables, or alternatively a polynomial in the skew-symmetric $2n \times 2n$ matrices $A = (A_{ij})$. It has degree n. We can consider $\mathrm{Pf}(A)$ as a map

$$\mathrm{Pf}\colon \mathfrak{so}_{2n} \to \mathbb{R},$$

from the space of skew-symmetric matrices. For an $A \in \mathfrak{so}_{2n}$, we let

$$\omega(A) = \sum_{i<j} A_{ij}\, e_i \wedge e_j \in \Lambda^2\big(\mathbb{R}^{2n}\big)$$

and define $\mathrm{Pf}(A)$ by the equation

$$\omega(A) \wedge \ldots \wedge \omega(A) = n!\mathrm{Pf}(A)\mathrm{vol},$$

where $\mathrm{vol} = e_1 \wedge e_2 \wedge \ldots \wedge e_{2n}$. For the block matrix

$$A = \mathrm{diag}\left(\begin{pmatrix} 0 & a_1 \\ -a_1 & 0 \end{pmatrix}, \begin{pmatrix} 0 & a_2 \\ -a_2 & 0 \end{pmatrix}, \ldots, \begin{pmatrix} 0 & a_n \\ -a_n & 0 \end{pmatrix}\right)$$

a simple calculation gives

(4) $$\omega(A) = a_1\, e_1 \wedge e_2 + a_2\, e_3 \wedge e_4 + \cdots + a_n\, e_{2n-1} \wedge e_{2n},$$

and one sees that

$$\omega(A) \wedge \ldots \wedge \omega(A) = n!(a_1 \ldots a_n)\, \mathrm{vol}.$$

It follows that $\mathrm{Pf}(A) = a_1 \ldots a_n$ and $\mathrm{Pf}(A)^2 = \det(A)$ in this case.

Theorem B.5 *If $A \in \mathfrak{so}_{2n}$ and B is an arbitrary matrix then*

(i) $\mathrm{Pf}(A)^2 = \det(A)$

(ii) $\mathrm{Pf}(BAB^t) = \mathrm{Pf}(A)\det(B)$.

Proof. Since A is a real matrix and $AA^t = A^t A$, A is *normal* when considered as element in $M_{2n}(\mathbb{C})$. The spectral theorem ensures the existence of an orthonormal basis of eigenvectors $e_1, \ldots, e_{2n}$ in $\mathbb{C}^{2n}$ and eigenvalues $\lambda_i \in \mathbb{C}$, $A e_i = \lambda_i e_i$. By conjugation we see that the complex conjugate vector $\bar{e}_i$ is an eigenvector with eigenvalue $\overline{\lambda}_i$. We claim that the basis can be chosen so that $e_{2j} = \bar{e}_{2j-1}$ $(1 \leq j \leq n)$. This is easy if all eigenvalues are zero, i.e. $A = 0$, so we may assume e_1 to be an eigenvector with non-zero eigenvalue λ_1. By skew-symmetry

$$\lambda_1 = \langle Ae_1, e_1 \rangle = \langle -e_1, Ae_1 \rangle = -\overline{\lambda}_1,$$

so λ_1 is purely imaginary. Pick $e_2 = \bar{e}_1$, and note that e_2 has eigenvalue $\lambda_2 = \overline{\lambda}_1 \neq \lambda_1$. Hence e_2 is orthogonal to e_1 in $\mathbb{C}^{2n}$. Note that the orthogonal complement $\mathrm{Span}(e_1, \bar{e}_1)^{\perp}$ is invariant under A and also under complex conjugation. This makes is possible to repeat the process in this subspace.

Having arranged that $e_{2j} = \bar{e}_{2j-1}$, we obtain an orthonormal basis for $\mathbb{R}^{2n}$ consisting of the vectors

$$v_j = \frac{1}{\sqrt{2}}(e_{2j-1} + e_{2j}), \quad w_j = \frac{1}{\sqrt{2}\sqrt{-1}}(e_{2j-1} - e_{2j}) \quad (1 \leq j \leq n).$$

Moreover for $a_j \in \mathbb{R}$ given by $\lambda_{2j-1} = \sqrt{-1}a_j$ we have $Av_j = -a_j w_j$ and $Aw_j = a_j v_j$. This proves the existence of $g \in O_{2n}$ such that

$$gAg^{-1} = \mathrm{diag}\left(\begin{pmatrix} 0 & a_1 \\ -a_1 & 0 \end{pmatrix}, \ldots, \begin{pmatrix} 0 & a_n \\ -a_n & 0 \end{pmatrix}\right).$$

In particular

$$\mathrm{Pf}(gAg^{-1})^2 = (a_1 \ldots a_n)^2 = \det(gAg^{-1}) = \det(A).$$

Since $g^{-1} = g^t$, assertion (ii) implies that $\mathrm{Pf}(A)^2 = \det(A)$ for all A.

In order to prove (ii) we consider the elements $f_i = Be_i \in \mathbb{R}^{2n}$. Since $f_i = \sum B_{\nu i} e_{\nu}$ we have that

$$\tau = \sum A_{ij} f_i \wedge f_j = \sum B_{\nu i} A_{ij} B_{\mu j} e_{\nu} \wedge e_{\mu} = \sum (BAB^t)_{\nu\mu} e_{\nu} \wedge e_{\mu},$$

so that $\tau = \omega(BAB^t)$. Hence

$$\omega(BAB^t) \wedge \ldots \wedge \omega(BAB^t) = \tau \wedge \ldots \wedge \tau = n! \, \mathrm{Pf}(A) f_1 \wedge \ldots \wedge f_{2n}.$$

By Theorem 2.18 the map

$$\Lambda^{2n}(B) \colon \Lambda^{2n}(\mathbb{R}^{2n}) \to \Lambda^{2n}(\mathbb{R}^{2n})$$

is multiplication by $\det{(B)}$, so that

$$f_1 \wedge \ldots \wedge f_{2n} = \det(B) e_1 \wedge \ldots \wedge e_{2n}.$$

This gives (ii).   □

Let $\mathfrak{su}_n \subset M_n(\mathbb{C})$ denote the subset of skew-hermitian matrices. The realification map from $M_n(\mathbb{C})$ to $M_{2n}(\mathbb{R})$ induces a map from $\mathfrak{su}_n$ to $\mathfrak{so}_{2n}$, denoted $A \mapsto A_{\mathbf{R}}$, and we have

Theorem B.6 $\mathrm{Pf}(A_{\mathbf{R}}) = \left(-\sqrt{-1}\right)^n \det(A).$

Proof. Since a skew-hermitian matrix has an orthonormal basis of eigenvectors, we may assume A is diagonal, $A = \mathrm{diag}\left(\sqrt{-1}a_1, \ldots, \sqrt{-1}a_n\right)$ with $a_i \in \mathbb{R}$. Multiplication by $\sqrt{-1}a_i$ on $\mathbb{C}$ corresponds to the matrix

$$\begin{pmatrix} 0 & -a_i \\ a_i & 0 \end{pmatrix} \in \mathfrak{so}_2$$

so by (4) $\mathrm{Pf}(A_{\mathbf{R}}) = (-1)^n a_1 \ldots a_n$. On the other hand

$$\det(A) = \left(\sqrt{-1}\right)^n a_1 \ldots a_n$$

and the result follows.   □

C. PROOF OF LEMMAS 12.12 AND 12.13

In the proof of Lemma 12.13 the diffeomorphism of (iv) is obtained by applying
Lemma 12.12 to a new Morse function $F: M \to \mathbb{R}$, which coincides with f outside
small neighborhoods of the critical points p_i. The sets U_i and V_i from (ii) and
(iii) together with F in a neighborhood of p_i are constructed by means of the
following lemma applied to $f \circ k_i^{-1}$, where k_i is a C^∞-chart around p_i given by
Theorem 12.6. Without loss of generality we may assume $f \circ k_i^{-1}$ to be in the
standard form of Example 12.5.

Lemma C.1 *Let $W \subseteq \mathbb{R}^n$ be an open neighborhood of the origin in $\mathbb{R}^n$ and let
$f: W \to \mathbb{R}$ be the function*

$$f(x) = a - \sum_{i=1}^{\lambda} x_i^2 + \sum_{i=\lambda+1}^{n} x_i^2,$$

where $a \in \mathbb{R}$, $\lambda \in \mathbb{Z}$ and $0 \le \lambda \le n$. Choose $\epsilon > 0$ such that W contains the set

$$E = \left\{ x \in \mathbb{R}^n \;\middle|\; \sum_{i=1}^{\lambda} x_i^2 + 2 \sum_{i=\lambda+1}^{n} x_i^2 \le 2\epsilon \right\}.$$

*Then there exists a Morse function $F: W \to \mathbb{R}$ and contractible open sets $U \subseteq \mathbb{R}^n$,
$V \subseteq \mathbb{R}^{n-\lambda+1}$ that satisfy:*

   (i) $F(x) = f(x)$ when $x \in W - E$.
   (ii) *The only critical point of F in W is 0 and $F(0) < a - \epsilon$.*
   (iii) $F^{-1}((-\infty, a + \epsilon)) = f^{-1}((-\infty, a + \epsilon))$.
   (iv) $F^{-1}((-\infty, a - \epsilon)) = f^{-1}((-\infty, a - \epsilon)) \cup U$.
   (v) $f^{-1}((-\infty, a - \epsilon)) \cap U$ *is diffeomorphic with $S^{\lambda-1} \times V$.*

Proof. We introduce the notation $\xi = \sum_{i=1}^{\lambda} x_i^2$ and $\eta = \sum_{i=\lambda+1}^{n} x_i^2$. Then

(1) $$f(x) = a - \xi + \eta$$

and we define $F \in C^\infty(W, \mathbb{R})$ to be

(2) $$F(x) = a - \xi + \eta - \mu(\xi + 2\eta),$$

where $\mu \in C^\infty(\mathbb{R}, \mathbb{R})$ is chosen to have the properties:

   (a) $-1 < \mu'(t) \le 0$ for all $t \in \mathbb{R}$.
   (b) $\mu(t) = 0$ when $t \ge 2\epsilon$.
   (c) μ is constant on an open interval around 0 with value $\mu(0) > \epsilon$.

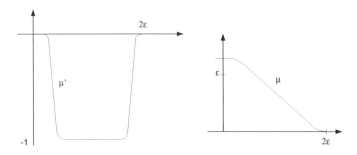

Figure 1

If $x \in W - E$, then $\xi + 2\eta > 2\epsilon$ and (b) implies that $\mu(\xi + 2\eta) = 0$. This gives (i). A simple calculation gives:

$$(3) \qquad \frac{\partial F}{\partial x_i} = \begin{cases} 2x_i(-1 - \mu'(\xi + 2\eta)) & \text{if } 1 \leq i \leq \lambda \\ 2x_i(1 - 2\mu'(\xi + 2\eta)) & \text{if } \lambda + 1 \leq i \leq n. \end{cases}$$

By (a) we have

$$(4) \qquad -1 - \mu'(\xi + 2\eta) < 0, \ 1 - 2\mu'(\xi + 2\eta) > 0.$$

It follows from (3) and (4) that 0 is the only critical point of F. By (c), F coincides with $f - \mu(0)$ on a neighborhood of 0 where $\mu(0) > \epsilon$. This shows that F is a Morse function and that (ii) is satisfied.

Since $\mu(t) \geq 0$ by (a) and (b), the formulas (1) and (2) show that $F(x) \leq f(x)$ for all $x \in W$. Hence

$$(5) \qquad \begin{aligned} f^{-1}((-\infty, a + \epsilon)) &\subseteq F^{-1}((-\infty, a + \epsilon)) \\ f^{-1}((-\infty, a - \epsilon)) &\subseteq F^{-1}((-\infty, a - \epsilon)). \end{aligned}$$

If (iii) were false, there would exist an $x \in W$ with $F(x) < a + \epsilon$ and $f(x) \geq a + \epsilon$. By Equations (1), (2) and (b) we conclude that $\xi + 2\eta < 2\epsilon$. This implies $\eta < \epsilon$, which contradicts (1), because $f(x) \leq a + \eta < a + \epsilon$. This proves (iii). Analogously one sees that (iv) is satisfied for the open set U defined by

$$(6) \qquad U = \{x \in W \mid \xi + 2\eta < 2\epsilon \text{ and } F(x) < a - \epsilon\}.$$

We shall show that U is contractible. By (ii) we have $0 \in U$. Let $\tilde{x} \in U$ be a point of the form $\tilde{x} = (x_1, \ldots, x_\lambda, 0, \ldots, 0)$ and let $\varphi_1 \colon [0, 1] \to \mathbb{R}$ be the function $\varphi_1(t) = F(t\tilde{x})$. By Equations (3) and (4) we have

$$\varphi_1'(t) = \sum_{i=1}^{\lambda} x_i \frac{\partial F}{\partial x_i}(t\tilde{x}) \leq 0,$$

so that φ_1 is decreasing. If $0 \leq t \leq 1$ then

$$F(t\tilde{x}) = \varphi_1(t) \leq \varphi_1(0) = F(0) < a - \epsilon.$$

Hence U contains the line segment with endpoints $\tilde{x}$ and 0. Consider an arbitrary $x \in U$ and let $\tilde{x} = (x_1, \ldots, x_\lambda, 0, \ldots, 0)$. Let φ_2 be

$$\varphi_2 \colon [0,1] \to \mathbb{R}; \quad \varphi_2(t) = F(tx + (1-t)\tilde{x}) = F(x_1, \ldots, x_\lambda, tx_{\lambda+1}, \ldots, tx_n).$$

It is increasing, because the formulas (3) and (4) give

$$\varphi_2'(t) = \sum_{i=\lambda+1}^{n} x_i \frac{\partial F}{\partial x_i}(tx + (1-t)\tilde{x}) \geq 0.$$

If $0 \leq t \leq 1$ then $F(tx + (1-t)\tilde{x}) = \varphi_2(t) \leq \varphi_2(1) = F(x) < a - \epsilon$ from which we conclude that U contains the line segment with endpoints x and $\tilde{x}$. Example 6.5 and Lemma 6.2 imply the contractibility of U.

It remains to construct V and prove (v). Set $B = f^{-1}((-\infty, a - \epsilon)) \cap U$. By Equations (5), (6) and (1) we have

$$B = \{x \in W \mid \xi + 2\eta < 2\epsilon \text{ and } f(x) < a - \epsilon\}$$
$$= \{x \in W \mid \epsilon < \xi < 2\epsilon \text{ and } \eta < \min(\xi - \epsilon, \epsilon - \xi/2)\}.$$

If $\lambda = 0$ then $B = \emptyset$ and (v) is true (with an arbitrarily chosen V). If $\lambda > 0$ we define the open set $V \subset \mathbb{R}^{n-\lambda+1}$ by

$$V = \{(s, x_{\lambda+1}, \ldots, x_n) \mid \sqrt{\epsilon} < s < \sqrt{2\epsilon} \text{ and } \eta < \min(s^2 - \epsilon, \epsilon - s^2/2)\}.$$

To see that V is contractible note that if $q = (s, x_{\lambda+1}, \ldots, x_n) \in V$, then V contains the line segments from q to $\tilde{q} = (s, 0, \ldots, 0)$ and from $\tilde{q}$ to $(s_0, 0, \ldots, 0)$, where $s_0 = \frac{1}{2}(\sqrt{\epsilon} + \sqrt{2\epsilon})$. Define finally a diffeomorphism

$$\Psi \colon S^{\lambda-1} \times V \longrightarrow B; \quad \Psi(y, s, x_{\lambda+1}, \ldots, x_n) = (s \cdot y, x_{\lambda+1}, \ldots, x_n),$$

where $y \in S^{\lambda-1} \subseteq \mathbb{R}^\lambda$ and $(s, x_{\lambda+1}, \ldots, x_n) \in V$. $\square$

Some of the sets introduced in the proof above are indicated in Figure 2. We note that Figure 2 only displays a quarter of the constructed sets; it should be reflected in both the $\sqrt{\xi}$ axis and the $\sqrt{\eta}$ axis for $n = 2$, and rotated correspondingly for

$n > 2$. In particular E is represented by the quarter ellipse between $\sqrt{\epsilon}$ and $\sqrt{2\epsilon}$.

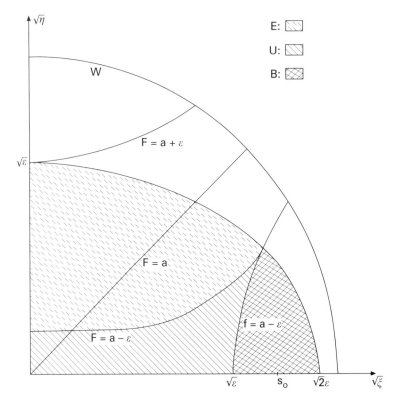

Figure 2 (U hatched; $f^{-1}((-\infty, a - \varepsilon)) \cap U$ double-hatched)

Proof of Lemma 12.13.   By Theorem 12.6 we can choose a smooth chart $h_i \colon \tilde{W}_i \to W_i \subseteq \mathbb{R}^n$ with $p_i \in \tilde{W}_i$, $h_i(p) = 0$ and such that

$$f \circ h_i^{-1}(x) = a - \sum_{j=1}^{\lambda_i} x_j^2 + \sum_{j=\lambda_i+1}^{n} x_j^2 \quad \text{for } x \in W_i.$$

The coordinate patches $\tilde{W}_i$ ($1 \le i \le r$) can be assumed to be mutually disjoint. Choose $\epsilon > 0$ such that a is the only critical value in $[a - \epsilon, a + \epsilon]$ and such that all W_i contain the closed ball $\sqrt{2\epsilon}D^n$. We apply Lemma C.1 to $f \circ h_i^{-1} \colon W_i \to \mathbb{R}$ and this ϵ. We obtain new Morse functions $F_i \colon W_i \to \mathbb{R}$ with $1 \le i \le r$, and an open set $U_i \subseteq \tilde{W}_i$ such that $h_i(U_i)$ is the open contractible subset of $\mathbb{R}^n$ given by Lemma C.1. Thus we have satisfied Lemma 12.13.(i) and (ii), and Lemma 12.13.(iii) follows from assertion (v) of Lemma C.1. By assertion (i) we get a Morse function

$$F \colon M \to \mathbb{R}; \quad F(q) = \begin{cases} F_i \circ h_i(q) & \text{if } q \in \tilde{W}_i \\ f(q) & \text{if } q \notin \bigcup_{i=1}^{r} \tilde{W}_i \end{cases}$$

and by Lemma C.1.(iii) and (iv),

$$F^{-1}((-\infty, a + \epsilon)) = M(a + \epsilon)$$
$$F^{-1}((-\infty, a - \epsilon)) = M(a - \epsilon) \cup U_1 \cup ... \cup U_r.$$

We know from Lemma C.1.(ii) that F has the same critical points as f and furthermore that $F(p_i) < a - \epsilon$ $(1 \leq i \leq r)$. If p is one of the other critical points, then

$$F(p) = f(p) \notin [a - \epsilon, a + \epsilon],$$

and hence $[a - \epsilon, a + \epsilon]$ does not contain any critical value of F. Hence assertion (iv) of Lemma 12.13 follows from Lemma 12.12 applied to F. $\square$

Lemma 12.12 is a consequence of the following theorem, which will be proved later in this appendix.

Theorem C.2 *Let N^n be a smooth manifold of dimension $n \geq 1$ and $f : N \to \mathbb{R}$ a smooth function without any critical points. Let J be an open interval $J \subseteq \mathbb{R}$ with $f(N) \subseteq J$ and such that $f^{-1}([a, b])$ is compact for every bounded closed interval $[a, b] \subset J$. There exists a compact smooth $(n-1)$-dimensional manifold Q^{n-1} and a diffeomorphism*

$$\Phi : Q \times J \to N$$

such that $f \circ \Phi : Q \times J \to J$ is the projection onto J.

Proof of Lemma 12.12.   Choose $c_1 < a_1$ and $c_2 > a_2$, so that the open interval $J = (c_1, c_2)$ does not contain any critical values of f. Since M is compact, we can apply Theorem C.2 to $N = f^{-1}(J)$. We thus have a compact smooth manifold Q and a diffeomorphism $\Phi : Q \times J \to N$ such that $f \circ \Phi(q, t) = t$ for $q \in Q$, $t \in J$. Consider a strictly increasing diffeomorphism $\rho : J \to J$, which is the identity map outside of a closed bounded subinterval of J. Via ρ we can construct the diffeomorphism

$$\Psi_\rho : M \to M; \quad \Psi_\rho(p) = \begin{cases} \Phi \circ (id_Q \times \rho) \circ \Phi^{-1}(p) & \text{if } p \in N \\ p & \text{if } p \notin N. \end{cases}$$

If $a \in J$ then Ψ_ρ maps $M(a)$ diffeomorphically onto $M(\rho(a))$. It suffices to choose ρ so that $\rho(a_1) = a_2$. One may choose

(7)
$$\rho(t) = t + \int_{c_1}^{t} g(x)dx,$$

where $g \in C_c^\infty(\mathbb{R}, \mathbb{R})$ satisfies the conditions:

$$\text{supp}(g) \subseteq J, \ g(x) > \ 1 \text{ for } x \in \mathbb{R}, \ \int_{c_1}^{a_1} g(x)dx = a_2 - a_1, \ \int_{c_1}^{c_2} g(x)dx = 0.$$

Now g can be constructed easily via Corollary A.4. See Figure 3 below. $\square$

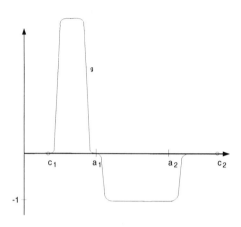

Figure 3

Remark C.3 In the proof above, we proved more than claimed in Lemma 12.12. Indeed, we found a diffeomorphism Ψ_ρ of M to itself for which

$$\Psi_\rho(M(a_1)) = M(a_2).$$

Let $\rho_s\colon J \to J$ be the map $\rho_s(t) = s\rho(t) + (1-s)t$, where $s \in [0,1]$. Then $\Psi_{\rho_s}(p)$ is smooth as a function of both s and p. Moreover, every Ψ_{ρ_s} is a diffeomorphism of M onto itself. This gives a so-called *isotopy* from $\Psi_{\rho_0} = \mathrm{id}_M$ to $\Psi_{\rho_1} = \Psi_\rho$.

It remains to prove Theorem C.2. We first prove a few lemmas.

Lemma C.4 *There exists a smooth tangent vector field X on N^n such that* $d_p f(X(p)) = 1$ *for all $p \in M$.*

Proof. We use Lemma 12.8 to find a gradient-like vector field Y on M. Now $\rho(p) = d_p f(Y(p)) > 0$ and we can choose $X(p) = \rho(p)^{-1} Y(p)$. $\square$

We shall investigate the integral curves $\alpha\colon I \to N$ for X. They are smooth on open intervals $I \subset \mathbb{R}$ and satisfy

(8) $$\alpha'(t) = X(\alpha(t)).$$

By Lemma C.4 and (8) we have $\frac{d}{dt} f \circ \alpha(t) = 1$, which gives $f \circ \alpha(t) = t + c$ for a constant c.

Lemma C.5 *Assume $p_0 \in N$ with $f(p_0) = t_0 \in J$. Then*

  (i) *$f^{-1}(t_0)$ is a compact $(n-1)$-dimensional smooth submanifold of N.*
  (ii) *There exists an open neighborhood $W_0 \subseteq f^{-1}(t_0)$ of p_0, a $\delta > 0$ with $(t_0 - \delta, t_0 + \delta) \subseteq J$ and a diffeomorphism*

$$\Phi_0 : W_0 \times (t_0 - \delta, t_0 + \delta) \to W,$$

    where W is an open neighborhood of p_0 in N such that the following conditions are satisfied:

  (a) *$\Phi_0(p, t_0) = p$ for all $p \in W_0$.*
  (b) *$f \circ \Phi_0$ is the projection onto $(t_0 - \delta, t_0 + \delta)$.*
  (c) *For fixed $p \in W_0$, the function $\Phi_0(p, t)$ is an integral curve of X. (We call W a product neighborhood of p_0.)*

Proof. Since t_0 is a regular value, $f^{-1}(t_0)$ is a smooth submanifold of dimension $n - 1$ (see Exercise 9.6). It is compact by the assumptions of Theorem C.2. Choose a C^∞-chart $h : U \to U'$ on N with $h(p_0) = 0$ and

$$h\big(U \cap f^{-1}(t_0)\big) = U' \cap (\{0\} \times \mathbb{R}^{n-1}).$$

Let us write $h_*(X_{|U}) = (F_1, \ldots, F_n)$ where $F_j \in C^\infty(U', \mathbb{R})$. Then $\alpha : I \to U$ is an integral curve of X precisely when $h \circ \alpha(t) = (x_1(t), \ldots, x_n(t))$ satisfies the system of differential equations

$$(9) \qquad x_j'(t) = F_j(x_1(t), \ldots, x_n(t)), \quad 1 \leq j \leq n.$$

For $y = (y_2, \ldots, y_n) \in \mathbb{R}^{n-1}$ with $(0, y) \in U'$ there exists a uniquely determined solution $x(t) : I(y) \to U'$ for (9), which is defined on a open interval $I(y)$ around t_0 with boundary condition $x(t_0) = (0, y)$.

The general theory of ordinary differential equations shows that the solution is smooth as a function of both t and y. Specifically there exists an open ball D in $\mathbb{R}^{n-1}$ with center at 0, a $\delta > 0$, and a smooth function $x : D \times (t_0 - \delta, t_0 + \delta) \to U'$ such that

  (α) If $y \in D$ then the map $t \to x(y, t)$ is a solution of (9).
  (β) If $y \in D$ then $x(y, t_0) = (0, y)$.

Then $W_0 = h^{-1}(D)$ is an open neighborhood around p_0 in $f^{-1}(t_0)$, and we can define a smooth map

$$\Phi_0 : W_0 \times (t_0 - \delta, t_0 + \delta) \to N; \quad \Phi_0(p, t) = h^{-1}(x(h(p), t)).$$

Now (α) implies (c) and (β) implies (a), while (b) follows from the remarks preceeding the lemma. By (a) the differential of Φ_0 at (p_0, t_0) is the identity on $T_{p_0} f^{-1}(t_0)$. By (c) $X(p_0)$ is the image of $\frac{\partial}{\partial t}$.

Hence $D_{(p_0, t_0)} \Phi_0$ is an isomorphism and the inverse function theorem implies that Φ_0 is a local diffeomorphism around (p_0, t_0). For suitable sizing of W_0 and δ, Φ_0 becomes a diffeomorphism. $\qquad\square$

Lemma C.6 *Let p_0 be a point of N such that $f(p_0) = t_0$. There exists a uniquely determined integral curve $\alpha: J \to N$ of X with $\alpha(t_0) = p_0$. Moreover, $f \circ \alpha(t) = t$ ($t \in J$).*

Proof. The existence and uniqueness theorem for the system of equations in (9) shows that for suitable open interval I around t_0 there exists a uniquely determined integral curve $\alpha: I \to N$ with $\alpha(t_0) = p_0$. The remark prior to Lemma C.5 shows that $f \circ \alpha(t) = t$ if $t \in I$. Hence $I \subseteq J$.

Assume that I has a right endpoint $t_1 \in J$. Since $f^{-1}([t_0, t_1])$ is compact, we can find a sequence s_m in $[t_0, t_1)$ that converges towards t_1, such that the sequence $(\alpha(s_m))$ in N converges towards a point $p_1 \in f^{-1}(t_1)$. We can apply Lemma C.5 to find a product neighborhood W_1 of p_1. There will be some s_m with $\alpha(s_m) \in W_1$, but now we are in a position to apply the uniqueness of the integral curve, which starts at $\alpha(s_m)$, to extend α slightly past t_1. The analogous extension can be performed if I has a left endpoint $t_2 \in J$. In total we see, again via local uniqueness, that α can be extended uniquely to all of J. $\qquad\square$

The image $\alpha(J)$ is a 1-dimensional smooth submanifold of N and $\alpha(J)$ intersects every fiber $f^{-1}(t)$ in exactly one point. For two points $t_1, t_2 \in J$ we can define

$$\varphi_{t_1, t_2}: f^{-1}(t_1) \to f^{-1}(t_2)$$

by mapping $p \in f^{-1}(t_1)$ to the point of intersection between $f^{-1}(t_2)$ and the integral curve through p. We have

Lemma C.7 *The maps φ_{t_1, t_2} are diffeomorphisms.*

Proof. It is sufficient to show the smoothness of φ_{t_1, t_2}, as $\varphi_{t_2, t_1} = \varphi_{t_1, t_2}^{-1}$. We may assume that $t_2 > t_1$. An arbitrary $p_1 \in f^{-1}(t_1)$ determines an integral curve $\alpha: J \to N$ with $\alpha(t_1) = p_1$.

We can find a subdivision $t_1 = s_0 < s_1 < s_2 < \ldots < s_{k-1} < s_k = t_2$ and product neighborhoods W_i, $1 \le i \le k$, of the type of Lemma C.5, such that

$$\alpha([s_{i-1}, s_i]) \subseteq W_i \quad (1 \le i \le k).$$

For each of these we have a diffeomorphism, analogous to Φ_0 in Lemma C.5. It follows that

$$\varphi_{s_{i-1}, s_i}: f^{-1}(s_{i-1}) \to f^{-1}(s_i)$$

is smooth in a neighborhood of $\alpha(s_{i-1})$. Note that $\varphi_{s_{i-1}, s_i}(\alpha(s_{i-1})) = \alpha(s_i)$. Since

$$\varphi_{t_1, t_2} = \varphi_{s_{k-1}, s_k} \circ \varphi_{s_{k-2}, s_{k-1}} \circ \ldots \circ \varphi_{s_1, s_2} \circ \varphi_{s_0, s_1},$$

we can conclude that φ_{t_1, t_2} is smooth at p_1. $\qquad\square$

Lemma C.8 *Let* $t_0 \in J$. *Define* $\pi\colon N \to f^{-1}(t_0)$ *by* $\pi(p) = \varphi_{t,t_0}(p)$ *where* $p \in f^{-1}(t)$. *Then* π *is smooth.*

Proof. Let W_0, W and Φ_0 be given as in Lemma C.5. Then $\pi_{|W} = pr_{W_0} \circ \Phi_0^{-1}$, and π is smooth on W. A $p_1 \in N$ with $f(p_1) = t_1$ enables us to define $\pi_1\colon N \to f^{-1}(t_1)$ by $\pi_1(p) = \varphi_{t,t_1}(p)$. Then π_1 is smooth on a product neighborhood W_1, of p_1 and since $\pi = \varphi_{t_1,t_0} \circ \pi_1$, π will be smooth on W_1; cf. Lemma C.7.   □

Proof of Theorem C.2. Let $Q = f^{-1}(t_0)$. By Lemma C.8

$$\Psi\colon N \to Q \times J; \quad \Psi(p) = (\pi(p), f(p))$$

is smooth. Let $p \in N$. Consider the differential

$$D_p\Psi\colon T_pN \to T_{\pi(p)}Q \times \mathbb{R}.$$

It follows from Lemmas C.7 and C.8 that the subspace $T_p f^{-1}(f(p)) \subseteq T_pN$ is mapped isomorphically onto $T_{\pi(p)}Q \times \{0\}$ and that $D_p\Psi(X(p)) = (0,1)$. Hence $D_p\Psi$ is an isomorphism. Since Ψ is bijective by construction, we can conclude that Ψ is a diffeomorphism. The assertion follows by letting Φ be the inverse diffeomorphism.   □

D. EXERCISES

1.1. Perform the calculations of Theorem 1.7.

1.2. Let $W \subseteq \mathbb{R}^3$ be the open set

$$W = \{(x_1, x_2, x_3) \in \mathbb{R}^3 \mid \text{ either } x_3 \neq 0 \text{ or } x_1^2 + x_2^2 < 1\}.$$

Prove the existence and uniqueness of a function $F \in C^\infty(W, \mathbb{R})$ such that $\operatorname{grad}(F)$ is the vector field considered in Example 1.8 and $F(0) = 0$.
Find a simple expression for F valid when $x_1^2 + x_2^2 < 1$.
(Hint: First note that F is constant on the open disc in the x_1, x_2-plane bounded by the unit circle S. Then integrate along lines parallel to the x_3–axis.)

2.1. Prove the formula in Remark 2.10.

2.2. Find an $\omega \in \operatorname{Alt}^2 \mathbb{R}^4$ such that $\omega \wedge \omega \neq 0$.

2.3. Show that there exist isomorphisms

$$\mathbb{R}^3 \xrightarrow{i} \operatorname{Alt}^1 \mathbb{R}^3, \qquad \mathbb{R}^3 \xrightarrow{j} \operatorname{Alt}^2 \mathbb{R}^3$$

given by

$$i(v)(w) = \langle v, w \rangle, \qquad j(v)(w_1, w_2) = \det(v, w_1, w_2),$$

where $\langle \, , \, \rangle$ is the usual inner product. Show that for $v_1, v_2 \in \mathbb{R}^3$, we have

$$i(v_1) \wedge i(v_2) = j(v_1 \times v_2).$$

2.4. Let V be a finite-dimensional vector space over $\mathbb{R}$ with inner product $\langle \, , \, \rangle$, and let

$$i: V \to V^* = \operatorname{Alt}^1(V)$$

be the $\mathbb{R}$-linear map given by

$$i(v)(\omega) = \langle \omega, v \rangle.$$

Show that if $\{b_1, \dots, b_n\}$ is an orthonormal basis of V, then

$$i(b_k) = b_k^*,$$

where $\{b_1^*, \dots, b_k^*\}$ is the dual basis. Conclude that i is an isomorphism.

2.5. Assumptions as in Exercise 2.4. Show the existence of an inner product on $\operatorname{Alt}^p(V)$ such that

$$\langle \omega_1 \wedge \dots \wedge \omega_p, \tau_1 \wedge \dots \wedge \tau_p \rangle = \det(\langle \omega_i, \tau_j \rangle),$$

whenever $\omega_i, \tau_j \in \text{Alt}^1(V)$, and

$$\langle \omega, \tau \rangle = \langle i^{-1}(\omega), i^{-1}(\tau) \rangle.$$

Let $\{b_1, \ldots, b_n\}$ be an orthonormal basis of V, and let $\beta_j = i(b_j)$. Show that

$$\{\beta_{\sigma(1)} \wedge \ldots \wedge \beta_{\sigma(p)} \,|\, \sigma \in S(p, n-p)\}$$

is an orthonormal basis of $\text{Alt}^p(V)$.

2.6. Suppose $\omega \in \text{Alt}^p(V)$. Let $v_1, \ldots, v_p$ be vectors in V and let $A = (a_{ij})$ be a $p \times p$ matrix. Show that for $w_i = \sum_{j=1}^{p} a_{ij} v_j$ $(1 \le i \le p)$ we have

$$\omega(w_1, \ldots, w_p) = \det A \, \omega(v_1, \ldots, v_p).$$

(Try $p = 2$ first.)

2.7. Show for $f: V \to W$ that

$$\text{Alt}^{p+q}(f)(\omega_1 \wedge \omega_2) = \text{Alt}^p(f)(\omega_1) \wedge \text{Alt}^q(f)(\omega_2),$$

where $\omega_1 \in \text{Alt}^p(W)$, $\omega_2 \in \text{Alt}^q(W)$.

2.8. Show that the set

$$\{f \in \text{End}(V) \mid \exists g \in GL(V) : gfg^{-1} \text{ a diagonal matrix}\}$$

is everywhere dense in $\text{End}(V)$, assuming that V is a finite-dimensional complex vector space.

2.9. Let V be an n-dimensional vector space with inner product $\langle , \rangle$. From Exercise 2.5 we obtain an inner product on $\text{Alt}^p(V)$ for all p, in particular on $\text{Alt}^n(V)$.

A volume element of V is a unit vector $\text{vol} \in \text{Alt}^n(V)$. Hodge's star operator

$$*: \text{Alt}^p(V) \to \text{Alt}^{n-p}(V)$$

is defined by the equation $\langle *\omega, \tau \rangle \text{vol} = \omega \wedge \tau$. Show that $*$ is well-defined and linear.

Let $\{e_1, \ldots, e_n\}$ be an orthonormal basis of V with $\text{vol}(e_1, \ldots, e_n) = 1$ and $\{\epsilon_1, \ldots, \epsilon_n\}$ the dual orthonormal basis of $\text{Alt}^1(V)$. Show that

$$*(\epsilon_1 \wedge \ldots \wedge \epsilon_p) = \epsilon_{p+1} \wedge \ldots \wedge \epsilon_n$$

and in general that

$$*(\epsilon_{\sigma(1)} \wedge \ldots \wedge \epsilon_{\sigma(p)}) = \text{sign}(\sigma)\, \epsilon_{\sigma(p+1)} \wedge \ldots \wedge \epsilon_{\sigma(n)}$$

with $\sigma \in S(p, n-p)$. Show that $* \circ * = (-1)^{p(n-p)}$ on $\text{Alt}^p(V)$.

2.10. Let V be a 4-dimensional vector space and $\{\epsilon_1, \ldots, \epsilon_4\}$ a basis of $\mathrm{Alt}^1(V)$. Let $A = (a_{ij})$ be a skew-symmetric matrix and define

$$\alpha = \sum_{i<j} a_{ij}\,\epsilon_i \wedge \epsilon_j.$$

Show that

$$\alpha \wedge \alpha = 0 \iff \det(A) = 0.$$

Say $\alpha \wedge \alpha = \lambda.\epsilon_1 \wedge \epsilon_2 \wedge \epsilon_3 \wedge \epsilon_4$. What is the relation between λ and $\det(A)$?

2.11. Let V be an n-dimensional vector space with inner product $\langle\,,\,\rangle$ and volume element $\mathrm{vol} \in \mathrm{Alt}^n(V)$, as in Exercise 2.9. Let $v \in \mathrm{Alt}^1(V)$ and

$$F_v\colon \mathrm{Alt}^p(V) \to \mathrm{Alt}^{p+1}(V)$$

be the map

$$F_v(\omega) = v \wedge \omega.$$

Show that the map

$$F_v^* = (-1)^{np} * \circ\, F_v \circ *\colon \mathrm{Alt}^{p+1}(V) \to \mathrm{Alt}^p(V)$$

is adjoint to F_v, i.e. $\langle F_v\omega, \tau\rangle = \langle \omega, F_v^*\tau\rangle$. Let $\{e_1, \ldots, e_n\}$ be an orthonormal basis of V with $\mathrm{vol}(\epsilon_1, \ldots, \epsilon_n) = 1$ and $\{\epsilon_1, \ldots, \epsilon_n\}$ the dual (orthonormal) basis of $\mathrm{Alt}^1(V)$; see Exercise 2.5. Show that

$$F_v^*(\epsilon_1 \wedge \ldots \wedge \epsilon_{p+1}) = \sum_{i=1}^{p+1} (-1)^{i+1}\langle v, \epsilon_i\rangle\, \epsilon_1 \wedge \ldots \wedge \hat\epsilon_i \wedge \ldots \wedge \epsilon_{p+1}.$$

Show that $F_v F_v^* + F_v^* F_v\colon \mathrm{Alt}^p(V) \to \mathrm{Alt}^p(V)$ is multiplication by $\|v\|^2$. (Hint: Suppose that $v = \lambda.\epsilon_1$ and show that the general case follows from the special case.)

2.12. Let V be an n-dimensional vector space. Show for a linear map $f\colon V \to V$ the existence of a number $d(f)$ such that

$$\mathrm{Alt}^n(f)(\omega) = d(f)\omega$$

for $\omega \in \mathrm{Alt}^n(V)$. Verify the product rule

$$d(g \circ f) = d(g)d(f)$$

for linear maps $f, g\colon V \to V$ using the functoriality of Alt^n.
Prove that $d(f) = \det(f)$.
(Hint: Pick a basis $e_1, \ldots, e_n$ for V, let $\epsilon_1, \ldots, \epsilon_n$ be the dual basis for $\mathrm{Alt}^n(V)$ and evaluate $\mathrm{Alt}^n(f)(\epsilon_1 \wedge \ldots \wedge \epsilon_n)$ on $(e_1, \ldots, e_n)$ in terms of the matrix for f with respect to the chosen basis.)

3.1. Show for an open set in $\mathbb{R}^2$ that the de Rham complex

$$0 \to \Omega^0(U) \to \Omega^1(U) \to \Omega^2(U) \to 0$$

is isomorphic to the the the complex

$$0 \to C^\infty(U, \mathbb{R}) \overset{\mathrm{grad}}{\to} C^\infty(U, \mathbb{R}^2) \overset{\mathrm{rot}}{\to} C^\infty(U, \mathbb{R}) \to 0.$$

Analogously, show that for an open set in $\mathbb{R}^3$ the de Rham complex is isomorphic to

$$0 \to C^\infty(U, \mathbb{R}) \overset{\mathrm{grad}}{\to} C^\infty(U, \mathbb{R}^3) \overset{\mathrm{rot}}{\to} C^\infty(U, \mathbb{R}^3) \overset{\mathrm{div}}{\to} C^\infty(U, \mathbb{R}) \to 0$$

defined in Chapter 1.

3.2. Let $U \subseteq \mathbb{R}^n$ be an open set and $dx_1, \ldots, dx_n$ the usual constant 1-forms $(dx_i = \epsilon_i)$. Let $\mathrm{vol} = dx_1 \wedge \ldots \wedge dx_n \in \Omega^n(U)$.

Use $*\colon \mathrm{Alt}^p(\mathbb{R}^n) \to \mathrm{Alt}^{n-p}(\mathbb{R}^n)$ (from Exercise 2.9) to define a linear operator (Hodge's star operator)

$$*\colon \Omega^p(U) \to \Omega^{n-p}(U)$$

and show that $*(dx_1 \wedge \ldots \wedge dx_p) = dx_{p+1} \wedge \ldots \wedge dx_n$ and $* \circ * = (-1)^{n(n-p)}$. Define $d^*\colon \Omega^p(U) \to \Omega^{p-1}(U)$ by

$$d^*(\omega) = (-1)^{np+n-1} * \circ d \circ *(\omega).$$

Show that $d^* \circ d^* = 0$.

Verify the formula

$$d^*(f \, dx_1 \wedge \ldots \wedge dx_p) = \sum_{j=1}^{p} (-1)^j \frac{\partial f}{\partial x_j} dx_1 \wedge \ldots \wedge \widehat{dx_j} \wedge \ldots \wedge dx_p$$

and more generally for $1 \le i_1 < i_2 < \ldots < i_p \le n$ that

$$d^*\left(f \, dx_{i_1} \wedge \ldots \wedge dx_{i_p}\right) = \sum_{\nu=1}^{p} (-1)^\nu \frac{\partial f}{\partial x_{i_\nu}} dx_{i_1} \wedge \ldots \wedge \widehat{dx_{i_\nu}} \wedge \ldots \wedge dx_{i_p}.$$

3.3. With the notation of Exercise 3.2, the Laplace operator $\Delta\colon \Omega^p(U) \to \Omega^p(U)$ is defined by

$$\Delta = d \circ d^* + d^* \circ d.$$

Let $f \in \Omega^0(U)$. Show that $\Delta(f \, dx_1 \wedge \ldots \wedge dx_p) = \Delta(f) dx_1 \wedge \ldots \wedge dx_p$ where

$$-\Delta(f) = \frac{\partial^2 f}{\partial x_1^2} + \cdots + \frac{\partial^2 f}{\partial x_n^2}.$$

(Hint: Try the case $p = 1, n = 2$ first. What can one say about $\Delta(f \cdot dx_I)$ where $I = (i_1, \ldots, i_p)$?)

A p-form $\omega \in \Omega^p(U)$ is said to be harmonic if $\Delta(\omega) = 0$. Show that

$$*: \Omega^p(U) \to \Omega^{n-p}(U)$$

maps harmonic forms into harmonic forms.

3.4. Let $\mathrm{Alt}^p(\mathbb{R}^m, \mathbb{C})$ be the $\mathbb{C}$-vector space of alternating $\mathbb{R}$-multilinear maps

$$\omega: \mathbb{R}^n \times \cdots \times \mathbb{R}^n \to \mathbb{C}$$

(p factors). Note that ω can be written uniquely

$$\omega = \mathrm{Re}\,\omega + i\,\mathrm{Im}\,\omega,$$

where $\mathrm{Re}\,\omega \in \mathrm{Alt}^p(\mathbb{R}^m)$, $\mathrm{Im}\,\omega \in \mathrm{Alt}^p(\mathbb{R}^n)$.

Extend the wedge product to a $\mathbb{C}$-bilinear map

$$\mathrm{Alt}^p(\mathbb{R}^n, \mathbb{C}) \times \mathrm{Alt}^q(\mathbb{R}^n, \mathbb{C}) \xrightarrow{\wedge} \mathrm{Alt}^{p+q}(\mathbb{R}^n, \mathbb{C})$$

and show that we obtain a graded anti-commutative $\mathbb{C}$-algebra $\mathrm{Alt}^*(\mathbb{R}^n, \mathbb{C})$.

3.5. Introduce $\mathbb{C}$-valued differential p-forms on an open set $U \subseteq \mathbb{R}^n$ by setting (see Exercise 3.4)

$$\Omega^p(U, \mathbb{C}) = C^\infty(U, \mathrm{Alt}^p(\mathbb{R}^n, \mathbb{C})).$$

Note that $\omega \in \Omega^p(U, \mathbb{C})$ can be written uniquely

$$\omega = \mathrm{Re}\,\omega + i\,\mathrm{Im}\,\omega,$$

where $\mathrm{Re}\,\omega \in \Omega^p(U)$. Extend d to a $\mathbb{C}$-linear operator

$$d: \Omega^p(U, \mathbb{C}) \to \Omega^{p+1}(U, \mathbb{C})$$

and show that Theorem 3.7 holds for $\mathbb{C}$-valued differential forms.

Generalize Theorem 3.12 to the case of $\mathbb{C}$-valued differential forms

3.6. Take $U = \mathbb{C} - \{0\} = \mathbb{R}^2 - \{0\}$ in Exercise 3.5 and let $z \in \Omega^0(U, \mathbb{C})$ be the inclusion map $U \to \mathbb{C}$. Write $x = \mathrm{Re}\,z$, $y = \mathrm{Im}\,z$.

Show that

$$\mathrm{Re}\left(z^{-1}dz\right) = d\log r,$$

where $r: U \to \mathbb{R}$ is defined by $r(z) = |z| = \sqrt{x^2 + y^2}$. Show that

$$\mathrm{Im}\left(z^{-1}dz\right) = \frac{-y}{x^2 + y^2}dx + \frac{x}{x^2 + y^2}dy.$$

(Observe that this is the 1-form corresponding to the vector field of Example 1.2.)

3.7. Prove for the complex exponential map $\exp: \mathbb{C} \to \mathbb{C}^*$ that

$$d_z\exp = \exp(z)dz \quad \text{and} \quad \exp^*(z^{-1}dz) = dz.$$

4.1. Consider a commutative diagram of vector spaces and linear maps with exact rows

$$
\begin{array}{ccccccccc}
A_1 & \longrightarrow & A_2 & \longrightarrow & A_3 & \longrightarrow & A_4 & \longrightarrow & A_5 \\
\downarrow{\scriptstyle f_1} & & \downarrow{\scriptstyle f_2} & & \downarrow{\scriptstyle f_3} & & \downarrow{\scriptstyle f_4} & & \downarrow{\scriptstyle f_5} \\
B_1 & \longrightarrow & B_2 & \longrightarrow & B_3 & \longrightarrow & B_4 & \longrightarrow & B_5
\end{array}
$$

Suppose that f_4 is injective, f_1 is surjective and f_2 is injective. Show that f_3 is injective. Suppose that f_2 is surjective, f_4 is surjective and f_5 is injective. Show that f_3 is surjective. In particular we have that if f_1, f_2, f_4 and f_5 are isomorphisms, then f_3 is an isomorphism. (This assertion is called the 5-lemma.)

4.2. Consider the following commutative diagram

$$
\begin{array}{ccccccccc}
0 & \longrightarrow & A_1 & \longrightarrow & A_2 & \longrightarrow & A_3 & \longrightarrow & 0 \\
& & \downarrow{\scriptstyle f_1} & & \downarrow{\scriptstyle f_2} & & \downarrow{\scriptstyle f_3} & & \\
0 & \longrightarrow & B_1 & \longrightarrow & B_2 & \longrightarrow & B_3 & \longrightarrow & 0
\end{array}
$$

where the rows are exact sequences. Show that there exists a exact sequence

$$0 \to \operatorname{Ker} f_1 \to \operatorname{Ker} f_2 \to \operatorname{Ker} f_3 \to$$
$$\to \operatorname{Cok} f_1 \to \operatorname{Cok} f_2 \to \operatorname{Cok} f_3 \to 0.$$

(Hint: Try the long exact cohomology sequence).

4.3. In the commutative diagram

$$
\begin{array}{ccccccccc}
& 0 & & 0 & & 0 & & 0 & \\
& \downarrow & & \downarrow & & \downarrow & & \downarrow & \\
0 \longrightarrow & A^{0,0} & \longrightarrow & A^{1,0} & \longrightarrow & A^{2,0} & \longrightarrow & A^{3,0} & \longrightarrow \cdots \\
& \downarrow & & \downarrow & & \downarrow & & \downarrow & \\
0 \longrightarrow & A^{0,1} & \longrightarrow & A^{1,1} & \longrightarrow & A^{2,1} & \longrightarrow & A^{3,1} & \longrightarrow \cdots \\
& \downarrow & & \downarrow & & \downarrow & & \downarrow & \\
0 \longrightarrow & A^{0,2} & \longrightarrow & A^{1,2} & \longrightarrow & A^{2,2} & \longrightarrow & A^{3,2} & \longrightarrow \cdots \\
& \downarrow & & \downarrow & & \downarrow & & \downarrow & \\
0 \longrightarrow & A^{0,3} & \longrightarrow & A^{1,3} & \longrightarrow & A^{2,3} & \longrightarrow & A^{3,3} & \longrightarrow \cdots \\
& \downarrow & & \downarrow & & \downarrow & & \downarrow & \\
& \vdots & & \vdots & & \vdots & & \vdots &
\end{array}
$$

the horizontal $(A^{*,q})$ and the vertical $(A^{p,*})$ are chain complexes where $A^{p,q} = 0$ if either $p < 0$ or $q < 0$. Suppose that

$$H^p(A^{*,q}) = 0 \text{ for } q \neq 0 \text{ and all } p$$
$$H^q(A^{p,*}) = 0 \text{ for } p \neq 0 \text{ and all } q.$$

Construct isomorphisms $H^p(A^{*,0}) \to H^p(A^{0,*})$ for all p.

4.4. Let $0 \to A^0 \xrightarrow{d^0} A^1 \xrightarrow{d^1} \cdots \xrightarrow{d^{n-1}} A^n \to 0$ be a chain complex and assume that $\dim_{\mathbb{R}} A^i < \infty$. The *Euler characteristic* is defined by

$$\chi(A^*) = \sum_{i=0}^{n} (-1)^i \dim A^i.$$

Show that $\chi(A^*) = 0$ if A^* is exact. Show that the sequence

$$0 \to H^i(A^*) \to A^i/\text{Im } d^{i-1} \xrightarrow{d^i} \text{Im } d^i \to 0$$

is exact and conclude that

$$\dim_{\mathbb{R}} A^i - \dim_{\mathbb{R}} \text{Im } d^{i-1} = \dim_{\mathbb{R}} H^i(A^*) + \dim_{\mathbb{R}} \text{Im } d^i.$$

Show that $\chi(A^*) = \sum_{i=0}^{n} (-1)^i \dim_{\mathbb{R}} H^i(A^*)$.

4.5. Associate to two composable linear maps

$$f: V_1 \to V_2, \qquad g: V_2 \to V_3$$

an exact sequence

$$0 \to \text{Ker}(f) \to \text{Ker}(g \circ f) \to \text{Ker}(g) \to$$
$$\to \text{Cok}(f) \to \text{Cok}(g \circ f) \to \text{Cok}(g) \to 0.$$

5.1. Adopt the notation of Example 5.4. A point $(x, y) \in U_1$ can be uniquely described in terms of polar coordinates $(r, \theta) \in (0, \infty) \times (0, 2\pi)$. Let $\arg_1 \in \Omega^0(U_1)$ be the function mapping (x, y) into $\theta \in (0, 2\pi)$ (why is $\arg_1$ smooth?).

Define similarly $\arg_2 \in \Omega^0(U_2)$ using polar coordinates with $\theta \in (-\pi, \pi)$ and prove the existence of a closed 1-form $\tau \in \Omega^1(\mathbb{R}^2 - \{0\})$ such that

$$\tau|_{U_\nu} = i_\nu^*(\tau) = d \arg_\nu \qquad (\nu = 1, 2).$$

Show that the connecting homomorphism

$$\partial^0: H^0(U_1 \cap U_2) \to H^1(\mathbb{R}^2 - \{0\})$$

carries the locally constant function with values $\{0, 2\pi\}$ on the upper and lower half-planes respectively into $[\tau]$.

5.2. Show that the 1-forms $\tau \in \Omega^1(\mathbb{R}^2 - \{0\})$ of Exercise 5.1 and $\text{Im}(z^{-1} dz)$ of Exercise 3.6 are the same.

5.3. Can $\mathbb{R}^2$ be written as $\mathbb{R}^2 = U \cup V$ where U, V are open connected sets such that $U \cap V$ is disconnected?

5.4. (Phragmen–Brouwer property of $\mathbb{R}^n$) Suppose $p \neq q$ in $\mathbb{R}^n$. A closed set $A \subseteq \mathbb{R}^n$ is said to separate p from q, when p and q belong to two different connected components of $\mathbb{R}^n - A$.

Let A and B be two disjoint closed subsets of $\mathbb{R}^n$. Given two distinct points p and q in $\mathbb{R}^n - (A \cup B)$. Show that if neither A nor B separates p from q, then $A \cup B$ does not separate p from q. (Apply Theorem 5.2 to $U_1 = \mathbb{R}^n - A$, $U_2 = \mathbb{R}^n - B$.)

6.1. Show that "homotopy equivalence" is an equivalence relation in the class of topological spaces.

6.2. Show that all continuous maps $f: U \to V$ that are homotopic to a constant map induce the 0-map $f^*: H^p(V) \to H^p(U)$ for $p > 0$.

6.3. Let $p_1, \ldots, p_k$ be k different points in $\mathbb{R}^n$, $n \geq 2$. Show that

$$H^d(\mathbb{R}^n - \{p_1, \ldots, p_k\}) \cong \begin{cases} \mathbb{R}^k & \text{for } d = n - 1 \\ \mathbb{R} & \text{for } d = 0 \\ 0 & \text{otherwise.} \end{cases}$$

6.4. Suppose $f, g: X \to S^n$ are two continuous maps, such that $f(x)$ and $g(x)$ are never antipodal. Show that $f \simeq g$.
Show that every non-surjective continuous map $f: X \to S^n$ is homotopic to a constant map.

6.5. Show that S^{n-1} is homotopy equivalent to $\mathbb{R}^n - \{0\}$. Show that two continuous maps

$$f_0, f_1: \mathbb{R}^n - \{0\} \to \mathbb{R}^n - \{0\}$$

are homotopic if and only if their restrictions to S^{n-1} are homotopic.

6.6. Show that S^{n-1} is not contractible.

7.1. Show that $\mathbb{R}^n$ does not contain a subset homeomorphic to D^m when $m > n$.

7.2. Let $\Sigma \subseteq \mathbb{R}^n$ be homeomorpic to S^k ($1 \leq k \leq n - 2$). Show that

$$H^p(\mathbb{R}^n - \Sigma) \cong \begin{cases} \mathbb{R} & \text{for } p = 0, n - k - 1, n - 1 \\ 0 & \text{otherwise.} \end{cases}$$

7.3. Show that there is no continuous map $g: D^n \to S^{n-1}$ with $g|_{S^{n-1}} \simeq \text{id}_{S^{n-1}}$.

7.4. Let $f: D^n \to \mathbb{R}^n$ be a continuous map, and let $r \in (0, 1)$ be given. Suppose for all $x \in S^{n-1}$ that $\|f(x) - x\| \leq 1 - r$. Show that $\text{Im } f(D^n)$ contains the closed disc with radius r and center 0.
(Hint: Modify the proof of Brouwer's fixed point theorem and use Exercises 6.4 and 7.3.)

7.5. Assume given two injective continuous maps $\alpha, \beta: [0, 1] \to D^2$ such that

$$\alpha(0) = (-1, 0), \quad \alpha(1) = (1, 0)$$
$$\beta(0) = (0, -1), \quad \beta(1) = (0, 1).$$

Prove that the curves α and β intersect (apply both parts of Theorem 7.10).

8.1. Fill in the details of Remark 8.2.

8.2. Let $\varphi: N \to M$ be a continuous map from a smooth manifold N to a smooth submanifold M of $\mathbb{R}^k$. Let $i: M \to \mathbb{R}^k$ be the inclusion. Show that φ is smooth if and only if $i \circ \varphi$ is smooth.

8.3. Suppose that $M \subseteq \mathbb{R}^k$ (with the induced topology from $\mathbb{R}^k$) is an n-dimensional topological manifold. Include M in $\mathbb{R}^{k+n}$. Show that M is locally flat in $\mathbb{R}^{k+n}$.

8.4. Set $\mathbf{T}^n = \mathbb{R}^n / \mathbb{Z}^n$, i.e. the set of cosets for the subgroup $\mathbb{Z}^n$ of $\mathbb{R}^n$ with respect to vector addition. Let $\pi \colon \mathbb{R}^n \to \mathbf{T}^n$ be the canonical map and equip $\mathbf{T}^n$ with the quotient topology (i.e. $W \subseteq \mathbf{T}^n$ is open if and only if $\pi^{-1}(W)$ is open in $\mathbb{R}^n$).

Show that $\mathbf{T}^n$ is a compact topological manifold of dimension n (the n-dimensional torus). Construct a differentiable structure on $\mathbf{T}^n$, such that π becomes smooth and every $p \in \mathbb{R}^n$ has an open neighborhood that is mapped diffeomorphically onto an open set in $\mathbf{T}^n$ by π. Prove that $\mathbf{T}^1$ is diffeomorphic to S^1.

8.5. Define $\hat{A} \colon \mathbb{R}^2 \to \mathbb{R}^2$ by $\hat{A}(x,y) = \left(x + \frac{1}{2}, -y\right)$. Show that there exists a smooth map $A \colon \mathbf{T}^2 \to \mathbf{T}^2$ satisfying $A \circ \pi = \pi \circ \hat{A}$. (Consult Exercise 8.4.) Show that A is a diffeomorphism, that $A = A^{-1}$ and that $A(q) \neq q$ for all $q \in \mathbf{T}^2$.

Let $\mathbf{K}^2$ be the set of pairs $\{q, A(q)\}$, $q \in \mathbf{T}^2$. Show that $\mathbf{K}^2$ with the quotient topology from $\mathbf{T}^2$ is a 2-dimensional topological manifold (Klein's bottle).

Construct a differentiable structure on $\mathbf{K}^2$.

8.6. Let $p_0 \in S^n$ be the "north pole" $p_0 = (0, \ldots, 0, 1)$. Show that $S^n - \{p_0\}$ is diffeomorphic to $\mathbb{R}^n$ under stereographic projection, i.e. the map $S^n - \{p_0\} \to \mathbb{R}^n$ that carries $p \in S^n$ into the point of intersection between the line through p_0 and p and the equatorial hyperplane $\mathbb{R}^n \subseteq \mathbb{R}^{n+1}$.

9.1. Let $M \subseteq \mathbb{R}^l$ be a differentiable submanifold and assume the points $p \in \mathbb{R}^l$ and $p_0 \in M$ are such that $\|p - p_0\| \leq \|p - q\|$ for all $q \in M$. Show that $p - p_0 \in T_{p_0} M^\perp$.

9.2. A smooth map $\varphi \colon M^m \to N^n$ between smooth manifolds is called *immersive* at $p \in M$, when

$$D_p \varphi \colon T_p M \to T_q N, \quad q = \varphi(p)$$

is *injective*. Show that there exist smooth charts (U, h) in M with $p \in U$, $h(p) = 0$, and (V, k) in N with $q \in V$, $k(q) = 0$, such that

$$k \circ \varphi \circ h^{-1}(x_1, \ldots, x_m) = (x_1, \ldots, x_m, 0, \ldots, 0)$$

in a neighborhood of 0.

(Hint: Reduce the problem to the case where $\varphi \colon W \to \mathbb{R}^n$ is smooth on an open neighborhood W in $\mathbb{R}^m$ of 0 with $\varphi(0) = 0$, and

$$\left(\frac{\partial \varphi_i(0)}{\partial x_j} \right)_{1 \leq i, j \leq m}$$

is an invertible $m \times m$ matrix. Apply the inverse function theorem to

$$F: W \times \mathbb{R}^{n-m} \to \mathbb{R}^n;$$
$$F(x_1, \ldots, x_n) = (\varphi_1(x_1, \ldots, x_m), \ldots, \varphi_m(x_1, \ldots, x_m), x_{m+1}, \ldots, x_n).)$$

9.3. The smooth map $\varphi: M^m \to N^n$ from Exercise 9.2 is an *immersion*, when it is immersive at every $p \in M$. We say that φ is *closed* if $\varphi(A)$ is closed in N^n for every closed set $A \subseteq M^m$. Show that an injective closed immersion is a smooth embedding.

9.4. Let ω be an irrational real number. Using the notation of Exercise 8.4, define the map $\alpha: \mathbb{R} \to \mathbf{T}^2$ by $\alpha(t) = \pi(t, \omega t)$. Show that α is an injective immersion and that the image $\alpha(\mathbb{R})$ is everywhere dense in $\mathbf{T}^2$. Conclude that α is not a smooth embedding.
(Hint: The additive group $\mathbb{Z} + \mathbb{Z}\omega \subseteq \mathbb{R}$ is dense in $\mathbb{R}$.)

9.5. A smooth map $\varphi: M^m \to N^n$ between smooth manifolds is called *submersive* at $p \in M$, when

$$D_p\varphi: T_pM \to T_qN, \quad q = \varphi(p)$$

is surjective. Show that there exist smooth charts (U, h) in M with $p \in U$, $h(p) = 0$, and (V, k) in N with $q \in V$, $k(q) = 0$, such that

$$k \circ \varphi \circ h^{-1}(x_1, \ldots, x_m) = (x_1, \ldots, x_n).$$

(Hint: Imitate Exercise 9.2.)

9.6. Suppose $\varphi: M^m \to N^n$ is a smooth map between smooth manifolds. Let $q \in N^n$ and assume that φ is submersive at every point of the fiber $\varphi^{-1}(q)$ (see Exercise 9.5). Show that $\varphi^{-1}(q)$ is an $(m-n)$-dimensional differentiable submanifold of M. Note that the result holds for all non-empty fibers $\varphi^{-1}(q)$, when φ is a *submersion*, i.e. φ is submersive at every $p \in M$.

9.7. Construct a smooth embedding of the n-dimensional torus $\mathbf{T}^n$ in $\mathbb{R}^{n+1}$.

9.8. Let a_1, a_2, a_3 be three distinct real numbers and define $f: \mathbb{R}^3 \to \mathbb{R}^4$ by

$$f(x_1, x_2, x_3) = \left(x_2 x_3, x_1 x_3, x_1 x_2, a_1 x_1^2 + a_2 x_2^2 + a_3 x_3^2\right).$$

The restriction $f_{|S^2}$ takes the same values at antipodal points and therefore f induces a map $\tilde{f}: \mathbb{RP}^2 \to \mathbb{R}^4$. Show that $\tilde{f}$ is a smooth embedding.

9.9. In the vector space $M = M_n(\mathbb{R})$ of real-valued $n \times n$ matrices we have the subspace of symmetric matrices S_n. Define a smooth map $\varphi: M \to S_n$ by

$$\varphi(A) = A^t A,$$

where A^t is the transpose of A. Note that the pre-image $\varphi^{-1}(I)$ of the identity matrix is exactly the set of orthogonal matrices O_n. Show that for $A \in M$ and $B \in M$ we have

$$D_A\varphi(B) = B^t A + A^t B.$$

(Hint: Use the curve $A + sB$, $s \in \mathbb{R}$, through A.)

Apply Exercise 9.6 to show that O_n is a differentiable submanifold of $M_n(\mathbb{R})$.

9.10. A *Lie group* G is a smooth manifold, which is also a group, such that both

$$\mu: G \times G \to G; \quad \mu(g_1, g_2) = g_1 g_2$$

and

$$i: G \to G; \quad i(g) = g^{-1}$$

are smooth. Show that the group O_n of orthogonal $n \times n$ matrices is a a Lie group. (Apply Exercise 9.9.)

9.11. Let $\varphi: M \to N$ be a smooth map between smooth manifolds. Show that

$$\varphi^*: \Omega^*(N) \to \Omega^*(M)$$

is a chain map.

9.12. The usual inner product on $\mathbb{R}^n$ induces an inner product on $\mathrm{Alt}^n(\mathbb{R}^n)$ (see Exercise 2.5). Show that $\omega \in \mathrm{Alt}^n(\mathbb{R}^n)$ is a unit vector if and only if $\omega(v_1, \ldots, v_n) = \pm 1$ for every orthonormal basis $\{v_1, \ldots, v_n\}$ of $\mathbb{R}^n$.

9.13. Show that Klein's bottle (Exercise 8.5) is non-orientable.

9.14. Let M^n be a Riemannian manifold and $f \in C^\infty(M, \mathbb{R})$. Define the gradient vector field $\mathrm{grad} f$ on M by demanding that $\mathrm{grad}_p f \in T_p M$ satisfies

$$\langle \mathrm{grad}_p f, v \rangle_p = d_p f(v)$$

for all $v \in T_p M$. Show that for a local parametrization, $h: W \to M$, we have

$$\mathrm{grad}_{h(x)} f = \sum_{j=1}^{n} a_j(x) \frac{\partial}{\partial x_j},$$

where $a_j \in C^\infty(W, \mathbb{R})$, $1 \le j \le n$ is determined by the set of linear equations

$$\sum_{i=1}^{n} g_{ij}(x) a_j(x) = \frac{\partial f}{\partial x_i}(x) \quad (1 \le i \le n).$$

Show that the map $\mathrm{grad} f: M \to TM$ is smooth.

Let $p \in M$ with $\mathrm{grad}_p f \ne 0$. Set $c = f(p)$. Show that $f^{-1}(c)$ in a neighborhood of p is an $(n-1)$-dimensional smooth submanifold, and that $\mathrm{grad}_p f$ is a normal vector to $f^{-1}(c)$ at p.

9.15.   Let M be a smooth n-dimensional manifold and let $\hat{M}$ denote the set of
pairs (p, o_p), where $p \in M$ and o_p is either of the two orientations of $T_p M$.
The projection $\pi: \hat{M} \to M$ sends (p, o_p) to p.

For an open oriented set $W \subseteq M$ with orientation form $\omega \in \Omega^n(W)$ we let
$\hat{W} \subseteq \hat{M}$ be the set of pairs (p, o_p), where $p \in W$ and o_p is the orientation
of $T_p M$ determined by $\omega_p \in \mathrm{Alt}^n T_p M$.

Show that $\hat{M}$ has a topology such that $\hat{W}$ is open and π maps $\hat{W}$
homeomorphically onto W for every open oriented set $W \subseteq M$. Note
that $\hat{M}$ is a topological manifold.

Show that $\hat{M}$ has a uniquely determined differentiable structure such that
π maps $\hat{W}$ diffeomorphically onto W for every oriented open set $W \subseteq M$.
Show that $\hat{M}$ has a canonical orientation. The pair consisting of $\hat{M}$ and
π is called the *oriented double covering* of M.

9.16.   Let M be a connected smooth manifold. Show that $\hat{M}$ (see Exercise 9.15)
consists of at most two connected components, and that M is orientable
if and only if $\hat{M}$ is not connected.

9.17.   Let $V \subseteq \mathbb{R}^{n+k}$ be an open tubular neighborhood of the smooth submanifold
$M^n \subseteq \mathbb{R}^{n+k}$ with the associated projection $r: V \to M$ (see Theorem 9.23).
Define a smooth map $f : V - M \to \mathbb{R}$ by

$$f(x) = \|x - r(x)\| = \min_{y \in M} \|x - y\|.$$

Show that f is a solution to the differential equation

$$\sum_{j=1}^{n+k} \left(\frac{\partial f}{\partial x_j} \right)^2 = 1$$

(the eikonal equation from geometrical optics).

Suppose that $k = 1$, and that M^n is oriented by the Gauss map Y. We
can define the signed distance from M, $\varphi: V \to \mathbb{R}$ by requiring

$$\varphi(x) Y(r(x)) = x - r(x) \qquad (x \in V).$$

Show that φ is a smooth solution to the eikonal equation

$$\sum_{j=1}^{n+1} \left(\frac{\partial \varphi}{\partial x_j} \right)^2 = 1.$$

9.18.   Let $\pi: \hat{M} \to M$ be the oriented double covering of Exercise 9.15. Let
$A: \hat{M} \to \hat{M}$ be the map that for $p \in M$ interchanges the two points in
$\pi^{-1}(p)$. Show that A is a diffeomorphism of order 2 and that

$$\Omega^r(\hat{M}) = \Omega^r(\hat{M})_+ \oplus \Omega^r(\hat{M})_-,$$

where $\Omega^r(\hat{M})_\pm$ is the eigenspace associated to ± 1 for the isomorphism

$$A^* \colon \Omega^r(\hat{M}) \to \Omega^r(\hat{M}).$$

Show that the de Rham complex $(\Omega^*(\hat{M}), d)$ decomposes into the direct sum of two subcomplexes

$$(\Omega^*(\hat{M})_+, d) \quad \text{and} \quad (\Omega^*(\hat{M})_-, d).$$

Show that π^* maps the de Rham complex $(\Omega^*(M), d)$ isomorphically onto $(\Omega^*(\hat{M})_+, d)$. Show that for every $k \in \mathbb{Z}$ we have that

$$H^k(\pi) \colon H^k(M) \to H^k(\hat{M})$$

maps $H^k(M)$ isomorphically onto the $(+1)$-eigenspace in $H^k(\hat{M})$ of $H^k(A)$.

10.1. Let $\pi \colon \mathbb{R}^2 \to \mathbf{T}^2$ be as in Exercise 8.4, and let

$$U_1 = \pi(\mathbb{R} \times (0,1)), \quad U_2 = \pi\left(\mathbb{R} \times \left(-\tfrac{1}{2}, -\tfrac{1}{2}\right)\right).$$

Show that U_1 and U_2 are diffeomorphic to $S^1 \times \mathbb{R}$, and that $U_1 \cap U_2$ has 2 connected components, which are both diffeomorphic to $S^1 \times \mathbb{R}$. Note that $U_1 \cup U_2 = \mathbf{T}^2$.

Use the exact Mayer–Vietoris sequence and Corollary 10.14 to show that

$$H^0(\mathbf{T}^2) \cong H^2(\mathbf{T}^2) \cong \mathbb{R} \quad \text{and} \quad H^1(\mathbf{T}^2) \cong \mathbb{R}^2.$$

10.2. In the notation of Exercise 10.1 we have smooth manifolds

$$C_1 = \pi(\mathbb{R} \times \{a\}), \quad C_2 = \pi(\{b\} \times \mathbb{R}) \quad (a, b \in \mathbb{R})$$

of $\mathbf{T}^2$ which are diffeomorphic to S^1. They are given the orientations induced by $\mathbb{R}$. Show that the map

$$\Omega^1(\mathbf{T}^2) \to \mathbb{R}^2; \quad \omega \mapsto \left(\int_{C_1} \omega, \int_{C_2} \omega\right)$$

induces an isomorphism $H^1(\mathbf{T}^2) \to \mathbb{R}^2$. Show that this isomorphism is independent of a and b.

10.3. Using the notation of Exercise 8.4, we have smooth submanifolds in the n-dimensional torus $\mathbf{T}^n = \mathbb{R}^n/\mathbb{Z}^n$, which are diffeomorphic to S^1

$$C_j = \{\pi(0, \ldots, 0, s, 0, \ldots, 0) \mid s \in \mathbb{R}\},$$

where s is in the j-th place, $1 \le j \le n$. They are given the canonical orientation. Let $\omega \in \Omega^1(\mathbf{T}^n)$ be a closed 1-form with

$$\int_{C_j} \omega = 0 \quad \text{for } 1 \le j \le n.$$

Show that ω is exact.

(Hint: Find an $f \in C^\infty(\mathbf{R}^n, \mathbf{R})$ such that $df = \pi^*(\omega)$, and show that f is periodic with period 1 in all n variables.)

Also show that the map

$$\Omega^1(\mathbf{T}^n) \to \mathbf{R}^n; \quad \omega \mapsto \left(\int_{C_1} \omega, \ldots, \int_{C_n} \omega \right)$$

induces an isomorphism $H^1(\mathbf{T}^n) \to \mathbf{R}^n$.)

10.4. Show that for every connected compact, non-orientable smooth n-dimensional manifold M we have that $H^n(M) = 0$.

(Hint: Use Exercise 9.18.)

10.5. Calculate the de Rham cohomology of Klein's bottle.

(Hint: The oriented double covering can be identified with the map $\mathbf{T}^2 \to \mathbf{K}^2$ from Exercise 8.5.)

10.6. (Partial integration). Let R be a compact domain with smooth boundary in an oriented n-dimensional smooth manifold M. Show that for for $\omega \in \Omega^{p-1}(M)$, $\tau \in \Omega^{n-p}(M)$ we have

$$\int_R d\omega \wedge \tau = \int_{\partial R} \omega \wedge \tau + (-1)^p \int_R \omega \wedge d\tau.$$

10.7. (Divergence theorem) Suppose that R is a compact domain with smooth boundary in $\mathbf{R}^3$. Let $N: \partial R \to S^2$ be the outward directed Gauss map and let $F \in C^\infty(U, \mathbf{R}^3)$ be a smooth vector field on an open set $U \subseteq \mathbf{R}^3$ with $R \subseteq U$. Show that

$$\int_R \operatorname{div} F \, d\mu_{\mathbf{R}^3} = \int_{\partial R} \langle F, N \rangle d\mu_{\partial R}.$$

(Hint: Consider $\omega \in \Omega^2(U)$ given by $\omega_p(w_1, w_2) = \det(F(p), w_1, w_2)$.)

10.8. (Classical Stokes) Let $S \subseteq \mathbf{R}^3$ be a regular surface, oriented by the Gauss map $N: S \to S^2$, and let $R \subseteq S$ be a compact domain with smooth boundary ∂R. Along ∂R we have a unit vector field V pointing in the positive direction. Let $F \in C^\infty(U, \mathbf{R}^3)$ be a smooth vector field on an open set $U \subseteq \mathbf{R}^3$ with $S \subseteq U$. Show that

$$\int_R \langle \operatorname{rot} F, N \rangle d\mu_S = \int_{\partial R} \langle F, V \rangle d\mu_{\partial R}.$$

10.9. Let M^n be a Riemannian manifold and $\pi: \hat{M} \to M$ the oriented double covering from Exercise 9.15. Let $\hat{M}$ be given a Riemannian metric such that

$$D_q \pi: T_q \hat{M} \to T_{\pi(q)} M$$

is an isometry for all $q \in \hat{M}$.

Show (by means of Riesz' representation theorem) that there exists a positive measure μ_M on M such that for all $f \in C_c^0(M, \mathbb{R})$

$$\int_M f \, d\mu_M = \frac{1}{2} \int_{\hat{M}} f \circ \pi \operatorname{vol}_{\hat{M}}.$$

Assume that $f \in C_c^0(M, \mathbb{R})$ has support contained in an oriented open subset $W \subseteq M$. Show that

$$\int_M f \, d\mu_M = \int_W f \operatorname{vol}_W.$$

10.10. Define $i_j \colon \mathbb{R}^{n-1} \to \mathbb{R}^n$ $(1 \leq j \leq n)$ by

$$i_j(x_1, \ldots, x_{n-1}) = (x_1, \ldots, 0, \ldots, x_{n-1})$$

with 0 at the j-th entry. Let

$$Y_n = \{x \in \mathbb{R}^n \mid x_i \geq 0, \ 1 \leq i \leq n\}.$$

and $\omega \in \Omega^{n-1}(\mathbb{R}^n)$ with $\operatorname{supp}(\omega) \cap Y_n$ compact. Show that we have

$$\int_{Y_n} d\omega = \sum_{j=1}^{n} (-1)^j \int_{Y_{n-1}} i_j^*(\omega).$$

10.11. Let $\omega \in \Omega^r(M^n)$. Suppose that

$$\int_\Sigma \omega = 0,$$

for every oriented smooth manifold $\Sigma \subseteq M^n$ that is diffeomorphic to S^r. Show that $d\omega = 0$.

10.12. Let P be a smooth manifold and

$$\tilde{d} \colon \Omega^{n-1}(P) \to \Omega^n(P),$$

a linear operator such that

$$\int_R \tilde{d}\omega = \int_{\partial R} \omega$$

for every $\omega \in \Omega^{n-1}(P)$ and every compact domain with smooth boundary R in an n-dimensional oriented submanifold $M^n \subseteq P$. Show that $\tilde{d} = d$.

10.13. Let M be a smooth manifold. A piecewise C^1-curve on M is a continuous parametrized curve $\alpha \colon [a, b] \to M$ for which there exists a partition

$$a = t_0 < t_1 < \ldots < t_{k-1} < t_k = b,$$

such that the restrictions $\alpha_{|[t_{i-1},t_i]}$ $(1 \le i \le k)$ are continuously differentiable (from one side at the endpoints). For $\omega \in \Omega^1(M)$ we define the path integral of ω along α by

$$\int_\alpha \omega = \sum_{i=1}^k \int_{t_{i-1}}^{t_i} \omega\big(\alpha'(t)\big)\,dt.$$

Suppose that

$$\int_\alpha \omega = 0$$

for every closed piecewise C^1-curve α on M. Show that ω is exact.

10.14. (Green's identity) Show for real-valued functions on an open set $U \subseteq \mathbb{R}^n$ that

$$d(f \wedge *dg) = (\langle \mathrm{grad}f, \mathrm{grad}g \rangle - f\Delta g)dx_1 \wedge \ldots \wedge dx_n$$

(see Exercises 3.2 and 3.3 for definitions of $*$ and Δ).
Conclude for a compact domain with smooth boundary $R \subseteq U$ that

$$\int_{\partial R} f \wedge *dg = \int_{\partial R} \langle \mathrm{grad}f, \mathrm{grad}g \rangle d\mu_n - \int_R f\Delta g\, d\mu_n,$$

where μ_n is the Lebesgue measure on $\mathbb{R}^n$. Derive the identity

$$\int_R f\Delta g\, d\mu_n - \int_R g\Delta f\, d\mu_n = \int_{\partial R} g \wedge *df - \int_{\partial R} f \wedge *dg.$$

10.15. Define $\rho \colon \mathbb{R}^n - \{0\} \to \mathbb{R}$ for $n \ge 2$ by

$$\rho(x) = \phi(\|x\|)$$

where $\phi(r)$ for $r > 0$ is given by

$$\phi(r) = \begin{cases} \log r & \text{if } n = 2 \\ \frac{1}{2-n}r^{2-n} & \text{if } n \ge 3. \end{cases}$$

Show for the closed form ω defined by equation (19) of Example 9.18 that $*d\rho = \omega$, and conclude that ρ is harmonic (i.e. $\Delta\rho = 0$). Apply the last identity of Exercise 10.14 to $f = \rho$, $g \in C_c^\infty(\mathbb{R}^n, \mathbb{R})$ and $R_\epsilon = \{x \in \mathbb{R}^n \mid \epsilon \le \|x\| \le a\}$, where $\epsilon > 0$ and a is suitably large, to obtain

$$\int_{\mathbb{R}^n - \epsilon\mathring{D}^n} \rho\Delta g = \int_{\epsilon S^{n-1}} \rho \wedge * dg - \int_{\epsilon S^{n-1}} g \wedge \omega,$$

where $\epsilon\mathring{D}^n$ is the open disc of radius ϵ around 0 and ϵS^{n-1} is its boundary sphere. Show that

$$\lim_{\epsilon \to 0+} \int_{\epsilon S^{n-1}} \rho \wedge *dg = 0 \quad \text{and} \quad \lim_{\epsilon \to 0+} \int_{\epsilon S^{n-1}} g \wedge \omega = \mathrm{Vol}\big(S^{n-1}\big)g(0).$$

(Hint: The pull-back of ω to ϵS^{n-1} is $\epsilon^{1-n}\,\mathrm{Vol}_{\epsilon S^{n-1}}$.) Conclude that

$$g(0) = \frac{-1}{\mathrm{Vol}(S^{n-1})} \int_{\mathbb{R}^n} \rho\Delta g\, d\mu_n$$

for every $g \in C_c^\infty(\mathbb{R}^n, \mathbb{C})$, where the right hand side is the Lebesgue integral.

11.1. Generalize the concept of *degree* to the case of continuous maps in such a way that Corollary 11.2 holds for f continuous.
Show that Corollary 11.3, the statement $\deg(f) \in \mathbb{Z}$ of Theorem 11.9, Corollary 11.10 and Proposition 11.11 generalize to the case of continuous maps.

11.2. Prove the formula of Theorem 11.9 under the following assumptions:
   (i) f is continuous.
   (ii) f is smooth outside a closed set $A \subseteq N$.
   (iii) $p \in M - f(A)$ is a regular value of f restricted to $N - A$.
(Hint: Lemma 11.8 holds with $U \subseteq M - f(A)$, $V_i \subseteq N - A$. Construct a homotopy F from f to a smooth map such that F_t $(0 \leq t \leq 1)$ is the identity near the points $q_i \in f^{-1}(p)$.)

11.3. Let N^n be an oriented closed (i.e. compact) manifold. Show that every integer occurs as the degree of some smooth map $N^n \to S^n$ (Use Exercise 11.2.)

11.4. (The fundamental theorem of algebra) Suppose

$$P(z) = z^n + Q(z) = z^n + \sum_{j=0}^{n-1} a_j z^j$$

is a complex polynomial of degree $n \geq 1$ without any complex root. This leads to a contradiction as follows: Regarding S^1 as the unit circle in $\mathbb{C}$ define for any $r \geq 0$ a smooth map

$$f_r \colon S^1 \to S^1; \quad f_r(w) = \frac{P(rw)}{|P(rw)|}.$$

Show that $\deg(f_r)$ is independent of r and then (by taking $r = 0$) that $\deg(f_r) = 0$. Prove for r chosen suitably large that the expression

$$F(w, t) = \frac{(rw)^n + tQ(rw)}{|(rw)^n + tQ(rw)|}, \quad 0 \leq t \leq 1$$

defines a homotopy $S^1 \times [0, 1] \to S^1$ between f_r and the map $w \mapsto w^n$ and conclude from this that $\deg(f_r) = n$.

11.5. Let $\Sigma \subseteq \mathbb{R}^n$ be a smooth submanifold diffeomorphic to S^{n-1}, $n \geq 2$, and let U_1, U_2 be the open sets given by the Jordan–Brouwer separation theorem 7.10. For $x_0 \in \mathbb{R}^n - \Sigma$ define

$$F \colon \Sigma \to S^{n-1}; \quad F(x) = \frac{x - x_0}{\|x - x_0\|}.$$

Show that

$$\deg(F) = \begin{cases} 0 & \text{if } x_0 \in U_2 \\ \pm 1 & \text{if } x_0 \in U_1 \end{cases}$$

where the sign depends only on the orientation of Σ.

11.6. Prove the case $n < m$ of Theorem 11.5 (reduce to the case of equal dimensions).

Apply this to show that every map $f: N^n \to S^m$ from a smooth manifold of dimension $n < m$ is homotopic to a constant.

11.7. Let $U \subseteq \mathbb{R}^n$ be a bounded open set and A a compact subset of U. Prove the existence of a compact domain with smooth boundary R such that $A \subseteq R \subseteq U$.

(Hint: Try $R = \psi^{-1}([c, \infty))$ where ψ is given by Lemma A.7 and $c > 0$ is a regular value of ψ.)

11.8. Let B be the set of boundary points of a bounded open set $U \subseteq \mathbb{R}^n$ $(n \geq 2)$, and let $F: B \to \mathbb{R}^n - \{0\}$ be a continuous vector field on B without zeros. Show the existence of $X: U \cup B \to \mathbb{R}^n$ with the following properties:

(a) X is a continuous extension of F.

(b) X is smooth on U.

(c) X has only finitely many zeros in U.

(Hint: (a) and (b) can be achieved by the proof of Lemma 7.4, and (c) from the proof of Lemma 11.25.)

Prove that the integer

$$\gamma(F, U) = \sum_{p \in U, X(p) = 0} \iota(X; p)$$

is independent of the choice of X satisfying (a), (b), (c).

(Hint: Use Exercise 11.7 and Theorem 11.22.)

Find an example in $\mathbb{R}^2$ to show that $\gamma(F, U)$ actually depends on U and not only on B and F.

11.9. Suppose under the assumptions of Exercise 11.8 that $F: B \to \mathbb{R}^n - \{0\}$ is homotopic to $G: B \to \mathbb{R}^n - \{0\}$. Prove that $\gamma(F, U) = \gamma(G, U)$.

11.10. With U and B given as in Exercise 11.8 assume $F: U \cup B \to \mathbb{R}^n$ to be continuous without any zeros on B. Prove that if $\gamma(F_{|B}, U) \neq 0$, then F has at least one zero in U.

(Hint: Otherwise find X with properties (a) and (b) of Exercise 11.8 such that X has no zeros in U.)

11.11. Show that the condition defining a non-degenerate zero of a vector field is independent of the choice of chart.

11.12. Let $F \in C^\infty(\mathbb{R}^n, \mathbb{R}^n)$ and $G \in C^\infty(\mathbb{R}^m, \mathbb{R}^m)$ be vector fields both with the origin as the only zero. Show that $F \times G$ is a vector field on $\mathbb{R}^{n+m}$ with the same property, and that

$$\iota(F \times G; 0) = \iota(F; 0)\iota(G; 0).$$

12.1. Show that $f: \mathbb{R}^n \to \mathbb{R}$ given by

$$f(x) = \sum_{j=1}^{n} \sin^2\left(\pi x_j\right)$$

is a Morse function. Determine all the critical points and their indices. Show with the notation of Exercise 8.4 that there is a Morse function $\tilde{f}$ on T^n such that $f = \tilde{f} \circ \pi$. Prove that the number of critical points of index λ for $\tilde{f}$ is $\binom{n}{\lambda}$.

12.2. Let f be a Morse function on a Riemannian manifold. Prove that all zeros of $\mathrm{grad}(f)$ are non-degenerate (see Exercise 9.14).

12.3. Show for any smooth map $f: M^m \to \mathbb{R}^n$ and $p \in M$ that there is a map (non-linear in general)

$$d_p^2 f: \mathrm{Ker}\, D_p f \to \mathrm{Cok}\, D_p f$$

such that

$$d_p^2 f\left(\alpha'(0)\right) = (f \circ \alpha)''(0) + \mathrm{Im}\, D_p f$$

whenever $\alpha: (-\delta, \delta) \to M$ is smooth with $\alpha(0) = p$ and $\alpha'(0) \in \mathrm{Ker}\, D_p f$.

12.4. Let f be a Morse function on a closed connected manifold with at least two critical points of index 0. Prove that f has a critical point of index 1.

(Hint: With notation as in the proof of Theorem 12.16 study $\dim H^0(M(b_j))$ as a function of j.)

12.5. Is it possible for a Morse function on $\mathbb{R}^2$ to have only two critical points both of index 0?

12.6. With the notation and assumptions of Lemma 12.13 we put

$$W_\nu = M(a - \epsilon) \cup \bigcup_{i=1}^{\nu} U_i \quad (0 \le \nu \le r).$$

Show that W_ν has finite-dimensional de Rham cohomology and that one of the following two cases occurs for each ν, $1 \le \nu \le r$:

Case 1: $\dim H^p(W_\nu) - \dim H^p(W_{\nu-1}) = \begin{cases} 0 & \text{if } p \ne \lambda_\nu \\ 1 & \text{if } p = \lambda_\nu \end{cases}$

Case 2: $\dim H^p(W_\nu) - \dim H^p(W_{\nu-1}) = \begin{cases} 0 & \text{if } p \ne \lambda_\nu - 1 \\ -1 & \text{if } p = \lambda_\nu - 1. \end{cases}$

12.7. (The Morse inequalities) For a smooth manifold M^n with finite-dimensional de Rham cohomology we define the *Poincaré polynomial* by

$$P_M(t) = \sum_{j=0}^{n} b_j t^j,$$

where b_j is the j-th Betti number of M, $b_j = \dim_{\mathbf{R}} H^j(M)$. Let $f: M \to \mathbf{R}$ be a Morse function with c_λ critical points of index λ ($c_\lambda < \infty$). Define the polynomial

$$C_M(t) = C_{M,f}(t) = \sum_{\lambda=0}^{n} c_\lambda t^\lambda.$$

Prove for any closed manifold M that

$$C_M(t) - P_M(t) = (t+1)R_M(t),$$

where $R_M(t)$ is a polynomial with non-negative integral coefficients. (Hint: Prove by induction a similar statement for each set W_ν introduced in Exercise 12.6.)
Derive the Morse inequalities

$$\sum_{k=0}^{j} (-1)^{j-k} c_j \geq \sum_{k=0}^{j} (-1)^{j-k} b_j \quad (0 \leq j \leq n).$$

Observe that the Morse inequalities imply

$$c_j \geq b_j \quad (0 \leq j \leq n).$$

12.8. (Morse's lacunary principle) Continuing with the assumptions and notation of Exercise 12.7, suppose for each λ ($1 \leq \lambda \leq n$) that either $c_{\lambda-1} = 0$ or $c_\lambda = 0$. Prove that $b_j = c_j$ for every j.

12.9. Let $\mathrm{pr}_i: \mathbf{T}^n = \mathbf{R}^n/\mathbf{Z}^n \to \mathbf{R}/\mathbf{Z}$ be the i-th projection (see Exercise 8.4). Pick $\omega \in \Omega^1(\mathbf{R}/\mathbf{Z})$ representing a generator of $H^1(\mathbf{R}/\mathbf{Z}) \cong H^1(S^1) \cong \mathbf{R}$ and define $\omega_i = \mathrm{pr}_i^*(\omega) \in \Omega^1(\mathbf{T}^n)$. To an increasing sequence $I: 1 \leq i_1 < i_2 < \ldots < i_p \leq n$ we associate the closed p-form

$$\omega_I = \omega_{i_1} \wedge \ldots \wedge \omega_{i_p} \in \Omega^p(\mathbf{T}^n).$$

Prove that the resulting classes $[\omega_I] \in H^p(\mathbf{T}^n)$ are linearly independent. (Hint: Consider integrals of linear combinations $\sum_I a_I \omega_I$ over subtori.)
Prove with the help of Exercises 12.1 and 12.7 that

$$\dim H^p(\mathbf{T}^n) = \binom{n}{p},$$

and conclude that there is an isomorphism

$$H^*(\mathbf{T}^n) \cong \mathrm{Alt}^*(\mathbf{R}^n)$$

of graded algebras.

12.10. Let $\pi\colon \tilde{M}^n \to M^n$ be a d-fold covering of closed manifolds $\tilde{M}$ and M, i.e. M^n can be covered by open sets U with the property that $\pi^{-1}(U)$ is a disjoint union of d open sets $U_1, \dots, U_d$ such that $\pi_{|U_i}\colon U_i \to U$ is a diffeomorphism for $1 \le i \le d$. Show that $\chi(\tilde{M}) = d\chi(M)$.

12.11. Assume f and g are Morse functions on the closed manifolds M and N respectively. Show that a Morse function h on $M \times N$ can be defined by

$$h(p, q) = f(p) + g(q).$$

Describe the critical points and indices for h in terms of similar data for f and g. Derive the product formula for Euler characteristics

$$\chi(M \times N) = \chi(M)\chi(N).$$

13.1. A *symplectic space* (V, ω) is a real vector space equipped with an alternating 2-form $\omega \in \mathrm{Alt}^2(V)$. A linear subspace $W \subseteq V$ is said to be *non-degenerate* if for every $e \in W - \{0\}$ we can find $f \in W$ such that $\omega(e, f) \ne 0$. Assume from now on that V is non-degenerate of finite dimension.
Let $W \subseteq V$ be a non-degenerate subspace. Show that

$$W^{\perp} = \{x \in V \mid \omega(x, y) = 0 \text{ for every } y \in W\}$$

is a non-degenerate subspace with $W \oplus W^{\perp} = V$. Prove that V has a basis $\{e_1, f_1, e_2, f_2, \dots, e_n, f_n\}$ such that

$$\omega(e_i, e_j) = \omega(f_i, f_j) = 0, \quad \omega(e_i, f_j) = \begin{cases} 1 & \text{if } i = j \\ 0 & \text{if } i \ne j \end{cases}$$

This is called a *symplectic basis*. Note that $\dim V$ must be even.
(Hint: Pick $e_1 \ne 0$ arbitrarily. Then find f_1 and apply induction to $W^{\perp}$, where W is spanned by e_1 and f_1.)
Let $\omega_1, \tau_1, \omega_2, \tau_2, \dots, \omega_n, \tau_n$ be the dual basis of $\mathrm{Alt}^1(V) = V^*$ to a symplectic basis for V. Show that

$$\omega = \sum_{j=1}^{n} \omega_j \wedge \tau_j.$$

13.2. Let M^n be an oriented closed smooth manifold of dimension $n \equiv 2 \pmod 4$. Show that Poincaré duality organizes $H^{n/2}(M)$ as a non-degenerate symplectic space (see Exercise 13.1). Prove that $\chi(M)$ is even.

13.3. Consider a smooth map $f\colon N^n \to M^n$ between n-dimensional oriented smooth closed manifolds, where M is connected. Prove that if $H^n(f) \ne 0$ then $H^p(f)\colon H^p(M) \to H^p(N)$ is injective for every p.

13.4. Let (S^n, M^m) be a smooth compact manifold pair with $0 < m < n$ and $U = S^n - M^m$. Construct isomorphisms

$$H^p(U) \cong H^{n-p-1}(M)^* \quad (1 \leq p \leq n-2).$$

Show that $H^n(U) = 0$ and find short exact sequences

$$0 \to H^{n-1}(U) \to H^0(M)^* \to \mathbb{R} \to 0$$
$$0 \to \mathbb{R} \to H^0(U) \to H^{n-1}(M)^* \to 0.$$

13.5. Let $\pi: \hat{M} \to M$ be the oriented double covering constructed in Exercise 9.15 and define $A: \hat{M} \to \hat{M}$ as in Exercise 9.18. Find isomorphisms

$$H^p(M) \to \left(H_c^{n-p}(\hat{M})_{-1}\right)^*$$

where $H_c^q(\hat{M})_{-1}$ denotes the (-1)-eigenspace of A^* on $H_c^q(\hat{M})$.

13.6. Compute $H^n(M^n)$ for every smooth connected n-dimensional manifold M^n.

(Hint: Use Exercise 13.5. The answer depends on whether M is compact or not, and also whether M is orientable or not,)

13.7. Prove that $H_c^q(M)$ for every smooth manifold and every q is at most countably generated.

(Hint: Induction on open sets.)

13.8. Show that any de Rham cohomology space $H^p(M)$ is either finite-dimensional or isomorphic to a product $\prod_{n=1}^{\infty} \mathbb{R}$ of countably many copies of $\mathbb{R}$.

13.9. A compact set $K \subseteq \mathbb{R}^n$ is said to be *cellular* if $K = \bigcap_{j=1}^{\infty} D_j$, where each $D_j \subseteq \mathbb{R}^n$ is homeomorphic to D^n and $D_{j+1} \subseteq \mathring{D}_j$ for every j. Show for K cellular that

$$H^p(\mathbb{R}^n - K) \cong \begin{cases} \mathbb{R} & \text{if } p = 0, \, n-1 \\ 0 & \text{otherwise.} \end{cases}$$

13.10. For $K \subseteq \mathbb{R}^n$ compact denote by $\check{H}^0(K, \mathbb{R})$ the vector space of locally constant functions $K \to \mathbb{R}$. Construct an isomorphism

$$\Phi^1: \check{H}^0(K, \mathbb{R}) \to H_c^1(\mathbb{R}^n - K)$$

such that $\Phi^1(f) = [d\tilde{f}]$ where $\tilde{f} \in C_c^{\infty}(\mathbb{R}^n, \mathbb{R})$ is locally constant on an open set containing K and $\tilde{f}$ extends f. Find an isomorphism

$$\Phi: H^{n-1}(\mathbb{R}^n - K) \to \check{H}^0(K, \mathbb{R})^*$$

such that $\Phi([\omega])$ for $\omega \in \Omega^{n-1}(\mathbb{R}^n - K)$ can be evaluated on $f \in \check{H}^0(K, \mathbb{R})$ by the following procedure:

There are disjoint compact domains $R_1, \ldots, R_d$ with smooth boundary such that $K \subseteq \bigcup_{j=1}^{d} (R_j - \partial R_j)$ and f is constant on $K \cap R_j$ with value a_j. Then

$$\Phi([\omega])(f) = \sum_{j=1}^{d} a_j \int_{\partial R_j} \omega.$$

(Hint: Exercise 11.7 is needed.)

14.1.   A rational function

$$R(z) = \frac{P(z)}{Q(z)},$$

where P and Q are complex polynomials, is initially defined only on $\mathbb{C}$ with the roots of Q removed. Show that R extends to a smooth map of the Riemann sphere $\mathbb{C} \cup \{\infty\}$ to itself.

14.2.   The n-th *symmetric power* $SP^n(X)$ of a topological space X is the set of orbits under the action of the symmetric group $S(n)$ on $X^n = X \times \ldots \times X$ (n factors) with the quotient topology from X^n.
Show that $SP^n(S^2)$ is homeomorphic to $\mathbb{CP}^n$ by the map $f : (\mathbb{CP}^1)^n \to \mathbb{CP}^n$ taking

$$([\alpha_1, \beta_1], \ldots, [\alpha_n, \beta_n])$$

into $[a_0, a_1, \ldots, a_n] \in \mathbb{CP}^n$, where the a_k are determined by the identity

$$\prod_{j=1}^{n} (\beta_j z - \alpha_j) = \sum_{k=0}^{n} a_k z^k.$$

14.3.   Show that $H^p(S^2 \times S^4) \cong H^p(\mathbb{CP}^3)$ for every p, but that the graded algebras $H^*(S^2 \times S^4)$ and $H^*(\mathbb{CP}^3)$ are not isomorphic.

14.4.   Show that any continuous $f : \mathbb{CP}^m \to \mathbb{CP}^n$ induces the zero homomorphism

$$f^* : H^p(\mathbb{CP}^n) \to H^p(\mathbb{CP}^m)$$

for $p \neq 0$ when $m > n$.

14.5.   Prove in the following steps that $\pi : S^{2n+1} \to \mathbb{CP}^n$ is not homotopic to a constant. Suppose

$$F : S^{2n+1} \times [0, 1] \to \mathbb{CP}^n$$

is a homotopy from a constant F_0 to $F_1 = \pi$, and extend π continuously to $g : D^{2n+2} \to \mathbb{CP}^n$ by

(1) $g(tz) = F(z, t), \quad z \in S^{2n+1}, \quad t \in [0, 1].$

Define $h\colon D^{2n+2} \to \mathbb{CP}^{n+1}$ by

$$h(z_0, z_1, \ldots, z_n) = \left[z_0, z_1, \ldots, z_n, \left(1 - \sum_{j=0}^{n} |z_j|^2\right)^{1/2} \right]$$

and observe that h maps the open disc $\mathring{D}^{2n+2}$ bijectively onto $U_{n+1} = \mathbb{CP}^{n+1} - \mathbb{CP}^n$. Moreover $h_{|S^{2n+1}}$ is the composite of π with the inclusion $j\colon \mathbb{CP}^n \to \mathbb{CP}^{n+1}$. Find $f\colon \mathbb{CP}^{n+1} \to \mathbb{CP}^n$ so that $f \circ h = g$ and argue that f is continuous. Observe that $f \circ j = \mathrm{id}_{\mathbb{CP}^n}$, and pass to de Rham cohomology to obtain a contradiction.
This proves Hopf's result mentioned in Example 14.1.

14.6.   Given $z \in \mathbb{C}^{n+1} - \{0\}$, $p = \pi(z) \in \mathbb{CP}^n$, and two vectors $v_j \in T_z\mathbb{C}^{n+1} = \mathbb{C}^{n+1}$, $j = 1, 2$. Let $w_j = D_z\pi(v_j) \in T_p\mathbb{CP}^n$. Show that the hermitian inner product $\langle\!\langle \ , \ \rangle\!\rangle_p$ on $T_p\mathbb{CP}^n$ from Lemma 14.4 satisfies

$$\langle\!\langle w_1, w_2 \rangle\!\rangle_p = \langle v_1, v_2 \rangle - \frac{\langle v_1, z \rangle \langle z, v_2 \rangle}{\langle z, z \rangle},$$

where $\langle \ , \ \rangle$ denotes the usual hermitian inner product on $\mathbb{C}^{n+1}$.

14.7.   A *symplectic manifold* (M, ω) is a smooth manifold M equipped with a 2-form $\omega \in \Omega^2(M)$ satisfying the following conditions

   (i) $d\omega = 0$.
   (ii) (T_pM, ω_p) is a non-degenerate symplectic space for every $p \in M$ (see Exercise 13.1).

Show that $\mathbb{CP}^n$ admits the structure of a symplectic manifold.
Show that a symplectic manifold (M, ω) has even dimension $2m$ and that $\omega^m = \omega \wedge \ldots \wedge \omega$ is an orientation form on M.
Show for a closed symplectic manifold (M^{2m}, ω) that $H^*(M)$ contains a subalgebra isomorphic to $\mathbb{R}[c]/(c^{m+1})$, where $c = [\omega] \in H^2(M)$.

14.8.   (Grassmann manifolds) Fix integers $0 < m < n$ and let F denote either $\mathbb{R}$ or $\mathbb{C}$. Equip F^n with the usual inner product. Denote by $G_m(\mathsf{F}^n)$ the set of m–dimensional linear subspaces $V \subseteq \mathsf{F}^n$.
Show that $G_m(\mathsf{F}^n)$ can be identified with a compact subspace of the $n \times n$–matrices over F by associating to V the orthogonal projection on V. This makes $G_m(\mathsf{F}^n)$ a compact Hausdorff space (with countable basis for the topology).
Denote by $\mathrm{Fr}_m(\mathsf{F}^n)$ the set of $n \times m$ matrices over F of rank m. Observe that $\mathrm{Fr}_m(\mathsf{F}^n)$ can be identified with an open subset of F^{mn} and hence is a smooth manifold. The group $\mathrm{GL}_m(\mathsf{F})$ acts by right multiplication on $\mathrm{Fr}_m(\mathsf{F}^n)$. Show that the orbit space $\mathrm{Fr}_m(\mathsf{F}^n)/\mathrm{GL}_m(\mathsf{F})$ with quotient topology can be identified with $G_m(\mathsf{F}^n)$ by associating to $A \in \mathrm{Fr}_m(\mathsf{F}^n)$ the span of its column vectors denoted $[A] \in G_m(\mathsf{F}^n)$.

To an increasing set of indices I: $1 \leq i_1 < i_2 < \cdots < i_m \leq n$ and $A \in \mathrm{Fr}_m(\mathsf{F}^n)$ is associated the $m \times m$-submatrix A_I of rows numbered by I and the $(n-m) \times m$-matrix $A_{I'}$ consisting of the remaining rows. Show that

$$U_I = \{[A] \mid A \in \mathrm{Fr}_m(\mathsf{F}^n), \ A_I \text{ invertible}\}$$

is an open set in $G_m(\mathsf{F}^n)$ and define $h_I \colon U_I \ \rightarrow \ \mathrm{Mat}_{n-m,m}(\mathsf{F})$ into $(n-m) \times m$ matrices by

$$h_I([A]) = A_{I'} A_I^{-1}.$$

Show that h_I is a homeomorphism, and finally (identifying $\mathrm{Mat}_{n-m,m}(\mathsf{F})$ with $\mathsf{F}^{(n-m)m}$) that $\mathcal{H} = \{(U_I, h_I)\}$ is a smooth atlas on $G_m(\mathsf{F}^n)$. The resulting closed smooth manifold $G_m(\mathsf{F}^n)$ is the *Grassmann manifold* of m-dimensional subspaces of F^n.

15.1. Show that the pull-back $\eta^*(\tau_{S^2})$ by the Hopf fibration $\eta \colon S^3 \to S^2$ is a trivial vector bundle (see Example 14.10). Prove also that the tangent bundle τ_{S^3} is trivial.

15.2. Let G be any Lie group (see Exercise 9.10). Right translation by $g \in G$ is the diffeomorphism $R_g \colon G \to G$; $R_g(x) = xg$. Given a tangent vector $X_e \in T_e G$ at the neutral element e, define $X_g \in T_g G$ for $g \in G$ by $X_g = D_e R_g(X_e)$. Show that this extends X_e to a smooth vector field X on G and moreover that X is right-invariant in the sense that

$$X_{hg} = D_h R_g(X_h) \quad \text{for all } h, g \in G.$$

Construct a frame over G for the tangent bundle TG consisting of right-invariant vector fields.

15.3. Let ξ be a smooth real line bundle over $\mathbb{RP}^n$ with total space $S^n \times_{S^0} \mathbb{R}$, i.e. the orbit space of $S^n \times \mathbb{R}$ under the action of $S^0 = \{\pm 1\}$, where -1 acts by $(x, t) \mapsto (-x, -t)$. This is the *canonical line bundle* over $\mathbb{RP}^n$. Construct a smooth isomorphism of vector bundles

$$\tau_{\mathbb{RP}^n} \oplus \varepsilon^1 \cong (n+1)\xi,$$

where $k\xi$ denotes the k-fold direct sum $\xi \oplus \cdots \oplus \xi$.
(Hint: Both total spaces can be identified with $S^n \times_{S^0} \mathbb{R}^{n+1}$.)

15.4. Prove that the tangent bundle $\tau_{\mathbb{CP}^n}$ with the complex structure on $T_p \mathbb{CP}^n$ given by Lemma 14.4 is a smooth complex vector bundle over $\mathbb{CP}^n$. Construct a smooth isomorphism of complex vector bundles (similar to that in Exercise 15.3)

$$\tau_{\mathbb{CP}^n} \oplus \varepsilon^1 \cong (n+1)\overline{H}_n,$$

where $\overline{H}_n$ is conjugate to H_n, i.e. as a real vector bundle $\overline{H}_n$ is the same as H_n, but $z \in \mathbb{C}$ acts on $\overline{H}_n$ the same way as $\overline{z} \in \mathbb{C}$ acts on H_n.

15.5. Continuing Exercise 14.8, define the smooth map $\pi: \mathrm{Fr}_m(\mathbb{F}^n) \to G_m(\mathbb{F}^n)$ by $\pi(A) = [A]$.

Construct for each $I: 1 \leq i_1 < i_2 < \ldots < i_m$ a smooth section $S_I: U_I \to \pi^{-1}(U_I)$ such that $S_I([A]) = AA_I^{-1}$. Show that the map

$$k_I: U_I \times \mathrm{GL}_m(\mathbb{F}) \to \pi^{-1}(U_I); \quad k_I([A], Q) = S_I([A])Q$$

is a diffeomorphism (note that $\mathrm{GL}_m(\mathbb{F})$ is a Lie group). Define a smooth action of $\mathrm{GL}_m(\mathbb{F})$ on $\mathrm{Fr}_m(\mathbb{F}^n) \times \mathbb{F}^m$:

$$(A, x)Q = \left(AQ, Q^{-1}x\right)$$

and form the orbit space $E = \mathrm{Fr}(\mathbb{F}^n)_m \times_{\mathrm{GL}_m(\mathbb{F})} \mathbb{F}^m$ with the induced projection $\overline{\pi}: E \to G_m(\mathbb{F}^n)$, $\overline{\pi}([A, x]) = [A]$. Show that this is a smooth m-dimensional $\mathbb{F}$-vector bundle in such a way that the assignment

$$[A] \mapsto \left([S_I([A]), e_1], \ldots, [S_I([A]), e_n]\right)$$

defines a smooth frame over U_I. This is the *canonical vector bundle* $\gamma = \gamma_n^{\mathbb{F}^m}$ over $G_m(\mathbb{F}^n)$.

Give an identification of the fiber γ_V over $V \in G_m(\mathbb{F}^n)$ with V, which to $[A, x] \in \gamma_V$ (with $[A] = V$, $x \in \mathbb{F}^n$, a column vector) assigns $Ax \in V$. Establish a 1-1 correspondence between smooth sections $S: U \to \pi^{-1}(U)$ over a given open set $U \subseteq G_m(\mathbb{F}^n)$ and smooth frames $(s_1, ..., s_n)$ for γ over U such that $(s_1(p), ..., s_n(p))$ for $p \in V$ via the identification $\gamma_p \cong V$ corresponds to the column vectors of $S(p)$.

15.6. Show that the canonical vector bundle constructed in Exercise 15.5 in the case $m = 1$ can be identified (smoothly) with the canonical line bundle over $\mathbb{RP}^{n-1}$ or $\mathbb{CP}^{n-1}$ (Exercise 15.3 and Example 15.2).

15.7. Let ξ be an m-dimensional vector bundle over B admitting a complement η, i.e. $\xi \oplus \eta \cong \varepsilon^N$. Construct a vector bundle homomorphism $(f, \tilde{f})$ to the m-dimensional canonical vector bundle γ over the Grassmannian $G_m = G_m(\mathbb{R}^N)$, such that the fiber ξ_b maps isomorphically to a fiber in γ by including ξ_b in $\varepsilon_b^N \cong \mathbb{R}^N$. Conclude that $\xi \cong f^*(\gamma)$. Do the same for complex vector bundles.

15.8. Show that any vector bundle over a compact smooth manifold B is isomorphic to a smooth vector bundle. (Apply Exercise 15.7 and pick a homotopy from f to a smooth map.)

15.9. Let $\pi: E \to B$ be a smooth fiber bundle. For $p \in E$, $b = \pi(p)$ we have the fiber F_b through p. Observe that F_b is a smooth submanifold of E, so that $T_p F_b$ can be identified with a subspace of $T_p E$. Show that the union of these subspaces $T_p F_b$ as p runs through E is the total space of a smooth vector bundle τ^v over E (the tangent bundle along the fibers or *vertical tangent bundle*).

Show that the orthogonal complements $(T_pF_b)^\perp$ to T_pF_b in T_pE with respect to a Riemannian metric on E are the fibers of another smooth vector bundle τ^h over E (the normal bundle to the fibers or *horizontal tangent bundle*).

Find smooth vector bundle isomorphisms

$$\tau^v \oplus \tau^h \cong \tau_E , \quad \tau^h \cong \pi^*(\tau_B).$$

15.10. Prove Theorem 15.18 without the compactness assumption on B for smooth real vector bundles by embedding $E(\xi)$ in some Euclidean space (Theorem 8.11). Make use of Exercise 15.9, observing that $\tau^v \cong \pi^*(\xi)$.

Deduce the smooth complex case of Theorem 15.18 without the compactness assumption.

15.11. Let ξ be a smooth vector bundle with inner product over B (not assumed to be compact). Suppose $A \subseteq B$ is a closed set and $U \subseteq B$ is open with $A \subseteq U$. Let s be a continuous section in ξ over B which is smooth on U. Construct for any given continuous map $\epsilon: B \to (0, \infty)$ a smooth section $\tilde{s}$ over B that satisfies

(i) $\tilde{s}(b) = s(b)$ for $b \in A$.
(ii) $\|\tilde{s}(b) - s(b)\|_b \leq \epsilon(b)$ for every $b \in B$.

15.12. Let $\hat{f}: \xi \to \eta$ be a map of vector bundles over id_B, such that the rank $r = \mathrm{rk}(\hat{f}_b)$ of the induced linear map of fibers is independent of $b \in B$. Show that the subspaces

$$E(\mathrm{Im}(\hat{f})) = \bigcup_{b \in B} \mathrm{Im}(\hat{f}_b) \subseteq E(\eta)$$

$$E(\mathrm{Ker}(\hat{f})) = \bigcup_{b \in B} \mathrm{Ker}(\hat{f}_b) \subseteq E(\xi)$$

are total spaces of vector bundles $\mathrm{Im}(\hat{f})$ and $\mathrm{Ker}(\hat{f})$ over B (smooth if ξ, η and $\hat{f}$ are smooth).

15.13. Show for a map $\hat{f}: \xi \to \eta$ of vector bundles over id_B, that the function $b \mapsto \mathrm{rk}(\hat{f}_b)$ is lower semicontinuous on B.

15.14. Suppose $\hat{p}: \xi \to \xi$ is a vector bundle map over id_B satisfying $\hat{p} \circ \hat{p} = \hat{p}$. Show that $r = \mathrm{rk}(\hat{p}_b)$ is independent of $b \in B$ when B is connected. Prove that $\xi \cong \mathrm{Im}(\hat{p}) \oplus \mathrm{Ker}(\hat{p})$ (see Exercise 15.12).

16.1. Show for integers $n > 0$, $m > 0$ that

$$\mathbb{Z}/n\mathbb{Z} \otimes_{\mathbb{Z}} \mathbb{Z}/m\mathbb{Z} \cong \mathbb{Z}/d\mathbb{Z}$$

where d is the greatest common divisor of n and m.

16.2. Prove the isomorphisms (4) listed above Lemma 16.4.

16.3. Let V be a finite-dimensional complex vector space with Hermitian inner product $\langle \, , \, \rangle$. Construct a hermitian inner product on $\Lambda^k V$ satisfying

$$\langle v_1 \wedge \ldots \wedge v_k, w_1 \wedge \ldots \wedge w_k \rangle = \det \left(\langle v_\alpha, w_\beta \rangle \right).$$

16.4. Let F be a covariant functor from the category of finite-dimensional real vector spaces to itself. Assume F to be *smooth* in the sense that the maps induced by F

$$\mathrm{Hom}_{\mathbf{R}}(V, W) \to \mathrm{Hom}_{\mathbf{R}}(F(V), F(W))$$

are smooth.

Construct for a given smooth real vector bundle ξ over B another smooth real vector bundle $F(\xi)$ over B with fibers $F(\xi)_b = F(\xi_b)$, $b \in B$.

Show that this extends F to a covariant functor from the category of smooth real vector bundles over B and smooth homomorphisms over id_B to itself. (Note: Many variations are possible: $\mathbf{R}$ can be replaced by $\mathbf{C}$ in the source and/or target category, F can be contravariant, and smooth can be changed to continuous.)

16.5. Prove that the construction of Exercise 16.4 is compatible with pull-back, i.e. construct for $h \colon B' \to B$ a smooth vector bundle isomorphism

$$\psi_\xi \colon F(h^*(\xi)) \to h^*(F(\xi)).$$

Show that pull-back by h extends to a functor h^* from the category of vector bundles over B (see Exercise 16.4) to the corresponding category over B', and that the diagram

$$
\begin{array}{ccc}
F(h^*(\xi)) & \xrightarrow{\ \psi_\xi\ } & h^*(F(\xi)) \\
{\scriptstyle F(h^*(\hat{g}))}\downarrow & & \downarrow{\scriptstyle h^*(F(\hat{g}))} \\
F(h^*(\eta)) & \xrightarrow{\ \psi_\eta\ } & h^*(F(\eta))
\end{array}
$$

commutes for every $\hat{g} \colon \xi \to \eta$ over id_B.

16.6. Consider two smooth covariant functors F, G as in Exercise 16.4. Assume given (for each finite-dimensional vector space V) a linear map $\phi_V \colon F(V) \to G(V)$ such that the diagram

$$
\begin{array}{ccc}
F(V) & \xrightarrow{\ \phi_V\ } & G(V) \\
{\scriptstyle F(h)}\downarrow & & \downarrow{\scriptstyle G(h)} \\
F(W) & \xrightarrow{\ \phi_W\ } & G(W)
\end{array}
$$

is commutative for every $h \in \mathrm{Hom}_{\mathbf{R}}(V, W)$.

Construct for a given smooth real vector bundle over B a smooth bundle homomorphism $\phi_\xi \colon F(\xi) \to G(\xi)$ over id_B, and show that any smooth vector bundle homomorphism $\hat{g} \colon \xi \to \eta$ over id_B leads to a commutative diagram

$$
\begin{CD}
F(\xi) @>{\phi_\xi}>> G(\xi) \\
@V{F(\hat{g})}VV @VV{G(\hat{g})}V \\
F(\eta) @>{\phi_\eta}>> G(\eta)
\end{CD}
$$

Prove finally with the assumptions and notation of Exercise 16.5 that the diagrams

$$
\begin{CD}
F(h^*(\xi)) @>{\phi_{h^*(\xi)}}>> G(h^*(\xi)) \\
@V{\psi_\xi}VV @VV{\psi_\xi}V \\
h^*(F(\xi)) @>{h^*(\phi_\xi)}>> h^*(G(\xi))
\end{CD}
$$

commute.

16.7. Let V be a finite-dimensional real or complex vector space. Construct a linear map $e \colon \bigotimes^k V \to \bigotimes^k V$ such that

$$
e(v_1 \otimes \cdots \otimes v_k) = \frac{1}{k!} \sum_{\sigma \in S(k)} \mathrm{sign}(\sigma) v_{\sigma(1)} \otimes \cdots \otimes v_{\sigma(k)}.
$$

Show that $e \circ e = e$, and that the quotient homomorphism $\otimes^k V \to \Lambda^k V$ induces an isomorphism $\mathrm{Im}(e) \cong \Lambda^k V$.

Apply Exercise 15.14 to give an alternative construction of the vector bundles $\Lambda^k(\xi)$.

16.8. For finite-dimensional vector spaces V and W, construct a linear map

$$
\Lambda^k V \otimes \Lambda^l W \to \Lambda^{k+l}(V \oplus W)
$$

carrying $(v_1 \wedge \ldots \wedge v_k) \otimes (w_1 \wedge \ldots \wedge w_l)$ into $(v_1, 0) \wedge \ldots \wedge (v_k, 0) \wedge (0, w_1) \wedge \ldots \wedge (0, w_l)$. Show that these combine to give isomorphisms

$$
\bigoplus_{k=0}^{n} \Lambda^k V \otimes \Lambda^{n-k} W \cong \Lambda^n (V \oplus W).
$$

Extend this to vector bundle isomorphisms

$$
\bigoplus_{k=0}^{n} \Lambda^k \xi \otimes \Lambda^{n-k} \eta \cong \Lambda^n (\xi \oplus \eta).
$$

16.9. Prove Lemma 16.10.(i).

16.10. Finish the proof of Theorem 16.13.(iv).

16.11. Let ξ and η be smooth vector bundles over a compact smooth manifold M. Show that ξ and η are isomorphic if and only if $\Omega^0(\xi) \cong \Omega^0(\eta)$ as $\Omega^0(M)$-modules. Prove that every finitely generated projective $\Omega^0(M)$-module P is isomorphic to $\Omega^0(\eta)$ for some smooth vector bundle η over M. (Hint: Apply Exercise 15.14 with $\xi = \varepsilon^n$.)

16.12. Prove for any line bundle ξ (real or complex) that $\text{Hom}(\xi, \xi)$ is trivial. Prove for any real line bundle that $\xi \otimes \xi$ is trivial.

16.13. Let T^*M be the total space of the dual tangent bundle τ_M^* of a smooth manifold M^n, and $\pi: T^*M \to M$ the projection. For $q \in T^*M$ (i.e. a linear form on $T_{\pi(q)}M$) define

$$\theta_q \in \text{Alt}^1(T_q(T^*M)); \quad \theta_q(X) = q(D_q\pi(X)).$$

Show that this defines a differential 1-form θ on T^*M and moreover that θ can be given in local coordinates by the expression

$$\sum_{i=1}^{n} t_i dx_i,$$

where $x_1, \ldots, x_n$ are the coordinate functions on a chart $U \subseteq M$ and $t_1, \ldots, t_n$ are linear coordinates on T_p^*M, $p \in U$ with respect to the basis $d_p x_1, \ldots, d_p x_n$.
Show that $(T^*M, d\theta)$ is a symplectic manifold (see Exercise 14.7).

17.1. A *derivation* on an $\mathbb{R}$-algebra A is an $\mathbb{R}$-linear map $D: A \to A$ that satisfies the identity

$$D(xy) = (Dx)y + x(Dy).$$

These form an $\mathbb{R}$-vector space $\text{Der } A$. Show for any smooth manifold M^m that there is a linear isomorphism

$$\Omega^0(\tau_M) \cong \text{Der}\Omega^0(M)$$

which to a vector field X assigns the derivation L_X given by

$$L_X(f) = Xf = df(X).$$

(Hint: Derivations are local operators.)
Show that the commutator $[D_1, D_2] = D_1 \circ D_2 - D_2 \circ D_1$ of $D_1, D_2 \in \text{Der} A$ also belongs to $\text{Der} A$, and define the Lie bracket $[X, Y] \in \Omega^0(\tau_M)$ of smooth vector fields X, Y on M by the condition

$$L_{[X,Y]} = [L_X, L_Y].$$

Prove that $\Omega^0(\tau_M)$ is a *Lie algebra*, i.e. that the following conditions hold

(L1) $[\ ,\ ]$ is bilinear.
(L2) $[X, X] = 0$.
(L3) $[X, [Y, Z]] + [Y, [Z, X]] + [Z, [X, Y]] = 0$ (Jacobi identity).

Show for $X, Y \in \Omega^0(\tau_M)$ and $f, g \in \Omega^0(M)$ that

$$[fX, Y] = f[X, Y] - (Yf)X$$
$$[X, gY] = g[X, Y] + (Xg)Y.$$

Suppose X and Y are given on a chart $U \subseteq M$ by the expressions (see Remark 9.4)

$$X = \sum_{i=1}^{m} a_i \frac{\partial}{\partial x_i}, \qquad Y = \sum_{j=1}^{m} b_j \frac{\partial}{\partial x_j}.$$

Show that in these coordinates $[X, Y]$ is given on U by

$$[X, Y] = \sum_{k=1}^{m} \left(\sum_{i=1}^{m} a_i \frac{\partial b_k}{\partial x_i} - \sum_{j=1}^{m} b_j \frac{\partial a_k}{\partial x_j} \right) \frac{\partial}{\partial x_k}.$$

17.2. Prove for $\omega \in \Omega^p(M)$ and $X_j \in \Omega^0(\tau_M)$, $1 \leq j \leq p+1$, the formula

$$d\omega(X_1, \ldots, X_{p+1}) = \sum_{i=1}^{p+1} (-1)^{i-1} X_i \big(\omega(X_1, \ldots, \hat{X}_i, \ldots, X_{p+1}) \big)$$

$$+ \sum_{1 \leq i < j \leq p+1} (-1)^{i+j} \omega \big(([X_i, X_j], X_1, \ldots, \hat{X}_i, \ldots, \hat{X}_j, \ldots, X_{p+1}) \big).$$

In particular

$$d\omega(X, Y) = X(\omega(Y)) - Y(\omega X) - \omega([X, Y]).$$

for $\omega \in \Omega^1(M)$, $X, Y \in \Omega^0(\tau_M)$

17.3. Let M be a Riemannian manifold with metric $\langle \ , \ \rangle$. Prove the existence and uniqueness of a connection ∇ on τ_M such that $\nabla_X : \Omega^0(\tau) \to \Omega^0(\tau)$ satisfies the following two conditions for smooth vector fields X, Y, Z:

    (a) $X(\langle Y, Z \rangle) = \langle \nabla_X Y, Z \rangle + \langle Y, \nabla_X Z \rangle$
    (b) $\nabla_X Y - \nabla_Y X = [X, Y]$

Prove moreover that ∇ satisfies the Koszul identity

$$2\langle \nabla_X Y, Z \rangle = X\langle Y, Z \rangle + Y\langle Z, X \rangle - Z\langle X, Y \rangle$$
$$- \langle X, [Y, Z] \rangle + \langle Y, [Z, X] \rangle + \langle Z, [X, Y] \rangle.$$

(Hints: Let $A(X, Y, Z)$ be the 6-term expression above. To prove uniqueness, derive the Koszul identity by rewriting the six terms using the equations obtained from (a) and (b) by cyclic permutation of X, Y, Z. For existence verify first the identity

$$A(X, Y, fZ) = fA(X, Y, Z),$$

where $f \in \Omega^0(M)$, and construct $\nabla_X Y$ so that the Koszul identity holds.) This connection is the *Levi-Civita connection* on M.

17.4. Prove for $M^n \subseteq \mathbb{R}^{n+k}$ that the connection on τ_M constructed in Example 17.2 is the same as the Levi-Civita connection of Exercise 17.3, when M is given the Riemannian metric induced from $\mathbb{R}^{n+k}$.

17.5. Let $\bigtriangledown$ be any connection on the vector bundle ξ over M. Given two vector fields $X, Y \in \Omega^0(\tau_M)$ define the operator $R(X,Y): \Omega^0(\xi) \to \Omega^0(\xi)$ by

$$R(X,Y) = \nabla_X \circ \nabla_Y - \nabla_Y \circ \nabla_X - \nabla_{[X,Y]}.$$

Verify that $R(X,Y)$ is a $\Omega^0(M)$-module homomorphism, and prove that $R(X,Y) = F_{X,Y}^\nabla$.

(Hint: Work locally using the formula (17.4). Show by direct computation that the two operators agree on e_i.)

17.6. Let M^n be a Riemannian manifold and $\bigtriangledown$ the Levi-Civita connection on τ_M from Exercise 17.3. Define $R(X,Y): \Omega^0(\tau_M) \to \Omega^0(\tau_M)$ as in Exercise 17.5. Prove the Bianchi identity

$$R(X,Y)Z + R(Y,Z)X + R(Z,X)Y = 0$$

for $X, Y, Z \in \Omega^0(\tau_M)$.

Show that the value of $R(X,Y)Z$ at $p \in M$ depends only on $X_p, Y_p, Z_p \in T_pM$. Hence R defines for $X_p, Y_p \in T_pM$ a linear map $R(X_p, Y_p): T_pM \to T_pM$.

17.7. Let $h: U \to U' \subseteq \mathbb{R}^n$ be a chart in M^n and $\partial_i = \frac{\partial}{\partial x_i}$ the vector fields on U considered in Remark 17.4. The Christoffel symbols are the smooth functions Γ_{ij}^k on U' determined by

$$\nabla_{\partial_i} \partial_j = \sum_k \left(\Gamma_{ij}^k \circ h \right) \partial_k.$$

Prove the formula

$$\Gamma_{ij}^k = \frac{1}{2} \sum_l g^{kl} \left(\frac{\partial g_{jl}}{\partial x_i} + \frac{\partial g_{il}}{\partial x_j} - \frac{\partial g_{ij}}{\partial x_l} \right),$$

where (g_{ij}) is the matrix of coefficients to the first fundamental form (see below Definition 9.15) and where (g^{kl}) is the inverse matrix $(g_{ij})^{-1}$.

(Hint: Apply the Koszul identity (cf. Exercise 17.3) to $\partial_i, \partial_j, \partial_k$.)

Define functions R_{ijk}^m on U' by

$$R(\partial_j, \partial_k)\partial_i = \sum_m \left(R_{ijk}^m \circ h \right) \partial_m.$$

Show that

$$R_{ijk}^m = \frac{\partial \Gamma_{ki}^m}{\partial x_j} - \frac{\partial \Gamma_{ji}^m}{\partial x_k} + \sum_l \left(\Gamma_{ki}^l \Gamma_{jl}^m - \Gamma_{ji}^l \Gamma_{kl}^m \right).$$

17.8. Let ∇ be a connection on ξ and consider two frames $e_1,\ldots,e_k$ and $\tilde{e}_1,\ldots,\tilde{e}_k$ on the same open set U and the corresponding matrices $A = (A_{ij})$, $\tilde{A} = (\tilde{A}_{ij})$ of 1-forms on U with

$$\nabla(e_i) = \sum_{j=1}^{k} A_{ij} \otimes e_j, \quad \nabla(\tilde{e}_i) = \sum_{j=1}^{k} \tilde{A}_{ij} \otimes \tilde{e}_j.$$

Let $\Phi = (\phi_{jm})$ be the invertible matrix of smooth functions given by

$$\tilde{e}_j = \sum_{m=1}^{k} \phi_{jm} e_m.$$

Show that $\tilde{A} = (d\Phi)\Phi^{-1} + \Phi A \Phi^{-1}$. What is the relationship between $d\tilde{A} - \tilde{A} \wedge \tilde{A}$ and $dA - A \wedge A$?

17.9. Prove Lemma 17.5.

17.10. Prove formula (17.18).

17.11. Given ξ with connection ∇_ξ, a frame $e_1,\ldots,e_k$ over an open set U and the connection matrix $A = (A_{ij})$.
Prove that ξ^* with dual connection ∇_{ξ^*} and dual frame $e_1^*,\ldots,e_k^*$ has connection matrix $-A^t = (-A_{ji})$.
Do a similar calculation for $\nabla_{\xi\otimes\eta}$ and $\nabla_{\mathrm{Hom}(\xi,\eta)}$ when ∇_η has connection matrix $B = (B_{rs})$ w.r.t. a frame $f_1,\ldots,f_m$ over U.

17.12. For ξ with connection ∇_ξ, construct connections $\nabla = \nabla_{\Lambda^i(\xi)}$ on $\Lambda^i(\xi)$ for all i, such that $\nabla_{\Lambda^1(\xi)} = \nabla_\xi$ and

$$\nabla_X(s \wedge t) = \nabla_X s \wedge t + (-1)^i s \wedge \nabla_X t$$

for $s \in \Omega^0\big(\Lambda^i(\xi)\big)$, $t \in \Omega^0\big(\Lambda^j(\xi)\big)$ and $X \in T_p M$.

17.13. Let $f : M' \to M$ be any smooth map and ξ a smooth (real or complex) vector bundle over M. Construct an isomorphism

$$\Omega^0\big(M'\big) \otimes_{\Omega^0(M)} \Omega^0(\xi) \xrightarrow{\cong} \Omega^0(f^*(\xi)); \quad \phi \otimes s \mapsto \phi f^*(s),$$

where $\psi \in \Omega^0(M)$ acts on $\Omega^0(M')$ by multiplication by $f^*(\psi) \in \Omega^0(M')$. (Hint: First handle trivial vector bundles. In general pick a complement to ξ as in Exercise 15.10.)

18.1. Prove for the canonical line bundle H over $\mathbb{CP}^n$ that H^* is not isomorphic to H.

18.2. Show for complex line bundles ξ,η over M that

$$c_1(\xi \otimes \eta) = c_1(\xi) + c_1(\eta).$$

Use the splitting principle to derive the formula

$$c_k(\xi \otimes \eta) = \sum_{r=0}^{k} \binom{n-r}{k-r} c_r(\xi) c_1(\eta)^{k-r},$$

where η is a complex line bundle and ξ any n-dimensional complex vector bundle over M.

18.3. Show for a complex n-dimensional vector bundle ξ that $c_1(\Lambda^n(\xi)) = c_1(\xi)$. (Hint: Apply the splitting principle and Exercise 16.8.)

18.4. Show for the vector bundle $H^\perp$ over $\mathbb{CP}^n$ defined in Example 18.13 that

$$c_k(H^\perp) = (-1)^k c_1(H)^k.$$

Show that n is the smallest possible dimension of a complementary vector bundle over $\mathbb{CP}^n$ to H.

18.5. Show for the complex line bundles H and $H^\perp$ over $\mathbb{CP}^1$ that $H^\perp \cong H^*$. (Hint: Consider $H \otimes H^\perp$.)

18.6. Adopt the notation of Exercise 14.8. Show that any tangent vector to $G_m(F^n)$ at the "point" $V \subseteq F^n$ can be written $\alpha'(0)$, where $\alpha(t) = [a_1(t), \ldots, a_m(t)]$ and $a_j(t)$ is the j-th column of a smooth curve $A: (-\delta, \delta) \to \mathrm{Fr}_m(F^n)$ with $V = [A(0)]$. Prove that there is a well-defined $\mathbb{R}$-linear isomorphism

$$D_V : T_V G_m(F^n) \to \mathrm{Hom}_F(V, F^n/V)$$

such that $D_V(\alpha'(0)): V \to F^n/V$ sends $a_j(0)$ into $a'_j(0) + V$.
Let γ be the canonical vector bundle over $G_m(F^n)$ (see Exercise 15.5). Construct another vector bundle $\gamma^\perp$ whose fiber over V is the orthogonal complement to V in F^n, and show that $\gamma \oplus \gamma^\perp \cong \varepsilon^n$.
Prove that the maps D_V define a smooth vector bundle isomorphism

$$\tau_{G_m(F^n)} \cong \mathrm{Hom}_F(\gamma, \gamma^\perp)$$

(in particular $\tau_{G_m(\mathbb{C}^m)}$ has a natural complex structure).

18.7. (Plücker embedding) Prove that a smooth embedding Φ of $G_m(F^n)$ into the copy $G_1(\Lambda^m F^n)$ of $\mathbb{RP}^N$ or $\mathbb{CP}^N$, $N = \binom{n}{m} - 1$, can be defined by

$$\Phi(V) = [a_1 \wedge \ldots \wedge a_m],$$

where $a_1, \ldots, a_m$ is any basis of V.
(Hint: Tangent maps can be computed in terms of the identification $T_V G_m(F^n) \cong \mathrm{Hom}_F(V, F^n/V)$ from Exercise 18.6.)

19.1. Let M^n be a Riemannian manifold with Levi-Civita connection ∇ on τ_M. Let $e_1, \ldots, e_n$ be an orthonormal frame in τ_M over $U \subseteq M$ with connection matrix (A_{ij}), and denote the dual frame in τ_M^* over U by $\epsilon_1, \ldots, \epsilon_n$. Show that

$$d\epsilon_i(e_j, e_k) = -A_{ki}(e_j) + A_{ji}(e_k).$$

(Hint: Use Exercises 17.2 and 17.3.)

Conclude in the case $n = 2$ that the connection described in Example 19.5 is the Levi-Civita connection.

19.2. Let V be an n-dimensional $\mathbb{R}$-vectorspace. A multilinear map $F: V \times V \times V \times V \to \mathbb{R}$ is said to be *curvature-like* when the following identities are satisfied:

(a) $F(x, y, z, w) = -F(y, x, z, w)$.
(b) $F(x, y, z, w) + F(y, z, x, w) + F(z, x, y, w) = 0$.
(c) $F(x, y, z, w) = -F(x, y, w, z)$.

Prove that a curvature-like function on a tangent space $T_p M^n$, where M is Riemannian, is given by the expression

$$\langle R_p(X_p, Y_p, Z_p), W_p \rangle_p$$

(see Exercises 17.5 and 17.6). Prove that the conditions (a), (b), (c) imply

(d) $F(x, y, z, w) = F(z, w, x, y)$.

19.3. Show that the curvature-like functions on V defined in Exercise 19.2 form a vector-space of dimension $\frac{1}{12} n^2 (n^2 - 1)$.

19.4. Let F be curvature-like on V as defined in Exercise 19.2, and suppose V is equipped with an inner product $\langle \ , \ \rangle$. For a 2-dimensional subspace $\Pi \subseteq V$, show that the expression

$$K(\Pi) = -\frac{F(x, y, x, y)}{\langle x, x \rangle \langle y, y \rangle - \langle x, y \rangle^2},$$

where x, y is a basis for Π, depends only on Π. Show that the function K on the Grassmannian $G_2(V)$ determines F uniquely. Conclude that if K is constant with value k on $G_2(V)$, then

$$F(x, y, z, w) = k(\langle y, z \rangle \langle x, w \rangle - \langle x, z \rangle \langle y, w \rangle).$$

19.5. Let $\Pi \subseteq T_p M^n$ be a 2-dimensional subspace, where M is a Riemannian manifold. The *sectional curvature* of M at Π is the real number

$$K(p, \Pi) = -\frac{\langle R_p(X_p, Y_p) X_p, Y_p \rangle}{\langle X_p, X_p \rangle \langle Y_p, Y_p \rangle - \langle X_p, Y_p \rangle^2},$$

where X_p, Y_p is a basis for Π (see Exercises 19.2 and 19.4). Show that S^n ($n \geq 2$) with the standard Riemannian metric induced from

$\mathbb{R}^{n+1}$ has constant sectional curvature, i.e. that $K(p, \Pi)$ is independent of $p \in S^n$ and the plane $\Pi \subseteq T_p S^n$.

Show that $F^{\nabla}_{X_p, Y_p} = R_p(X_p, Y_p) \colon T_p S^n \to T_p S^n$ acts by

$$F^{\nabla}_{X_p, Y_p}(Z_p) = k(\langle Y_p, Z_p \rangle X_p - \langle X_p, Z_p \rangle Y_p),$$

where k is the sectional curvature (in fact $k = 1$, see Exercise 19.8).

19.6. Show for the connection described in Example 19.5 and Exercise 19.1 that the Gaussian curvature $\kappa \in \Omega^0(M^2)$ satisfies

$$\kappa(p) = K(p, T_p M^2).$$

19.7. Show with the notation of Exercise 17.7 that

$$K(p, \operatorname{Span}(\partial_1, \partial_2)) = -\frac{\sum_m R^m_{112}}{g_{11} g_{22} - g^2_{12}}.$$

Show for $\dim M = 2$ that this reduces to

$$K(p, T_p M^2) = -g^{-1}_{11} R^2_{112}.$$

(Hint: Use that $\langle R(\partial_1, \partial_2)\partial_1, \partial_1 \rangle = 0$ to eliminate R^1_{112} from the previous expression.)

Remark: Gaussian curvature $K(p)$ of a surface $S \subseteq \mathbb{R}^3$ as defined in Example 12.18 is given in local coordinates by the same expression; see e.g. [do Carmo] page 234.

19.8. Show that the constant k in Exercise 19.5 is independent of $n \geq 2$. Compute it for $n = 2$.

(Hint: Embed $S^n \subseteq S^{n+1}$ as an equatorial sphere and compare connection matrices by applying Exercise 19.1 to an orthonormal frame $e_1, \ldots, e_n, e_{n+1}$, where $e_{n+1} \in (T_p S^n)^{\perp}$ for $p \in S^n$.)

19.9. Prove that the Pfaffian polynomial in variables X_{ij}, $1 \leq i < j \leq 2n$ is given by

$$\operatorname{Pf}(X_{ij}) = \sum_{\sigma \in T(n)} \operatorname{sign}(\sigma) \prod_{\nu=1}^{n} X_{\sigma(2\nu-1)\sigma(2\nu)},$$

where

$$T(n) = \{\sigma \in S(2n) \mid \sigma(2\nu - 1) < \sigma(2\nu) \text{ for } 1 \leq \nu \leq n\}.$$

19.10. Let $\epsilon_1, \ldots, \epsilon_{2n}$ be the standard basis for $\operatorname{Alt}^1(\mathbb{R}^{2n})$. Show that

$$\operatorname{Pf}(\epsilon_i \wedge \epsilon_j) = 1 \cdot 3 \cdot 5 \cdots (2n - 1) \operatorname{vol}.$$

20.1.   Verify formula (3) in the proof of Theorem 20.1.

20.2.   Let M_1 and M_2 be smooth manifolds, and assume that at least one of them have finite-dimensional de Rham cohomology. Show that the maps

$$H^p(M_1) \otimes H^q(M_2) \to H^{p+q}(M_1 \times M_2)$$
$$a_1 \otimes a_2 \mapsto \mathrm{pr}^*_{M_1}(a_1) \cdot \mathrm{pr}^*_{M_2}(a_2)$$

combine to give isomorphisms

$$\bigoplus_{p+q=n} H^p(M_1) \otimes H^q(M_2) \cong H^n(M_1 \times M_2).$$

(Hint: Use Theorem 20.1.)

20.3.   Construct a smooth manifold structure on $\tilde{G}_2(\mathbb{R}^m)$ such that the map $\tilde{\phi}$ in diagram (20.9) is a smooth embedding and π_0 a submersion.

20.4.   Show that exponentiation of 2×2 matrices

$$\exp(A) = \sum_{n=0}^{\infty} \frac{1}{n!} A^n$$

can be used to define a homeomorphism $\psi: \mathbb{R}^2 \to Q$ by

$$\psi(\alpha, \beta) = \exp\begin{pmatrix} \alpha & \beta \\ \beta & -\alpha \end{pmatrix}.$$

Show that the space X defined in the proof of Proposition 20.6 can be identified with the total space of a certain complex line bundle over $\tilde{G}_2(\mathbb{R}^m)$. How is this line bundle related to the canonical real vector bundle γ_2 over $\tilde{G}_2(\mathbb{R}^m)$?

20.5.   A complex structure J on $\mathbb{R}^{2n}$ is a linear map $J: \mathbb{R}^{2n} \to \mathbb{R}^{2n}$ with $J^2 = -\mathrm{id}$. Given J, $\mathbb{R}^{2n}$ becomes a $\mathbb{C}$-vector-space by defining

$$(a + ib)x = ax + bJx.$$

Show that $\mathrm{GL}_{2n}(\mathbb{R})$ acts transitively by conjugation on the set of complex structures, and that the subgroup fixing J is isomorphic to $\mathrm{GL}_n(\mathbb{C})$.
Prove a similar statement for the subgroup $\mathrm{GL}^+_{2n}(\mathbb{R})$ of matrices with positive determinant and complex structures inducing the standard orientation.

21.1.   (The Gysin sequence) Let ξ be an m-dimensional real smooth vector bundle over a compact manifold M with Riemannian metric and $S(\xi)$ the unit sphere bundle. Construct an exact sequence

$$\cdots \to H^{p-1}(S(\xi)) \to H^{p-m}(M) \xrightarrow{e\cdot} H^p(M) \xrightarrow{\pi^*} H^p(S(\xi)) \to \cdots$$

where the labeled maps are multiplication by the Euler class e of ξ and pull-back by the projection.

(Hint: Consider the Mayer–Vietoris sequence for $S(\xi \oplus 1)$ covered by $U_\infty = S(\xi \oplus 1) - s_\infty(M)$ and $U_0 = S(\xi \oplus 1) - s_0(M)$, where s_0 is the zero section $M \to E = E(\xi) \subseteq S(\xi \oplus 1)$.)

21.2. Show for $n \geq 1$ that

$$H^p\left(V_2\left(\mathbf{R}^{2n+1}\right)\right) \cong \begin{cases} \mathbf{R} & \text{for } p = 0, 4n - 1 \\ 0 & \text{otherwise} \end{cases}$$

$$H^p\left(V_2\left(\mathbf{R}^{2n+2}\right)\right) \cong \begin{cases} \mathbf{R} & \text{for } p = 0, 2n, 2n + 1, 4n + 1 \\ 0 & \text{otherwise} \end{cases}$$

(Hint: Apply the Gysin sequence of Exercise 21.1 to the tangent bundle of a sphere.)

21.3. Let $M^n \subseteq \mathbf{R}^{n+k}$ be a compact orientable smooth submanifold with normal bundle ν. Show that ν is orientable with $\hat{e}(\nu) = 0$.
(Hint: Identify the total space $E(\nu)$ with a tubular neighborhood of M.)

21.4. Verify Theorem 21.13 directly for the sphere S^{2n} with the Levi-Civita connection on $\tau_{S^{2n}}$ by using the information given in Example 10.12 and Exercises 19.5 and 19.10.

21.5. Let ξ^{2m} be an even-dimensional smooth real vector bundle over M. Let τ^ν be the vertical tangent bundle from Exercise 15.9 of $S(\xi \oplus 1) \to M$. Show that the orientation of ξ induces a natural orientation of τ^ν.
Let F^∇ be the curvature of some metric connection on τ^ν. Show that

$$u = \frac{1}{2}\left[\text{Pf}\left(\frac{-F^\nabla}{2\pi}\right) - \pi^* s_\infty^* \text{Pf}\left(\frac{-F^\nabla}{2\pi}\right)\right]$$

is an orientation class.

21.6. Give an alternative proof of Theorem 21.13 beginning with the computation in Exercise 21.4. Construct u as in Exercise 21.5 and show that $s_0^*(u) = e(\xi)$.

References

[Bredon], G. E. Bredon: *Topology and Geometry*, Springer Verlag, New York 1993

[do Carmo], M. P. do Carmo: *Differential Geometry of Curves and Surfaces*, Prentice-Hall Inc., New Jersey 1976.

[Donaldson–Kronheimer], S. K. Donaldson and P. B. Kronheimer: *The Geometry of Four-Manifolds*, Oxford University Press, Oxford 1990.

[Freedman–Quinn], M. H. Freedman and F. Quinn: *Topology of 4-Manifolds*, Princeton University Press, New Jersey 1990.

[Hirsch], M. W. Hirsch: *Differential Topology*, Springer-Verlag, New York 1976.

[Lang], S. Lang: *Algebra*, Addison-Wesley, Massachusetts 1965.

[Massey], W. S. Massey: *Algebraic Topology: An Introduction*, Hartcourt, Brace & World Inc. 1967.

[Milnor], J. Milnor: *Morse Theory*, Princeton University Press, New Jersey 1963.

[Milnor–Stasheff], J. Milnor and J. Stasheff: *Characteristic classes*, Annals of Math. studies, No **76**, Princeton University Press, New Jersey 1974.

[Moise], E. E. Moise: *Geometric Topology in Dimensions 2 and 3*, Springer-Verlag, New York 1977.

[Rudin], W. Rudin: *Real and Complex Analysis*, McGraw-Hill, New York 1966.

[Rushing], T. B. Rushing: *Topological Embeddings*, Academic Press, New York 1973.

[Whitney], H. Whitney: *Geometric Integration Theory*, Princeton University Press, New York 1957.

Index